Medical genetics

Medical genetics

Ian D. Young

Department of Clinical Genetics, Leicester Royal Infirmary,
Leicester, UK

OXFORD

UNIVERSITY PRESS

OXFORD

UNIVERSITY PRESS

Great Clarendon Street, Oxford OX2 6DP

Oxford University Press is a department of the University of Oxford.
It furthers the University's objective of excellence in research, scholarship,
and education by publishing worldwide in

Oxford New York

Auckland Cape Town Dar es Salaam Hong Kong Karachi Kuala Lumpur
Madrid Melbourne Mexico City Nairobi New Delhi Taipei Toronto Shanghai

With offices in

Argentina Austria Brazil Chile Czech Republic France Greece Guatemala Hungary
Italy Japan South Korea Poland Portugal Singapore Switzerland Thailand Turkey
Ukraine Vietnam

Oxford is a registered trade mark of Oxford University Press
in the UK and in certain other countries

Published in the United States
by Oxford University Press Inc., New York

© Ian D. Young, 2005

A catalogue record for this title is available from the British Library
Data available

Library of Congress Cataloguing in Publication Data
(Data available)
ISBN 978-0-19-959461-0 (pbk)

10 9 8 7 6 5 4 3

Typeset by EXPO Holdings Sdn Bhd., Malaysia
Printed by CPI Group (UK) Ltd, Croydon, CR0 4YY

Dedication

To my friend and colleague, Professor Bob Mueller

Preface

As the impact of genetics in medicine has expanded, so has the choice of related texts for medical students. Most of these are excellent, often running into several editions, each longer and more elaborate than its predecessor with increasingly complex explanations of new concepts and discoveries. No wonder then that this burgeoning profusion of detailed 'introductory texts' should have spawned a new generation of short fact-filled guides designed as aides-memoire for the beleaguered student on the eve of the genetics examination. Somewhere between these two extremes there is a need for a book of intermediate length providing the basic information required for the medical syllabus in a comprehensible and digestible fashion, which is both scientifically sound and clinically relevant.

It is with precisely this goal in mind that this text has been prepared. It has been compiled to provide the basic knowledge and core clinical skills recommended by the American Society of Human Genetics Information and Education Committee and, in the United Kingdom, by the Joint Committee on Medical Genetics and the British Society of Human Genetics. It has also been designed to help students pass examinations, and—perhaps of greater note—it has been written with a keen awareness of the importance of maintaining the student's interest and respect with clear explanations of difficult concepts and regular reference to clinical applications.

The emphasis throughout is on the elucidation of basic principles followed by consideration of clinical relevance. The chapters have been arranged to provide a logical, stepwise introduction to the important concepts and themes of human genetics with an increasing focus on medical applications. The latter chapters are primarily clinically orientated and are designed to provide the clinical knowledge and skills expected of a medical student. To assist those involved in the teaching of medical genetics, the core curriculum recommendations of the American and British Societies of Human Genetics have been included as appendices. An attempt has also been made to stimulate and sustain the reader's interest by the inclusion in each chapter of a brief review of a relevant 'landmark' publication, together with a clinically or socially related vignette, relating for example to a famous figure in history, such as Abraham Lincoln, or to a society celebrity such as Dolly the sheep! Frequent reference is made to medical disorders caused by genetic abnormalities. A student who wishes to learn more about a particular condition can obtain further information using the disease reference number provided for OMIM, the online database on genetic disorders maintained by NIH (www3.ncbi.nlm.nih.gov/entrez/query.fcgi?db=OMIM). Finally, to ensure that the text is as sensitive as possible to the needs of the student, each chapter concludes with a brief selection of recommended reading followed by multiple choice questions of the type much beloved by devious examiners worldwide.

The challenges facing a student of medicine at all stages of his or her career are daunting. Anything that can ease this burden of acquiring knowledge and understanding is to be welcomed, and this book has been written with precisely this intention. Sadly, the publishers cannot be persuaded to offer a money-back guarantee in the unlikely event of a conscientious reader failing a crucial test of genetics knowledge, but it is hoped that everyone who consults this text will find it helpful, instructive, and user-friendly.

Finally, the author would like to extend sincere thanks to the many colleagues who have offered encouragement, provided illustrations, and corrected misconceptions. Any remaining errors are entirely the author's responsibility. The support and assistance of Catherine Barnes and Georgia Pinteau at OUP have been particularly appreciated, together with the secretarial expertise of Mrs. Diane Castledine.

Ian Young
July 2004

Contents

Gene structure and function

The human body contains around 75 trillion (75×10^{12}) cells, most of which have a nucleus bounded by a nuclear membrane and surrounded by cytoplasm. Within each nucleus there is a set of 46 chromosomes made up of **chromatin**. This consists of an extremely long strand of **deoxyribonucleic acid** (DNA) interwoven around structural proteins known as histones. The structure of chromosomes and how they behave during cell division are outlined in the next chapter. This chapter is concerned with the structure of DNA, how it replicates, and how the 35 000 **genes** present in each set of chromosomes are expressed to produce proteins.

DNA and the double helix

A **nucleic acid** consists of a chain of **nucleotides** linked together through an alternating series of sugar and phosphate residues from which nitrogenous bases protrude (Fig. 1.1). The sugar residues in DNA and **ribonucleic acid** (RNA) are 2-deoxyribose and ribose respectively. DNA contains four types of base consisting of two purines, adenine (A) and guanine (G), and two pyrimidines, cytosine (C) and thymine (T). The same bases are present in RNA with the exception that the pyrimidine uracil (U) replaces thymine (Table 1.1). Purines contain two organic rings, whereas pyrimidines contain only a single ring.

The carbons within an organic ring are numbered consecutively in a clockwise direction around the ring. At one end of a nucleic acid chain, the phosphate group is attached to the fifth carbon atom of the sugar moiety. This is referred to as the 5' (pronounced 'five prime') end. At the other end the phosphate is attached to the third

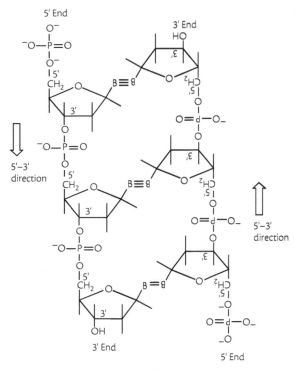

Fig. 1.1 Two antiparallel strands of DNA. B = base. Reproduced with permission from Elliott WH, Elliott DC (2001) *Biochemistry and molecular biology*, 2nd edn. Oxford University Press, Oxford.

TABLE 1.1	Nucleotide base pairing in DNA and RNA	
	Purines	**Pyrimidines**
DNA		
	Adenine (A) —	Thymine (T)
	Guanine (G) —	Cytosine (C)
RNA		
	Adenine (A) —	Uracil (U)
	Guanine (G) —	Cytosine (C)

carbon atom of the sugar moiety. This is referred to as the 3' ('three prime') end of the chain. The convention is to describe nucleic acid sequences in the 5' to 3' direction, beginning therefore with the nucleotide in which there is a free phosphate attached to the fifth carbon atom and ending with the nucleotide in which the third carbon atom is hydroxylated (Fig. 1.1) with no attached phosphate.

In 1953 Watson and Crick (Box 1.1) proposed that two DNA strands form a double helix by the formation of hydrogen bonds between the nitrogenous bases (Fig. 1.2).

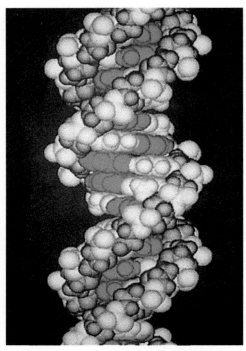

Fig. 1.2 Model of how the two strands of DNA form a double helix. Reproduced with permission from Lewin B (2000) *Genes VII*. Oxford University Press, New York.

Guanine always pairs with cytosine (G with C). Adenine pairs with thymine in DNA (A with T) and uracil in RNA (A with U). The paired strands are described as being **complementary**. One strand runs in a 5' to 3' direction, and the complementary strand runs in a 3' to 5' direction. Thus if one strand reads 5'-TAGCTG-3' then the complementary strand will read 3'-ATCGAC-5'. They are therefore said to be **antiparallel**.

Note that a pair of complementary bases, e.g. C with G, or A with T, is referred to as a **base pair** (bp); 1000 bp equal one **kilobase** (kb) and 1000 kilobases equal one **megabase** (Mb). The human genome consists of 2 sets of 23 chromosomes each made up of 3.2×10^9 base pairs. Note also that two adjacent bases on the same strand are depicted as CpG or ApG, with the intervening phosphate, represented as p, included to distinguish two adjacent bases from a complementary base pair.

Key Point

Genes are made up of DNA which consists of two antiparallel strands held together in a double helix by hydrogen bonds between complementary nitrogenous bases. Guanine always pairs with cytosine. Adenine always pairs with thymine.

BOX 1.1 LANDMARK PUBLICATION: THE STRUCTURE OF DNA

It is generally agreed that the discovery of the structure of DNA marks *the* seminal moment in the history of biology. It had been known for several years that genes were composed of DNA rather than protein and that DNA contained four nucleotides, adenine, cytosine, guanine and thymine. Various structures had been proposed including a triple-helical model in which the nucleotide bases pointed outwards. However, none of the proposed models satisfactorily accounted for all of DNA's observed physical characteristics.

In 1953 three papers appeared in a single issue of *Nature* providing support for a double helical structure. Two of these, one by Maurice Wilkins and colleagues and the other by Rosalind Franklin and Raymond Gosling, presented impressive evidence based on X-ray diffraction studies. In the third paper, James Watson and Francis Crick succinctly proposed that DNA is made up of two helical chains held together by hydrogen bonds between complementary purine and pyrimidine bases pointing inwards from a phosphate–sugar backbone. They observed that the most plausible three-dimensional structure would involve exclusive pairing of adenine with thymine and cytosine with guanine. They also observed, in a masterpiece of understatement, that their proposed structure 'suggests a possible copying mechanism for the genetic material'. The message of this manuscript was contained within less than a single published page and contained only a single illustration, consisting of a simple diagrammatic representation of the double helix as drawn by Francis Crick's wife. Unlike today, the paper was not peer reviewed and the order of authorship was decided by the toss of a coin.

In recognition of their achievements, Watson and Crick, together with Wilkins, were awarded the Nobel prize for medicine in 1962. There is a school of thought which maintains that Rosalind Franklin's contribution never received due recognition. Before publication Watson and Crick had seen her illuminating X-ray photographs of DNA, which may have helped guide them towards their brilliant, intuitive double-helix model. The issue of whether she would or should have joined the ranks of the Nobel laureates remains unresolved. According to the Nobel rules, no more than three individuals can share a prize and posthumous awards cannot be made. Sadly, Rosalind Franklin died at the age of 37 years in 1958, four years before the contribution of Watson, Crick, and Wilkins was duly acknowledged.

References

Watson JD, Crick FHC (1953) A structure for deoxyribose nucleic acid. *Nature*, **171**, 737–738.

Wilkins MHF, Stokes AR, Wilson HR (1953) Molecular structure of deoxypentose nucleic acids. *Nature*, **171**, 738–740.

Franklin RE, Gosling RG (1953) Molecular configuration in sodium thymonucleate. *Nature*, **171**, 740–742.

Maddox B (2002) *Rosalind Franklin. The dark lady of DNA.* HarperCollins, London.

DNA replication

Watson and Crick noted that the complementary structure of DNA was consistent with a plausible mechanism for DNA synthesis, which occurs during the S phase of the cell cycle (p. 32). At localized sites of replication, known as DNA forks, an enzyme appropriately known as a helicase unwinds the DNA helix. (Inherited defects in helicase enzymes cause a chromosome breakage disorder known as Bloom syndrome—p. 66.) The separated parental strands act as templates for the synthesis of complementary daughter strands by DNA polymerase, which incorporates the appropriate complementary bases (Fig. 1.2).

Replication proceeds in the 5′ to 3′ direction with new nucleotides being attached to the 3′ end. One strand, known as the *leading strand*, is formed continuously, moving in the direction of the fork (Fig. 1.3). The other strand, the *lagging strand*, is formed in 100–1000-nucleotide blocks known as **Okazaki fragments**. The synthesis of these fragments is initiated using an RNA primer. The fragments are then joined together by an enzyme known as a ligase. Because of this difference in the way that the DNA strands are synthesized, the process is described as being **semi-discontinuous**.

By the completion of DNA synthesis, each chromosome has divided into two chromatids (p. 26), each consisting of one original parent strand and one new complementary strand. This pattern of replication whereby one parent strand is retained in each daughter cell is described as **semi-conservative**.

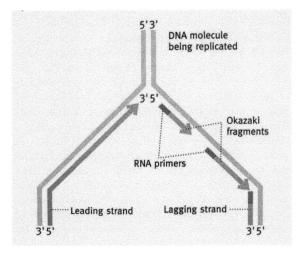

Fig. 1.3 Diagrammatic representation of DNA strand synthesis during DNA replication. The leading strand is synthesized continuously moving in the direction in which the fork is moving, i.e. upwards in this diagram. Reproduced with permission from Elliott WH, Elliott DC (2001) *Biochemistry and molecular biology*, 2nd edn. Oxford University Press, Oxford.

TABLE 1.2	Examples of repetitive DNA sequences
Microsatellites	Short tandem repeats. Also known as single sequence repeats. Contain multiple copies of bi-, tri-, or tetranucleotide repeat sequences
Minisatellites	Also known as variable number tandem repeats (VNTRs). Consist of multiple copies of a 10–100-bp core sequence
SINES	Short interspersed nuclear elements. 100–400 bp in length. Most common is the Alu repeat—see text
LINES	Long interspersed nuclear elements, 1–6 kb in length. Three LINE families make up >20% of total genome
Satellite DNA	Consists of large tandem repeats of 5–171 bp. α and β satellite DNA maintain function of the centromeres
Triplet repeats	Tandem repeats of three base pairs, usually CGG, CAG, or CTG. Can increase in number ('expand') to cause disorders which show anticipation (p 81).

The content of the human genome

Each set of chromosomes is made up of approximately 3.2×10^9 bp. Less than 10% of these encode proteins. Approximately 50% of the genome consists of different forms of repetitive DNA widely distributed across the genome (Table 1.2). This is thought to play a major role in the maintenance of chromosome structure.

The existence of these different forms of non-coding, 'junk' DNA has been exploited for a number of clinical applications. *Polymorphic* **microsatellites** formed the basis of the first genetic maps of the human genome. They are used in linkage analysis for gene tracking (p. 91). **Minisatellites** formed the basis of the original genetic fingerprinting studies (p. 142). FISH probes for α and β satellite DNA are used to detect the centromeres of specific chromosomes for rapid aneuploidy screening in prenatal diagnosis (p. 29).

The interspersed sequences, consisting of SINES and LINES (Table 1.2), are transposable elements, also known as **transposons.** These constitute 35–45% of the total genome. Originally transposons had the ability to move within the genome, but many have lost this ability as a result of acquired inactivating mutations. The most abundant form of interspersed sequences are known as *Alu* repeats, of which approximately 1 million copies are present in the human genome, occurring on average every 3 kb. These are 280–300 bp long

with a single site at the 170th nucleotide where they can be cleaved by the enzyme *Alu*I (hence the name *Alu* repeat). The high frequency of *Alu* repeats in the genome predisposes to unequal crossing-over in meiosis I, when chromosomes misalign resulting in deletions and duplications (p. 12). As an example, most of the deletion breakpoints in the low-density lipoprotein receptor gene (*LDLR*), which causes familial hypercholesterolaemia (p. 213), occur in *Alu* repeats.

Structure of a typical human gene

The Human Genome Project (Box 1.2) has revealed that the human genome contains 30 000–35 000 nuclear genes, indicating that an average-sized chromosome has 1300–1500 genes. Human genes show striking variation in size, ranging from around 1 kb for the genes coding for β-globin and insulin to 2.5 Mb for the dystrophin gene responsible for Duchenne muscular dystrophy (p. 109).

Almost all genes contain coding regions known as **exons**, which are expressed, with intervening sequences, known as **introns,** which are not expressed (Fig. 1.4). Introns are transcribed into primary RNA, also known as pre-mRNA, in the cell nucleus but are spliced out of the mature mRNA in the cytoplasm. The average number of exons per human gene is nine (so that the average number of introns is eight), but there is great

BOX 1.2 CASE CÉLÈBRE: THE HUMAN GENOME PROJECT

The Human Genome Project (HGP) was initiated in the USA in 1990 by the National Institutes of Health and the Department of Energy. The primary goal was to map and sequence the human genome by 2005. Secondary aspirations were to develop new mapping and sequencing technologies and to sequence the genomes of five other organisms (bacteria, yeast, roundworm, fruit fly, and mouse). Due recognition was given to prevailing public concern about the ethical and legal aspects of the project by allocating up to 5% of the total budget of US$3 billion to research into societal implications of a successful outcome.

Through collaboration with other national genome programmes in Canada, China, France, Germany, Japan, and the UK, and spurred on by the late intervention of a commercial competitor (Celera), the success of the project exceeded expectations. Under the auspices of the HGP, comprehensive genetic (linkage) and physical maps of the human genome were developed in the mid 1990s. These greatly expedited ongoing efforts to identify disease genes by positional cloning (p. 98). By 1996 yeast and bacterial genomes had been sequenced and attention turned to the challenges posed by trying to sequence the 3.2×10^9 bases in the human genome.

The strategy employed was to cut the human DNA into fragments of 100 000–200 000 bp which were then cloned into bacterial artificial chromosomes (BACs) and inserted into bacteria. The 20 000 BAC clones were then ordered and mapped before each clone was cut into small 2000-bp fragments for sequencing. The sequences for each BAC clone were determined by piecing together the sequences from all of the small fragments and in turn a draft sequence of the entire genome was assembled. This was published in *Nature* in 2001, in the same week that the private company, Celera, published its own draft sequence in *Science*.

Analysis of the draft sequence revealed that the genome contains approximately 30 000–40 000 coding genes. This was considerably less than had been expected, particularly when compared with the 18 500 present in a roundworm and the 28 000 in a mustard plant. Human complexity is now thought to be achieved by a much more elaborate set of protein products than is found in other species. The latest estimate of the total number of coding human genes, based on the final genome sequence completed in 2003, is around 30 000.

Completion of the human genome sequence was not achieved without controversy. Competition between laboratories in the publicly funded consortium was overshadowed by debate over patenting rights and the issue of whether private companies can claim ownership over a public property such as the structure of a human gene (see Box 10.2). These issues remain unresolved. What is clear is that the successful sequencing marks a major milestone in human biology, which offers enormous benefits in terms of understanding human susceptibility to disease and the development of new genetically targeted treatments.

References

International Human Genome Sequencing Consortium (2001) Initial sequencing and analysis of the human genome. *Nature*, **409**, 860–921.

Venter JC, Adams MD, Myers EW *et al.* (2001). The sequence of the human genome. *Science*, **291**, 1304–1351.

Web sites

The US National Human Genome Research Institute: http://www.nhgri.nih.gov/
The UK Human Genome Mapping Resource Centre: http://www.hgmp.mrc.ac.uk/GenomeWeb/

variation with the dystrophin gene containing 79 exons whereas the β-globin gene has only 3. The average size of an exon in a human gene is 145 bp.

In addition to exons and introns, each gene contains a closely adjacent upstream (5′) regulatory promoter region and other regulatory sequences including enhancers, silencers, and sometimes a locus control region. The promoter region contains specific sequences such as a TATA box, a CG box, and a CAAT box, which provide binding sites for transcription factors. The enhancer and silencer sequences fulfil a similar purpose, but are located at a greater distance from the coding sequences.

The first and last exons contain *untranslated regions*, known as the 5′ UTR and 3′ UTR respectively. The 5′ UTR marks the start of transcription and contains an initiator codon which indicates the site of the start of translation. The 3′ UTR contains a termination codon, which marks the end of translation, plus nucleotides which encode a sequence of adenosine residues known as the *poly(A) tail*.

Key Point

Most genes contain several exons, which are expressed, separated by intervening sequences known as introns, which are spliced out of mature mRNA.

Transcription and translation

Nuclear DNA contains the blueprint for the manufacture of polypeptides. The process linking DNA and its protein product is complex. For convenience this is usually considered as consisting of two stages, transcription and translation. **Transcription** involves the conversion of DNA to RNA. **Translation** refers to the manufacture of a polypeptide chain from an extranuclear RNA template.

The regulation of transcription

Transcription is initiated when proteins known as transcription factors bind to the promoter region and to other upstream regulatory regions such as the enhancers. It can also be initiated more autonomously in selected genes, such as α- and β-globin (p. 155), by activity at an upstream *locus control region* (LCR). The relevant transcription factors contain specific structural domains, such as leucine zippers and zinc fingers, which bind to the regulatory regions that become exposed in surface folds of the double helix through the process of **chromatin remodelling**. This involves a change in the structure of the basic unit of DNA packaging, known as a **nucleosome** (p. 25), resulting in strands of DNA which had previously been concealed becoming unfolded.

There is increasing evidence that changes in chromatin structure play a key role in gene expression. Chromatin exists in two forms, **euchromatin** and **heterochromatin**, which differ in that euchromatin is loosely packed whereas heterochromatin is tightly condensed. Euchromatin is transcriptionally active and replicates early in the S phase of the cell cycle (p. 32). Heterochromatin is transcriptionally inactive and replicates late in the S phase of the cell cycle.

Actively transcribed genes in euchromatin occur in clusters in a loop or 'domain' which is unfolded. The locus control region for the β-globin gene cluster acts by initiating chromatin unfolding in red blood cell precursors. This region remains tightly folded in other cells in which β-globin is not synthesized.

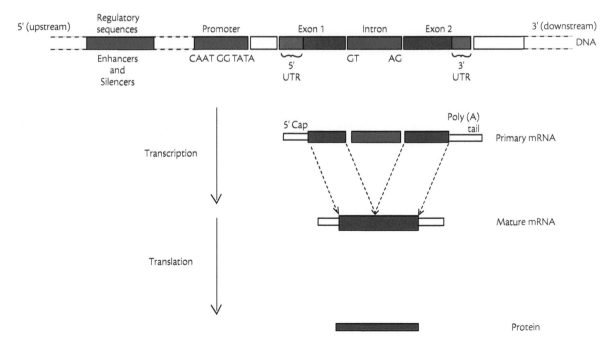

Fig. 1.4 The structure of a typical gene and the changes that occur in transcription and translation.

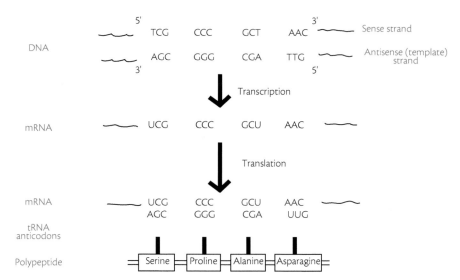

Fig. 1.5 Diagrammatic representation of how information in the sense strand of DNA is converted into a polypeptide chain. For the sake of simplicity, four codons are shown separated by a short gap.

Transcription—DNA to mRNA

Transcription commences at a transcriptional initiation site and involves the synthesis of a single strand of messenger RNA (mRNA) by RNA polymerase, proceeding in a 5′ to 3′ direction (Fig. 1.5). This 5′ to 3′ orientation refers to the coding or 'sense' strand. The DNA used as a template for this synthesis is the non-coding or 'antisense' strand. This is transcribed in a 3′ to 5′ direction. The net effect is the synthesis of a primary RNA transcript consisting of a strand of mRNA which is an exact replica of the sense strand, with the exception that uracil replaces thymine.

The primary transcript is then processed to produce the mature mRNA for transportation into the cytoplasm. Processing involves the splicing out of the non-coding introns, and the addition of a protective 5′ guanine nucleotide to form a 5′ cap and of adenine residues at the 3′ end to form a poly-A tail.

The splicing out of the introns requires recognition of the GT and AG dinucleotides which mark the beginning and end of most intron boundaries. These are known as the *splice donor* and *splice acceptor* sites respectively. Mutations which involve these dinucleotides, so-called *splice-site mutations*, can have a serious effect on protein structure and function.

Translation—mRNA to protein

The mRNA passes through the nuclear membrane to the ribosomes in the cytoplasm, where it provides a template for the synthesis of a polypeptide chain. Ribosomes are made up of a complex of protein and ribosomal RNA (rRNA) which catalyses the synthesis of the polypeptide. This is achieved by binding of mRNA to transfer RNA (tRNA) on the surface of the ribosome. Each tRNA contains a set of three nucleotide bases, referred to as an **anticodon**, which complement a set of three bases in mRNA known as a **codon**. Each codon is specific for an amino acid. Thus each tRNA conveys a specific amino acid residue to its position on the mRNA template, resulting in the sequential synthesis of the encoded polypeptide chain.

The 'degenerate' genetic code

A codon consists of three bases which code for a specific amino acid. As there are four bases, there are $4 \times 4 \times 4 = 64$ possible codons. These represent the genetic code (Fig. 1.6). The code is described as being *degenerate* because there are 64 codons but only 20 amino acids, so several different codons specify the same amino acid. Translation always begins at a codon for methionine (AUG), which is often then removed before the completion of synthesis of the polypeptide. This is referred to as the *initiator codon*, and it determines the reading frame of the mRNA (see p. 11). Three combinations (UAA, UAG, UGA) specify *stop codons*, also known as *nonsense* or *termination codons*. These signal the end of translation.

Protein localization

The initial destiny of a protein is determined by short peptide sequences in each polypeptide chain. A specific *signal peptide* of 20–30 amino acids binds to a signal recognition particle, which in turn binds to a receptor on the endoplasmic reticulum. This enables the signal

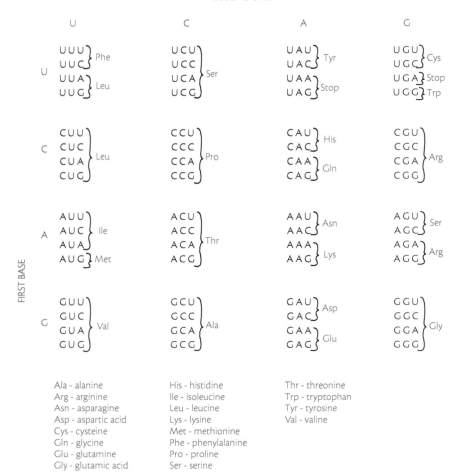

SECOND BASE

Fig. 1.6 The genetic code. Note that AGA, AGG, AUA, and UGA have a different interpretation in mitochondrial DNA, in which AGA and AGG encode STOP, AUA encodes methionine, and UGA encodes tryptophan.

Ala - alanine
Arg - arginine
Asn - asparagine
Asp - aspartic acid
Cys - cysteine
Gln - glycine
Glu - glutamine
Gly - glutamic acid

His - histidine
Ile - isoleucine
Leu - leucine
Lys - lysine
Met - methionine
Phe - phenylalanine
Pro - proline
Ser - serine

Thr - threonine
Trp - tryptophan
Tyr - tyrosine
Val - valine

peptide to cross into the lumen of the endoplasmic reticulum together with the rest of the translated polypeptide chain. The signal peptide is then cleaved out of the chain by signal peptidase. Other signal sequences, known as *stop transfer* and *start transfer* sequences, either prevent or facilitate transfer of the polypeptide across the cell membrane. Proteins destined for organelles within the cell, such as lysosomes and peroxisomes, require modification by addition of target sequences and transport to targeting signal receptors. Errors in these targeting systems account for a significant proportion of lysosomal and peroxisomal metabolic disorders (p. 214).

Post-translational modification

The addition of target sequences to facilitate localization is one example of how newly synthesized poly-

peptide chains can be modified, a process known as post-translational modification. Other examples include the formation of disulphide bonds, the cleavage of transport polypeptides, hydroxylation, and phosphorylation.

The formation of disulphide bonds is important for protein folding and the establishment of a three-dimensional protein structure. As an example, mutations in *FGFR3* that result in additional disulphide bonds between adjacent cell receptors lead to the lethal condition known as thanatophoric dysplasia (p. 172). Failure to cleave the transport peptide in procollagen to form mature collagen causes a severe joint hypermobility condition known as Ehlers–Danlos syndrome type VII (MIM 225410—'MIM' refers to 'OMIM' disease number, see Preface). Hydroxylation of lysine and proline is necessary for the formation of hydroxylysine and hydroxyproline, for which there are no genetic codes. Absence of the enzyme lysine hydroxylase results in another hypermobility syndrome (Ehlers–Danlos type VI—MIM 225400), which is also associated with ocular fragility. Phosphorylation is a key event in signal transduction, as discussed in Chapter 9.

> ### Key Point
>
> DNA is encoded through the processes of transcription, to produce mRNA, and translation, to produce protein. Proteins can undergo post-translational modification by addition of target sequences or by cleavage of transport polypeptides.

Mutations and their consequences

Mention has already been made of how errors at various stages of transcription and translation can result in clinical problems. In this section the full spectrum of mutations and their clinical consequences is reviewed.

A **mutation** can be defined as a structural change in genomic DNA which can be transmitted from a cell to its daughter cells. A mutation is described as **germline** if it is present in a gamete involved in a fertilization event leading to the conception of an individual in whom it is present in every cell. Alternatively, it is described as **somatic** if it occurs after conception and is therefore present in only a proportion of the body cells.

Structural classification of mutations

Structural changes in the genome can be subdivided into those that involve a change to all or part of a chromosome, as discussed in the next chapter, and those that involve a small segment of genomic DNA. This latter group can be considered under the following headings.

Point mutations

A point mutation involves a change to a single nucleotide and is also referred or as a nucleotide **substitution**. Substitution of a purine for a purine (changing A to G or G to A) or a pyrimidine for a pyrimidine (changing C to T or T to C) is known as a **transition**. Substitution of a purine for a pyrimidine, or vice versa, is known as a **transversion**. Approximately two thirds of substitutions are transitions and one third are transversions. The most common mutation observed in humans involves a transition in the C of a CpG dinucleotide pair to form TpG. This occurs much more often in the male germline than in the female germline, probably because the much greater number of cell divisions involved in spermatogenesis than in

oogenesis creates more opportunity for copy errors to occur in DNA replication.

Depending upon the effect on the protein product, nucleotide substitutions can be subdivided into silent, missense, nonsense, regulatory, and RNA processing mutations.

- *Silent mutations.* These are mutations which, because of the degeneracy of the genetic code, do not alter the amino acid being encoded. For example, valine is encoded by four different codons—GUU, GUA, GUC, and GUG (Fig. 1.6)—so any substitution of the third nucleotide will be silent.

- *Missense mutations.* A **missense** mutation results in the substitution of one amino acid for another. A *conservative* missense substitution has no effect on protein function, whereas a *non-conservative* missense mutation alters the function of the protein product, usually in a deleterious fashion. The common mutation that causes sickle-cell disease (p. 156) involves an A to T missense transversion, resulting in the substitution of glutamic acid by valine. Missense mutations represent 45–50% of all known pathogenic human mutations (Table 1.3).

- *Nonsense mutations.* A **nonsense** mutation creates a new stop codon (UAA, UAG, or UGA), thereby resulting in premature termination of translation. Nonsense mutations, also known as *premature termination mutations*, account for around 11% of all

TABLE 1.3 The spectrum of known pathogenic mutations in humans

Type of mutation	Proportion of total (expressed as %)
Point mutations	
Missense	47
Nonsense	11
Splice-site	10
Regulatory	1
Deletions and insertions	
Gross deletions	5
Small deletions	16
Gross insertions and duplications	1
Small insertions	6
Other rearrangements	3

Data obtained from the Human Gene Mutation Database (www.hgmd.org). Stenson PD, Ball EV, Mort M *et al.* (2003) The human gene mutation database (HGMD) : 2003 update. *Hum Mutat,* **21**, 577–581

known pathogenic mutations (Table 1.3). They usually result in rapid destruction of the mRNA product before translation by a process known as *nonsense-mediated mRNA decay*.

- *Regulatory mutations*. A regulatory mutation is one which involves the promoter or another regulatory sequence such as an enhancer, silencer, or locus control region. These are also known as *transcription mutations*. Most such mutations reduce mRNA production, but some can increase the level of transcrip-

tion, as occurs in hereditary persistence of fetal haemoglobin (p. 162).

- *RNA processing mutations*. These affect the processing of the primary RNA transcript to form mRNA, either by altering normal RNA splicing or by preventing either normal 5′ capping or 3′ polyadenylation. The so-called splice-site mutations usually occur in one of the GT or AG dinucleotides found at the beginning and end respectively of each intron. This results in skipping of an exon. Alternatively, a mutation may

A

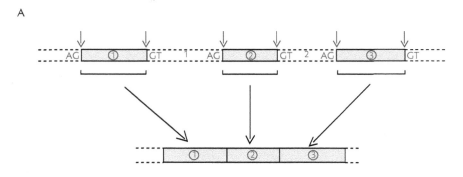

B

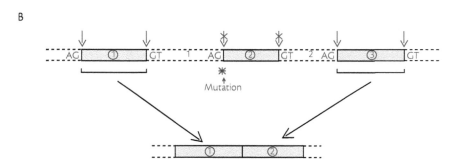

C

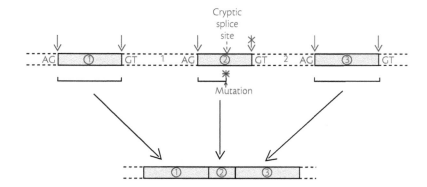

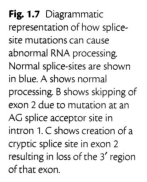

Fig. 1.7 Diagrammatic representation of how splice-site mutations can cause abnormal RNA processing. Normal splice-sites are shown in blue. A shows normal processing. B shows skipping of exon 2 due to mutation at an AG splice acceptor site in intron 1. C shows creation of a cryptic splice site in exon 2 resulting in loss of the 3′ region of that exon.

activate another normally silent (cryptic) splice site, leading to loss of part of an exon (Fig. 1.7). Splice-site mutations account for many cases of β-thalassaemia (p. 161).

Deletions and insertions

Deletions and insertions, which include duplications, can be subdivided on the basis of their size.

Small deletions and insertions

These involve loss (**deletion**) or gain (**insertion**) of a few nucleotides, probably as a result of *slippage* in DNA replication (Fig. 1.8). This is caused by mispairing between the complementary strands due to staggering of closely adjacent identical sequences.

If the number of nucleotides deleted or inserted in an exon is not a multiple of three, then the sequence of codons, known as the **reading frame**, is disrupted. This is referred to as a **frame-shift** and it usually results in a shortened (truncated) protein product. An *in-frame* deletion or insertion, i.e. one that does not disturb the reading frame, tends to have a less severe effect. One of the most common in-frame deletions encountered in western Europeans is the ΔF508 deletion which accounts for approximately 70% of all cystic fibrosis mutations in this population (p. 107).

Small deletions and insertions account for approximately 22% of all known mutations.

Large deletions and insertions

These range in size from 20 bp to 10 Mb, beyond which they become visible using a light microscope and are classified as chromosome abnormalities (Chapter 2).

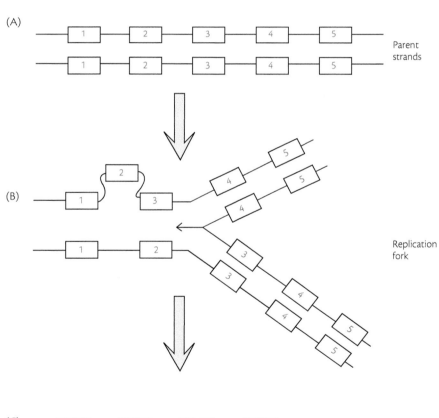

Fig. 1.8 Diagrammatic representation of how slipped strand mispairing ('slippage') during DNA replication can cause a deletion. Bubbling out in a parent strand (B) causes a deletion. Bubbling out in a daughter strand would lead to an insertion.

Large deletions and insertions account for approximately 5–6% of known pathogenic mutations. Most large deletions and insertions are caused by unequal crossing-over between homologous sequences. A small number of insertions are caused by retrotransposition.

Unequal crossing-over

Crossing-over between misaligned closely adjacent sequences which show close homology results in the formation of a deletion in one chromatid and a duplication in the other (Fig. 1.9). This mechanism accounts for the formation of haemoglobins Lepore and anti-Lepore (p. 155) and most α-thalassaemia mutations (p. 159). It is also thought to account for the generation of most of the chromosome microdeletion syndromes (p. 63) and for many cases of the Duchenne and Becker forms of muscular dystrophy (DMD and BMD—p. 109). Approximately two thirds of boys with DMD have a large

frame-shift deletion or duplication in the gene that encodes dystrophin on the X chromosome. In-frame deletions result in the much milder clinical features of BMD.

Retrotransposition

Although rare, a small number of large insertions have been shown to consist of transposable elements, SINES and LINES, which have moved from an inert region of the genome to become inserted into an exon elsewhere. *Alu* repeats have been identified as the cause of some cases of neurofibromatosis type 1 (p. 198). A truncated LINE sequence (Table 1.2) has been identified in several patients with haemophilia A.

Unstable trinucleotide ('triplet') repeat expansions

This unusual type of mutation was first identified in humans in the condition known as fragile X syndrome (p. 112). The mutation consists of an increase in the size

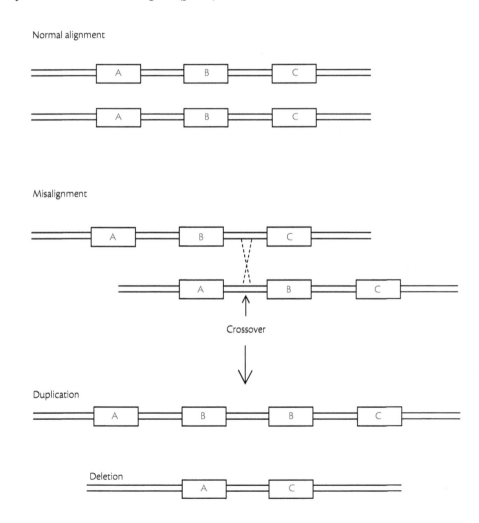

Fig. 1.9 Diagrammatic representation of how misalignment with 'unequal crossing-over' generates a deletion and a duplication.

of a triplet repeat, probably as a result of slippage (Fig. 1.8). When the mutation reaches a critical size, it either suppresses expression of the gene in which it is located, as in the fragile X syndrome, or it results in the production of an abnormal and potentially toxic protein, as in Huntington disease (p. 105). Trinucleotide repeats have been identified as the cause of approximately 30 disorders, most of which are extremely rare and involve the central nervous system. The more common trinucleotide repeat disorders are listed in Table 1.4. Other conditions caused by expansion of a triplet repeat include spinocerebellar ataxia types 1, 2, 3, 6, 7, 8, 10, and 12, Kennedy disease (MIM 313200), and a very rare form of progressive myoclonic epilepsy (MIM 254800).

Unstable, or *dynamic*, triplet repeat expansions can result in a disorder becoming more severe in successive generations. This is discussed further in the section on anticipation in Chapter 4.

Functional classification of mutations

Medical genetics is concerned with the functional consequences of human mutations. When considering the molecular pathogenesis of human mutations three functional effects can be recognized.

Loss-of-function

A mutation which results in reduced activity or quantity of the normal product is said to have a **loss-of-function** effect. Most mutations other than missense mutations result in a loss-of-function effect.

Loss-of-function mutations can be associated with either dominant or recessive inheritance. Most inborn errors of metabolism (Chapter 11) are caused by loss-of-function mutations which are harmless in the heterozygous (carrier) state, indicating that 50% of normal enzyme activity is usually sufficient for normal func-

tion. Thus these conditions show autosomal recessive inheritance (p. 73). However, in a few inborn errors, such as the porphyrias (p. 217), loss of 50% of enzyme activity has a deleterious effect, so inheritance is autosomal dominant (p. 71). In this situation the loss-of-function effect is referred to as **haploinsufficiency**.

Gain-of-function

A mutation which results in increased activity or quantity of the gene product is described as showing a **gain-of-function** effect. These are less common than loss-of-function mutations. Two examples considered elsewhere in this book are Huntington disease and achondroplasia/thanatophoric dysplasia. In Huntington disease (p. 105) the mutant protein forms cellular aggregates which have a neurotoxic effect. In achondroplasia and thanatophoric dysplasia (p. 172), mutations in *FGFR3* result in increased cell receptor activity leading respectively to mild and severe suppression of bone growth. Mutations exerting a gain-of-function effect usually cause conditions that show dominant inheritance.

Rarely, a different type of gain-of-function effect can occur when a mutation results in a gene product which has an entirely new activity. This is seen in acquired somatic mutations which contribute to cancer. These mutations often involve the formation of a new **chimeric** gene as a result of a translocation (p. 190).

Dominant negative effect

A mutation has a **dominant negative** effect if the product of the mutant allele interferes with the product of another normal allele, resulting in an overall adverse outcome. Dominant negative effects usually arise when the protein product has multiple components, i.e. it is *multimeric*, enabling the mutant protein to disrupt the function of the final product.

TABLE 1.4 Disorders caused by trinucleotide repeat expansions

Disorder	MIM	Inheritance	Repeat	Number of repeats		
				Normal	Mutable or premutation	Full mutation
Huntington's disease	143100	Autosomal dominant	CAG	10–25	26–35	36–120
Myotonic dystrophy	160900	Autosomal dominant	CTG	5–37	50–80 (mildly affected)	80–2000
Friedreich ataxia	229300	Autosomal recessive	GAA	6–34	36–100	100–1700
Fragile X syndrome	309550	X-linked	CGG	6–54	55–200	200–1000

Note that in many of the triplet repeat disorders there is an intermediate range of repeat sizes which result in either mild features or in a susceptibility to increase to a full mutation.

Osteogenesis imperfecta (MIM 166200), also known as brittle bone disease, provides an example of a condition which can result either from haploinsufficiency or from a dominant negative effect. Osteogenesis imperfecta is caused by mutations in one of the two genes *COL1A1* and *COL1A2*, which encode chains that combine as trimers to form type I collagen (Fig. 1.10). Mutations which result in failure of synthesis of a mutant chain, i.e. haploinsufficiency, cause mild osteogenesis imperfecta, with collagen that is reduced in quantity but is normal structurally. Mutations which result in a mutant chain that becomes incorporated in the mature collagen trimer cause a much more severe form of the disease by exerting a dominant negative effect, as illustrated in Fig. 1.11.

> **Key Point**
>
> Mutations can be classified on the basis of the structural change which occurs in the gene or on their functional effect. This can involve loss-of-function, gain-of-function, or a dominant negative disruption of a multimeric protein product.

Diagnostic molecular tests

Medical students are expected to have a working knowledge of common molecular analytical techniques and how these can be used to diagnose genetic disorders. The development of diagnostic laboratory tests for mutations in DNA dates back to 1975, when

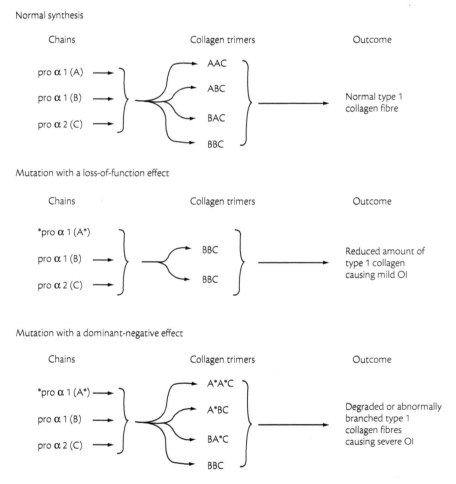

Fig. 1.10 Diagrammatic representation of different mutational mechanisms in osteogenesis imperfecta. The normal type I collagen molecule is a trimer made up of two proα1 chains and one proα2 chain. A mutant proα1 chain that becomes incorporated in the collagen trimers (dominant negative effect) has more severe consequences than a mutant chain which is not synthesized (haploinsufficiency or loss-of-function effect). Reproduced with permission from Suri M, Young ID (2004) *Genetics for pediatricians.* Remedica, London.

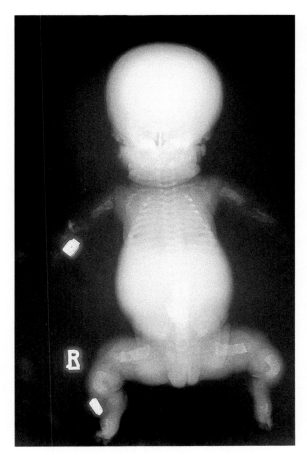

Fig. 1.11 Radiograph showing multiple fractures and abnormal modelling of the long bones in a baby with severe osteogenesis imperfecta caused by a mutation with a dominant negative effect.

Southern described a method, now referred to as **Southern blotting**, for transferring fragments of DNA on to a nylon filter, so that they can then be hybridized with complementary DNA or RNA sequences. Ten years later a method for amplifying DNA, known as the **polymerase chain reaction** (PCR), was developed and subsequently used to provide a rapid and sensitive test for the mutation that causes sickle-cell disease (p. 156).

Southern blotting and PCR remain the two standard procedures on which most DNA analysis is based. In addition, several newer techniques are now used routinely in diagnostic laboratories to screen or *scan* a gene in which a mutation is suspected.

Southern blotting

Southern blotting can be used to detect large deletions or insertions in a gene as well as large triplet repeat expansions. In the past it has also been used as an adjunct in linkage analysis.

The underlying principle is illustrated in Fig. 1.12. DNA from a patient is denatured by chemical or heat treatment and is then cleaved into a large number of small fragments using a **restriction enzyme**. Restriction enzymes, also known as restriction endonucleases, are derived from bacteria and cut DNA at specific restriction sites. The resulting fragments are separated out on an electrophoretic gel on which the small fragments move more quickly towards the anode. This gel is too fragile for further analysis so the DNA fragments are then blotted onto a nylon membrane known as a filter, and permanently fixed in position using ultraviolet light.

The next step involves the use of a **probe**, which has been modified (*labelled*) by addition of biotin or radioactive ^{32}P. A probe is a short double-stranded segment of DNA which is complementary to the gene that is being investigated. After denaturation to make it single-stranded, the labelled probe is added to the nylon filter where it hybridizes to complementary sequences. The DNA fragments, or *bands*, to which it has hybridized can be identified by autoradiography. If the patient's gene has a deletion then the probe will hybridize to a smaller band than normally. If the patient's gene contains an insertion or a triplet repeat expansion, then the probe will hybridize to a larger band than normally (Fig. 1.13).

If there is a mutation in a restriction site that is normally cleaved by the restriction enzyme, then this will result in hybridization of the probe to a different fragment. The presence or absence of a restriction site tightly linked to a disease gene is an example of the type of polymorphism (**restriction fragment length polymorphism—RFLP**) which can be used as a linked marker in clinical genetics for carrier detection, predictive testing, or prenatal diagnosis (p. 91).

Although as already indicated, 'Southern' was the name of the scientist who developed the technique, the scientific community subsequently attached the designations 'Northern' and 'Western' to two other analytical procedures.

◆ **Northern blotting** is almost identical to Southern blotting except that tissue mRNA rather than DNA is separated on the gel. After transfer to a filter the mRNA is hybridized to a cDNA probe. Successful hybridization indicates that the mRNA is being expressed in the relevant tissue. Close inspection of the gel indicates the size of the mRNA. The intensity of the hybridization signal gives an indication of the extent to which it is being expressed.

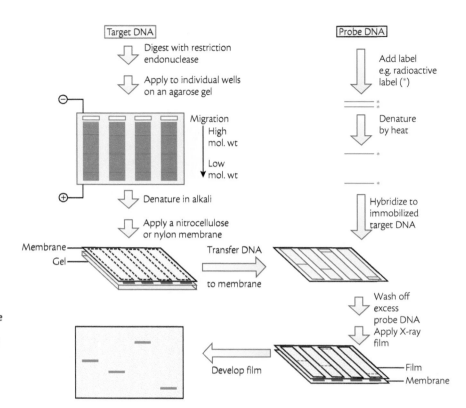

Fig. 1.12 Diagram showing the principle underlying Southern blot hybridization. Reproduced with permission from Strachan T, Read AP (2004) *Human molecular genetics 3.* Garland Science, London.

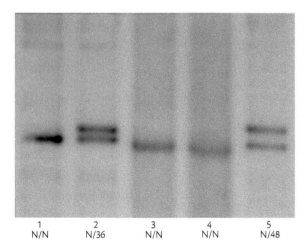

Fig. 1.13 Southern blot showing expansions in the gene which causes Huntington disease. Lanes 1, 3, and 4 show two normal alleles. Lane 2 shows one normal allele and a small expansion of 36 repeats. Lane 5 shows one normal allele and a full expansion of 48 alleles (see Table 5.3 for details of expansions that cause Huntington disease). Courtesy of Dr Judith Hudson, Department of Molecular Genetics, Leicester Royal Infirmary, Leicester.

♦ **Western blotting** is a technique for analysing protein expression in which proteins extracted from a tissue are fractionated on a gel according to size and then transferred on to a membrane. The protein of interest is then identified using a specific stain or antibody.

Polymerase chain reaction (PCR)

This technique enables a tiny quantity of a short sequence of DNA to be amplified by a factor of several million-fold, so that it can then be subjected to mutation analysis. The basic principle involves several cycles of denaturation, annealing of primers, and template extension, leading to an exponential increase in the amount of DNA being amplified (Fig. 1.14).

Each cycle starts with denaturation of the DNA by heating to form single-stranded DNA. Primers consisting of two short oligonucleotides each 15–30 nucleotides in length then anneal to their complementary sequences in the single-stranded DNA. A heat-resistant enzyme, *Taq* polymerase, then synthesizes new DNA strands from the end of the annealed primers using the

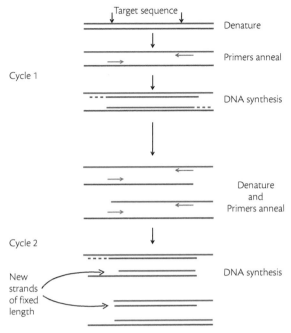

Cycle 1

Target sequence

Denature

Primers anneal

DNA synthesis

Denature
and
Primers anneal

Cycle 2

New
strands
of fixed
length

DNA synthesis

Fig. 1.14 Simplified diagrammatic representation of the PCR. By the end of the second cycle two new strands of fixed length, each containing the target sequence, will have been synthesized. By the end of the third cycle there will be six such strands and the number increases exponentially thereafter.

existing strands as templates. This completes the cycle, which is then repeated in an automated procedure. After 20–30 cycles the target sequence will have been amplified by a factor of over 1 million.

PCR has several advantages over Southern blotting (Table 1.5). In particular, much smaller amounts of DNA are required and the procedure can be completed overnight, in contrast to Southern blotting which takes at least 7 days. Technically PCR is less demanding than Southern blotting and no radiation is involved. However, a major disadvantage is that PCR cannot be used to amplify large alleles such as those containing an insertion or a large triplet repeat expansion. Small triplet repeat expansions, such as those which cause Huntington disease (p. 105) can be identified by PCR-based methods (Fig. 1.15), but large expansions, such as those which cause the fragile X syndrome (p. 112), can only be detected by Southern blotting.

> **Key Point**
>
> Most service laboratory molecular analysis is based on either the PCR or Southern blotting. PCR has the advantages that only a small amount of DNA is needed and a rapid result is obtained. Southern blotting has the advantages that large rearrangements can be detected and precise knowledge of the gene locus is not essential.

PCR-based mutation detection methods

Once a suitable sample of amplified DNA has been obtained, the products can be visualized by running them on a gel, followed by staining with ethidium bromide and inspection under ultraviolet light. Thus a small deletion, such as the common ΔF508 cystic fibrosis mutation (p. 108), will result in a product which is

TABLE 1.5	Advantages and disadvantages of the polymerase chain reaction and Southern blotting	
	Polymerase chain reaction	**Southern blotting**
Advantages	Only a small amount of DNA is needed	Can detect large rearrangements such as deletions, insertions, and large expansions
	No radiation involved	Knowledge of precise gene location is not necessary
	Rapid (<24 hours)	
	Automated and technically undemanding	
Disadvantages	Can only amplify short sequences	Requires several micrograms of DNA
	Therefore cannot be used to identify deletions or insertions > 1 kb	Takes 7–14 days
	High risk of contamination	Cannot detect point mutations, unless a restriction site is created or abolished
	Flanking DNA sequences must be known	

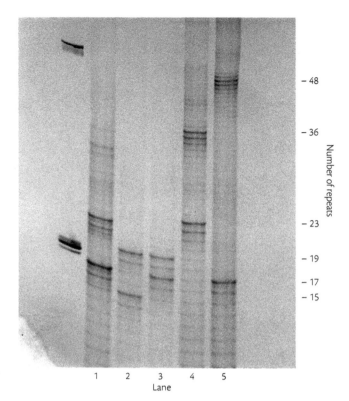

Fig. 1.15 PCR analysis of the gene which causes Huntington disease showing expansions ranging in size from 15 to 48 repeats. Courtesy of Dr Judith Hudson, Department of Molecular Genetics, Leicester Royal Infirmary, Leicester.

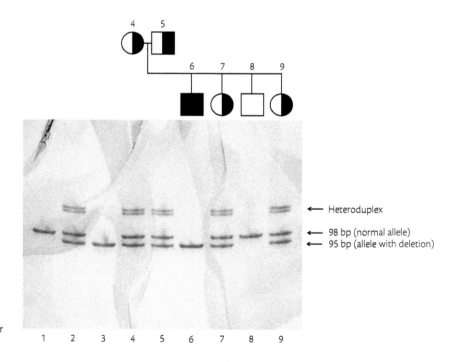

Fig. 1.16 PCR analysis of the common 3-bp deletion (ΔF 508) which causes cystic fibrosis. The analysis also detects a heteroduplex in heterozygotes. Courtesy of Dr Judith Hudson, Department of Molecular Genetics, Leicester Royal Infirmary, Leicester.

smaller than that of the corresponding normal allele (Fig. 1.16). Similarly, a small insertion or triplet repeat expansion will generate a bigger product than the normal allele. If it is known that a mutation alters a restriction site, as applies for the sickle-cell mutation (p. 157), then the amplified products can be digested with the appropriate restriction enzyme before being run and visualized on the gel.

PCR can be modified to amplify more than one exon in a gene using the technique known as *multiplex PCR*. This involves the use of several pairs of primers, each specific for a particular exon, in the reaction tube.

The primers are designed to ensure that the amplified products are of different lengths so that they will separate out clearly when run on a gel. Multiplex PCR has proved to be particularly useful for identifying deletions in boys with Duchenne muscular dystrophy (p. 111). A modified form of multiplex PCR known as *amplification refractory mutation system* (ARMS) multiplex can be used to screen a gene for several common mutations in one reaction. This involves the use of several pairs of primers, in which one primer is complementary to a common conserved sequence and the other is complementary to a short sequence of the gene

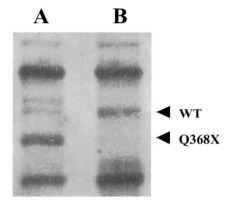

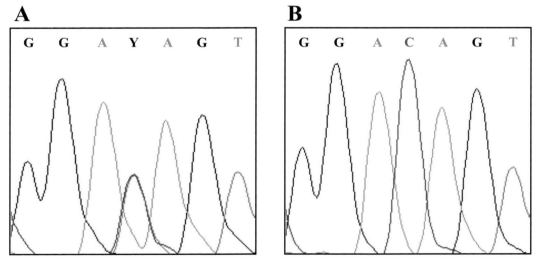

Fig. 1.17 SSCP analysis of the *GLC1A* gene in patients with primary open angle glaucoma. This figure illustrates the common heterozygous Q368X mutation (lane A), which causes incorporation of a premature termination codon and resultant truncation of the GLC1A protein. Lane B shows the banding expected with the homozygous wild-type (WT = normal) genotype. Direct sequence analysis confirms the mutation, arising from a C→T substitution (lower panel, indicated as Y = heterozygosity for C and T). Courtesy of Mrs Laura Baumber, Department of Genetics, Leicester University, Leicester.

which contains a common mutation. Amplification occurs only for a primer which anneals to a mutated sequence. ARMS multiplex is used to screen for the common cystic fibrosis mutations (p. 107) in affected individuals and in members of the general population.

PCR-based mutation scanning methods

Most single-gene disorders show marked mutational **heterogeneity**, indicating that they can be caused by many different mutations in the same gene. In cystic fibrosis, for example, over 1000 different mutations have been reported (p. 108). The ideal approach to the detection of an unknown mutation would be simply to sequence the gene, but as discussed below, this is expensive and time consuming. As a compromise, several relatively inexpensive methods have been developed to carry out an initial screen for the presence of a mutation.

Single-strand conformational polymorphism (SSCP) analysis

Single-stranded DNA folds to form a complex three-dimensional structure maintained by intramolecular bonds. Approximately 80% of all mutations alter the pattern of folding, i.e. the conformation, which in turn alters the mobility of the DNA strand on an electrophoretic gel. Thus a conformational change can be identified by running the denatured PCR product on a gel and comparing the band pattern with a normal control sample (Fig. 1.17). Any difference between the test and control samples suggests that a mutation is present in the amplified exon, also known as an *amplicon*, which can then be sequenced to determine the exact change.

Heteroduplex analysis

A heteroduplex consists of a hybrid segment of double-stranded DNA made up of two mismatched single strands. Heteroduplexes usually show an abnormal mobility on a polyacrylamide gel. Thus if PCR products from a suspected carrier are denatured and then allowed to reanneal so that the normal and mutant single strands form heteroduplexes, the presence of the heteroduplexes can be detected on the gel. The relevant exon can then be sequenced to identify the underlying mutation.

If DNA is amplified from a suspected homozygote with two identical mutations, or from a male with an X-linked recessive disorder, then a sample of normal PCR product must be added to the mixture to allow heteroduplexes to form.

Denaturing gradient gel electrophoresis (DGGE)

This technique is technically more demanding than SSCP and heteroduplex analysis but also has a higher sensitivity of over 90% for identifying mutations. The basic principle relies on the fact that normal and mutant sequences will show different band patterns on a gel. The PCR products are run on a special denaturing gradient gel which contains increasing concentrations of a denaturing chemical such as urea. When the double-stranded DNA denatures to form single strands, these stop at different points if the product has come from a heterozygote. Once again, the relevant exon can then be sequenced to identify the specific underlying mutation.

Denaturing high performance liquid chromatography (DHPLC)

This technique, which is also referred to as *WAVE* technology, is similar to DGGE in that it is based on the differential separation of mismatched heteroduplexes which form after the reannealing of normal and mutant DNA strands. PCR products are injected on to a column containing an increasing gradient of acetonitrite. The concentration of acetonitrite required to elute each homo- or heteroduplex varies according to the size and sequence of the fragment. DHPLC is an extremely sensitive method for detecting base substitutions and small deletions or insertions.

Key Point

Several PCR-based screening techniques have been developed which enable up to 90% of all mutations to be detected.

Sequencing

Sequencing can be viewed as the 'gold standard' method for detecting a mutation, but it is both time consuming and expensive. Increasingly, new automated methods are being used in service laboratories to identify rare mutations.

The standard technique for DNA sequencing is known as the *Sanger dideoxy sequencing method*. Frederick Sanger won two Nobel prizes, the first for developing protein sequencing and the second for developing DNA sequencing. The method relies on the fact that incorporation of a dideoxynucleotide (ddNTP) rather than a normal deoxynucleotide (dNTP) causes termination of chain synthesis because a ddNTP lacks a hydroxyl group and cannot form a phosphodiester bond.

In the dideoxy sequencing method, the sample of DNA to be sequenced is provided as a single strand. This acts as a template for the synthesis of a complementary strand using DNA polymerase and a mixture of dNTPs in four separate reactions in each of which a specific ddNTP is added (Fig. 1.18). In each reaction DNA synthesis ceases when a ddNTP is incorporated. As this is random, each reaction generates a series of different sized fragments. The fragments from each reaction are then run in separate wells on a long denaturing gel, enabling the position of each fragment to be compared and the sequence of the complementary strand to be read.

Automated sequencing now provides a much faster and cheaper alternative to the standard dideoxy method. This is based on the use of fluorescence labelling. Each of the four different nucleotides is labelled with a fluorochrome which emits a specific spectrum of light. The products of the sequencing reaction are subjected to electrophoresis, during which a monitor records the fluorescence signal generated by a beam of light projected on to each labelled nucleotide as it moves past a window. The emitted light is converted into an electronic signal which is computer analysed to produce a graph in which each of the four different nucleotides is indicated by a different coloured peak in a succession of intensity profiles (Fig. 1.19). This type of automated procedure played a major role in the sequencing of the human genome (Box 1.2).

DNA microarray ('chip') technology

A DNA microarray consists of a small device such as a coated glass microscope slide on to which large numbers of different DNA sequences have been positioned in the form of a grid. In one type of microarray tiny amounts of DNA are spotted onto the coated slide by an automated robot. In another type the different DNA sequences consist of oligonucleotides produced by *in situ* synthesis. This latter type is known as a *gene chip*.

Microarrays can be used to study the extent to which different genes are expressed in a particular tissue. mRNA from the tissue is converted to cDNA using reverse transcriptase and the cDNA is then labelled using different coloured fluorochromes. The tissue sample of labelled cDNA is then hybridized to the array and the pattern of hybridization is analysed by computer software. In this way an expression profile can be established for the relevant tissue.

Microarray technology is extremely powerful in that several hundred hybridizations can be analysed on a single slide, enabling the rapid analysis of gene expression in different tissues at different stages of development. Although still mainly a research tool, microarray technology will almost certainly be used in a service setting for multiple simultaneous mutation analysis, by spotting thousands of probes from different loci onto the array. Then PCR products from the relevant loci can be hybridized to the array, with the pattern of binding at each pair of spots giving a visual record of an individual's genotype at each locus.

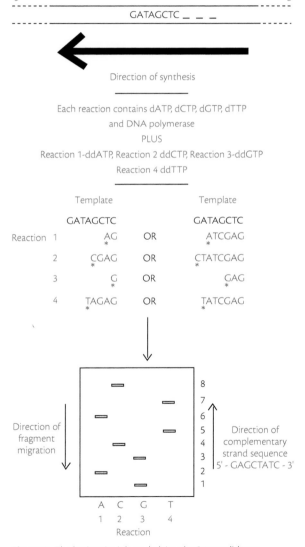

Fig. 1.18 The basic principle underlying the Sanger dideoxy sequencing method.

Summary

The human genome consists of a set of 23 chromosomes, containing 3.2×10^9 bp of DNA. The nucleotide bases in DNA are made up of two purines, adenine and guanine, and two pyrimidines, cytosine and thymine. In RNA thymine is replaced by uracil. Adenine always pairs with

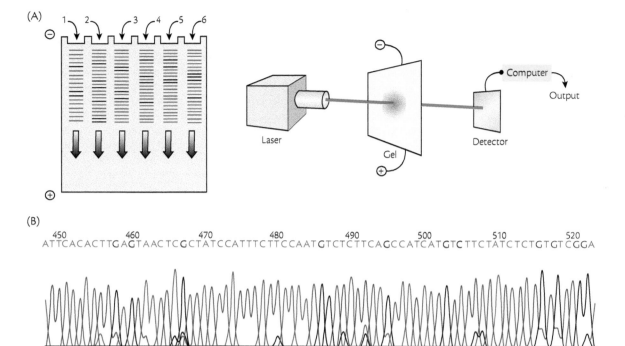

Fig. 1.19 (A) Automated DNA sequencing using fluorescent primers. The reaction products are loaded into single lanes of the electrophoresis gel or single gel capillaries. Four separate dyes are used as labels for the base specific reactions. During the electrophoresis run a laser beam is focused at a specific constant position on the gel. As the individual DNA fragments migrate past this position, the laser causes the dyes to fluoresce. Maximum fluorescence occurs at different wavelengths for the four dyes. The information is recorded electronically and the interpreted sequence is stored in a computer database. (B) Typical output of sequence data as a succession of dye-specific (base-specific) intensity profiles. Reproduced with permission from Strachan T, Read AP (2004) *Human molecular genetics 3*. Garland Science, London.

thymine in DNA and with uracil in RNA. Guanine always pairs with cytosine. A DNA molecule consists of two complementary strands in the form of a double helix held together by hydrogen bonds between the nitrogenous bases. DNA replication is semi-conservative in that one parent strand is retained in each daughter cell.

Approximately 50% of the genome consists of repetitive DNA including minisatellites, microsatellites, short and long interspersed nuclear elements, and triplet repeats. The genome also contains 30 000–35 000 genes made up of expressed sequences (known as exons), intervening sequences (known as introns), and regulatory sequences such as promotors, enhancers, and silencers. In transcription, DNA is converted to mRNA, which involves splicing out of the non-coding introns. In translation, mRNA acts as a template for the synthesis of a complementary polypeptide chain.

Mutations are heritable structural changes in DNA. These include point mutations involving a single nucleotide base pair and rearrangements such as deletions, insertions, and triplet repeat expansions. Mutations can also be classified on the basis of their functional consequences into loss-of-function, gain-of-function, and dominant negative.

Standard molecular diagnostic tests are based on the polymerase chain reaction (PCR) and Southern blotting. Newer PCR-based techniques have been developed for screening a gene for mutations. These include single-strand conformational polymorphism analysis, heteroduplex analysis, denaturing gradient gel electrophoresis and denaturing high performance liquid chromatography. New automated techniques have been developed to replace the older dideoxy method for gene sequencing.

Further reading

Epstein RJ (2002) *Human molecular biology.* Cambridge University Press, Cambridge.

Gardner A, Howell RT, Davies T (2000) *Biomedical sciences explained. Human genetics.* Arnold, London.

Lewin B (2000) *Genes VII.* Oxford University Press, New York.

Strachan T, Read AP (2004) *Human molecular genetics 3.* Garland Science, New York.

Multiple choice questions

1 In human DNA

  (a) cytosine always pairs with guanine

  (b) adenine always pairs with uracil

  (c) the leading strand forms continuously in replication

  (d) each chromatid consists of two new DNA strands

  (e) euchromatin is transcriptionally active

2 The sense strand of DNA reads CGGAACTCGCGC. Therefore the mRNA strand will read

  (a) CGGAACTCGCGC

  (b) CGGAACUCGCGC

  (c) GCCTTGAGCGCG

  (d) GCCUUGAGCGCG

  (e) GCGCGAGUUCCG

3 The following statements about mutations are correct:

  (a) an A to T substitution is a transversion

  (b) a missense mutation creates a stop codon

  (c) splice-site mutations can result in exon skipping

  (d) a 5-base insertion will disturb the reading frame

  (e) slippage can generate small deletions or insertions

4 Advantages of Southern blotting over the polymerase chain reaction (PCR) are that

  (a) it is technically easier

  (b) it can detect large rearrangements

  (c) no radiation is involved

  (d) it can be carried out overnight

  (e) large triplet repeat expansions can be detected

5 The following statements about molecular tests are correct:

  (a) Northern blotting involves analysis of RNA

  (b) Western blotting involves analysis of proteins

  (c) Southern blotting identifies all point mutations

  (d) the polymerase chain reaction requires only a small amount of DNA

  (e) single-strand conformational polymorphism (SSCP) analysis identifies all point mutations

Answers

1 (a) true—C pairs with G, A pairs with T

(b) false—adenine (A) pairs with thymine (T) in DNA

(c) true—the lagging strand forms in fragments

(d) false—each chromatid consists of one parent strand and one new strand

(e) true—heterochromatin is transcriptionally inactive

2 The correct answer is (b). mRNA is an exact replica of the sense strand with uracil (U) replacing thymine (T).

3 (a) true—substitution of a purine (A or G) for a pyrimidine (C or T) is known as a transversion

(b) false—a missense mutation creates a new amino acid

(c) true—if the mutation occurs at the beginning or end of an intron

(d) true—any deletion or insertion that is not a multiple of 3 will disturb the reading frame

(e) true—slippage can cause small deletions and insertions as well as triplet repeat expansions

4 (a) false—PCR is technically much easier

(b) true—this is its main advantage

(c) false—a radiolabelled probe is often used

(d) false—at least a week is required

(e) true—PCR can detect only small expansions

5 (a) true—Northern blotting tests for mRNA expression

(b) true—Western blotting tests for protein expression

(c) false—only point mutations which alter an RFLP are detected

(d) true—this is one of its main advantages

(e) false—SSCP analysis detects only around 80% of all mutations

Chromosomes and cell division

In the previous chapter the structure and function of the human genome were considered at the molecular level. This chapter describes the way in which the human genome is packaged and transmitted both from cell to cell and from generation to generation. **Chromosomes** are the packages of genetic material present within the cell nucleus, where they appear as thread-like structures when viewed using a powerful light microscope. The word chromosome is derived from the Greek for coloured body ('chroma soma'). Chromosomes are responsible for the transmission of DNA from parent cell to daughter cell through the process of mitosis and from parent to child through the process of meiosis. The chromosomal constitution of an organism is one of the main factors that determines its classification as a species. Primates such as chimpanzees and gorillas have 48 chromosomes in each cell, whereas humans have 46.

Cytogenetics is the name given to the study of chromosomes and their abnormalities. The human embryo is extremely sensitive to any abnormality that results in an increase or decrease in the correct amount of chromosome material in each cell. This is illustrated by the very large number of chromosomal abnormalities identified in humans, as outlined in the next chapter.

Chromosome structure

Chromosomes are made up of **chromatin,** which consists of an extremely long strand of DNA interwoven around structural proteins known as **histones.** The basic unit of packaging, known as a **nucleosome,** con-

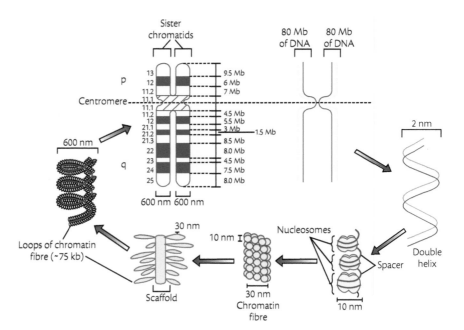

Fig. 2.1 Stages in the packaging of DNA to form chromosome number 17. Reproduced with permission from Strachan T, Read AP (2004) *Human molecular genetics* 3. Garland Science, London.

sists of a central core unit made up of 1.75 coils of the DNA strand around 8 histone proteins. Adjacent nucleosomes are linked together by a short 20–60-bp length of spacer DNA, giving an overall appearance similar to that of a string of beads (Fig. 2.1).

The next level of packaging (Table 2.1) consists of coiling of adjacent nucleosomes to form a chromatin fibre in the form of a superhelix 30 nm in diameter. Loops of chromatin fibre, each containing 20–100 kb of DNA, are attached to a central scaffold made up of nonhistone proteins. At these points of attachment there is an enzyme, topoisomerase II, which can induce and then repair double strand breaks, allowing loops of DNA to become further compacted around the central scaffold. Further coiling of the loop–scaffold complex results in the final appearance of a metaphase chromo-

some with an estimated packaging ratio of approximately 10 000 to 1. This elaborate supercoiling of DNA is sometimes referred to as the *solenoid* model of chromosome structure.

Chromosomes have three specific landmarks which are of particular importance: the centromere and the two terminal telomeres. The **centromere**, or primary constriction, is the point at which the sister chromatids are joined and at which microtubules attach to pull the sister chromatids apart in meiosis. In humans, centromeres consist of tandem head-to-tail repeats of 171 bp sequences classified as α satellites. The sequence differs from chromosome to chromosome, so specific centromeric DNA probes can be used for rapid FISH analysis (p. 29).

Centromeres contain a subdomain known as the *kinetochore*, which forms the site of attachment for the microtubules in the cell spindle during meiosis and mitosis. Rarely, autoantibodies can be generated against the kinetochore proteins resulting in a form of scleroderma known as the CREST syndrome (MIM 181750). (CREST stands for calcinosis, Raynaud phenomenon, esophageal involvement, sclerodactyly, and telangiectasia.) Kinetochore-specific autoantibodies are found in the serum of over 90% of patients with this condition.

Telomeres are specialized DNA-protein structures located at the end of each chromosome arm. Their function is to maintain the integrity of each chromo-

TABLE 2.1	Levels of DNA packaging to form a chromosome	
Level 1	Nucleosome	Consists of $1\frac{3}{4}$ coils of DNA strand around histone proteins
Level 2	Chromatin fibre	Superhelix made up of coiled nucleosomes
Level 3	Fibre–scaffold complex	Made up of loops of fibre around a central scaffold
Level 4	Chromosome	Formed by coiling of the fibre–scaffold complex

some and thus prevent random chromosomes fusing together. Human telomeres consist of a long (15 000–20 000 bp) tandem array of TTAGGG repeat sequences which are maintained by the enzyme telomerase. Normally telomerase is expressed only in embryonic cells; after birth, somatic cells gradually lose telomeric sequences during each cell cycle leading eventually to cell death. Activation of telomerase results in cell immortalization and is one of several genetic events seen in cells from malignant tumours.

The normal human chromosome complement

The normal human chromosome complement, or **karyotype**, consists of 46 chromosomes made up of 44 **autosomes** and 2 **sex chromosomes**, XX in the female and XY in the male (Box 2.1). Morphologically chromosomes are divided into **metacentric**, **submetacentric** and **acrocentric** depending upon the position of the centromere (Fig. 2.2). Acrocentric chromosomes sometimes have very small short arms consisting of stalk-like satellites. These contain several hundred copies of genes encoding ribosomal RNA. Unlike almost all other chromosome material, loss of these satellites, as can occur in a Robertsonian translocation (p. 40), is of no clinical significance. Note that these satellites have nothing to do with microsatellites (p. 143) or minisatellites (p. 143).

On the basis of individual length and the position of the centromere, chromosomes were originally divided into seven groups: A (1–3), B (4–5), C (6–12 and X), D (13–15), E (16–18), F (19–20), and G (21, 22, and Y). Subsequent developments in chromosome identification enabled individual chromosomes to be distinguished. These are numbered in decreasing order of

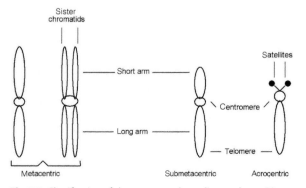

Fig. 2.2 Classification of chromosomes depending on the position of the centromeres.

BOX 2.1 **LANDMARK PUBLICATION: THE CHROMOSOME NUMBER OF MAN**

In 1956, Joe Hin Tjio and Albert Levan published a short paper that heralded a new era in the study of human chromosomes. Until that time it had proved very difficult to obtain satisfactory chromosome preparations for analysis and it was generally believed that the total number of chromosomes in each human cell was 48. Tjio and Levan modified the existing techniques by reducing the exposure of cells to hypotonic saline and adding colchicine to the culture medium. Hypotonic saline was known to have the extremely useful property of spreading the chromosomes, but overexposure had the undesirable effect of making the chromosome outlines blurred and therefore less distinguishable. The addition of colchicine resulted in an increase in the number of cells in metaphase, the period when chromosomes are maximally contracted and therefore most readily visible.

Using cultured human lung fibroblasts, Tjio and Levan were surprised to find that almost all of the 261 cells analysed had only 46 chromosomes. The quality of their chromosome preparations was such that they could classify each chromosome on the basis of the position of the centromere into M (median–submedian), S (subterminal), and T (nearly terminal). This corresponds to the classification still in existence today, i.e. metacentric, submetacentric, and acrocentric.

In their discussion the authors pointed out that colleagues elsewhere had observed chromosome counts of 46 in human preparations but had assumed that these must be incorrect in view of the prevailing dogma that the normal human chromosome count was 48. The authors concluded their paper with the rather modest statement that 'we do not wish to generalize our present findings into a statement that the chromosome number of man is 2n = 46, but it is hard to avoid the conclusion that this would be the most natural explanation of our observation'.

The technique described by Tjio and Levan remains the basis of standard chromosome analysis as carried out worldwide today.

Reference

Tjio JH, Levan A (1956) The chromosome number of man. *Hereditas*, **42**, 1–6.

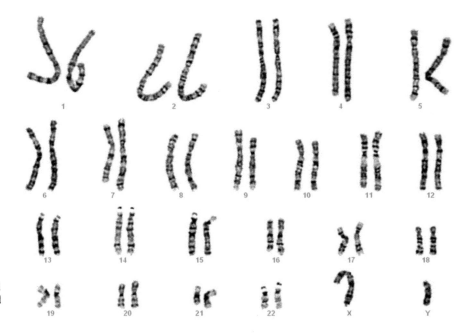

Fig. 2.3 A standard G-banded karyotype. Courtesy of Applied Imaging Corp.

size, and in a photograph are usually depicted as pairs (Fig. 2.3). The term **karyotype** can be used to describe either an individual's chromosome pattern or a photograph of their chromosomes. Generally chromosomes are shown as they appear in metaphase of mitosis, when they consist of two sister **chromatids** joined together at the centromere. These chromatids have arisen as a result of DNA synthesis before the cell enters the mitotic phase.

Normally half of an individual's chromosomes will have been inherited from each parent and half will be transmitted to each child. A single set of 23 chromosomes constitutes a **haploid** complement: the normal count of 46 chromosomes seen in all somatic cells represents the **diploid** complement. Pairs of chromosomes are referred to as **homologues**.

Chromosome nomenclature

A standardized system for naming chromosomes was established in 1971. Each chromosome has short (p) and long (q) arms separated by the centromere. Each arm is divided into regions numbered outwards from the centromere—p1, p2, etc.—and each region is subdivided into bands—e.g. p11, p12, etc.—again numbering outwards from the centromere. Thus p21 lies centromeric to p22 but telomeric to p12. The terms *proximal* and *distal* are sometimes used with reference to the location of a point on a chromosome. Proximal implies that the locus is in the half of the chromosome adja-

cent to the centromere, whereas distal indicates that it is located in the portion adjacent to the telomere.

> **Key Point**
>
> The normal human chromosome complement is made up of 46 chromosomes (46,XX in the female and 46,XY in the male), each of which consists of long and short arms separated by a centromere.

Methods of chromosome analysis

Routine analysis of human chromosomes can only be undertaken using dividing living cells. The most readily accessible source of such cells is venous blood, from which T lymphocytes can be extracted and subjected to a standard analytical procedure (Table 2.2). Other sources include fibroblasts cultured from skin and lymphocyte precursor cells from bone marrow. Cells from chorionic villi and fetal cells shed into the amniotic fluid, known as *amniocytes*, can be used to analyse the chromosomes of an unborn baby for the purpose of prenatal diagnosis.

A major step in the development of standard chromosome analysis, following its introduction in the early 1960s, was the discovery of special banding techniques enabling the reliable identification of individual chromosomes. The most common method in general

TABLE 2.2   Steps in routine chromosome analysis
Collect venous blood (2–5 ml) into lithium heparin
Add blood to culture medium containing phytohaemagglutinin, which stimulates T lymphocytes to divide.
Culture for 72–96 hours at 37°C
Add a synchronization agent such as thymidine after 48–72 hours
Add colchicine to arrest cells in metaphase
Add hypotonic saline to make cells swell
Fix cells with methanol and acetic acid
Drop cells on slide and add trypsin to denature protein
Stain chromosomes with Giemsa
Analyse metaphase spreads and issue report

TABLE 2.3   Chromosome nomenclature. See also Table 13.6 (p. 253).

Common abbreviations

p	short arm	i	isochromosome
q	long arm	ins	insertion
cen	centromere	inv	inversion
tel	telomere	r	ring
del	deletion	t	translocation
dup	duplication	rcp	reciprocal translocation
+	gain of a chromosome	rob	robertsonian translocation
–	loss of a chromosome	mos	mosaicism

Examples of formal karyotype reports

46,XX	Normal female
46,XY	Normal male
45,X	Female with Turner syndrome due to monosomy X
47,XY,+21	Male with Down syndrome due to trisomy 21
47,XXY	Male with Klinefelter syndrome
69,XXX	Triploidy
46,XX,del(5p)	Female with cri-du-chat syndrome
45,XX,der(14;21)(q10;q10)	Female with a balanced 14;21 Robertsonian translocation (by convention, the positions of the centromeres are designated as q10;q10)
46,XX,der(14;21)(q10;q10)+21	Female with an unbalanced 14;21 Robertsonian translocation resulting in Down syndrome

use involves the addition of trypsin to denature the protein content of each chromosome, followed by staining with a dye known as Giemsa. This binds with DNA to yield a specific pattern of light and dark bands. A well-stained slide of chromosomes in metaphase, commonly referred to as a *metaphase spread*, will enable chromosome analysis at a level of approximately 550 bands per cell. Other techniques which are used less frequently include Q (quinacrine) banding, R (reverse) banding, and C (centromeric) banding. These techniques are only used in special situations when cytogeneticists wish to study a particular region of interest.

Generally the total chromosome count is determined in a total of 10–15 cells, although if mosaicism is suspected the total number of cells counted is increased to at least 30. Detailed analysis of the banding pattern is undertaken in 3–5 cells. A formal report is then issued, which in normal individuals will read as 46,XX (female) or 46,XY (male) (Tables 2.3 and 13.6).

Fluorescence *in situ* hybridization

The technique of **fluorescence *in situ* hybridization** (FISH) involves the use of a DNA probe set or a single probe specific for an individual chromosome or chromosome region. This provides a very elegant and photogenic method for the rapid diagnosis of a suspected chromosome abnormality or rearrangement. Unlike conventional chromosome analysis, FISH can be used to analyse chromosomes in cells that are not dividing and even, in some circumstances, in cells that have died.

The basic principle involves adding a fluorescent dye to a DNA probe so that when the labelled probe, known as a *fluorochrome*, hybridizes to its complementary chromosome on a metaphase spread, it fluoresces when viewed under ultraviolet light. The most commonly used probes are single genomic sequences, which will hybridize to the region of a chromosome from which they have been derived. These unique sequence probes are extremely useful for detecting submicroscopic abnormalities such as microdeletions (Fig. 2.4).

Increasing use is being made of other types of probes in FISH analytical procedures (Table 2.4). Centromeric probes consist of α satellite DNA and can be used for the rapid prenatal diagnosis of autosomal and sex chromosomal trisomy syndromes. Subtelomeric probe sets have been developed to enable the detection of very subtle chromosome rearrangements and microdeletions/microduplications, which cannot be detected by conventional analysis (Fig. 2.5). This procedure detects

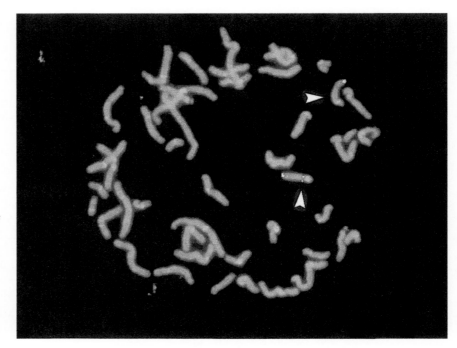

Fig. 2.4 FISH demonstration of a microdeletion. The green signal is generated by the chromosome 15 marker probe. The arrows point to the locus on chromosome 15 at which the red probe should hybridize. (Courtesy of Karen Marshall, Department of Cytogenetics, Leicester Royal Infirmary).

TABLE 2.4 Types of FISH probes used in chromosome analysis

Type of probe	Purpose
Unique sequence	To detect microdeletions and microduplications
Centromeric	For rapid prenatal diagnosis using cells in interphase
Telomeric	To detect cryptic rearrangements in children with unexplained mental retardation
Whole chromosome cocktails ('paints')	To identify chromosome material of unknown origin

abnormalities in around 5–10% of children with unexplained severe mental retardation. These subtle submicroscopic chromosome abnormalities are sometimes described as *cryptic* (Box 2.2). Finally, a selection of probes from a particular chromosome, referred to as a *chromosome paint*, can be used to clarify the nature of a subtle rearrangement, such as a translocation (Fig. 2.6), or an unidentified additional chromosome, such as a ring.

The great advantages of FISH over conventional chromosome analysis are that it can be used to study chromosomes in interphase, allowing a rapid result to be obtained, and that it can detect submicroscopic abnormalities, such as deletions, which are beyond the resolution of light microscopy. The main disadvantage of FISH is that it does not provide an overall analysis of each chromosome, so this procedure should be seen as an adjunct to conventional analysis rather than as a substitute or replacement for it.

Attempts have been made to get round this problem by using probes from each of the 24 different chromosomes (1–22, X, and Y) simultaneously, with each probe having its own specific spectrum of fluorescence wavelength. This technique, known as spectral karyotyping or M- (multicolour) FISH, yields beautiful photographs (Fig. 2.7) and is useful in cancer genetics, but does not as yet offer the degree of resolution provided by conventional chromosome analysis.

Comparative genomic hybridization (CGH)

This extension of FISH technology can be used to demonstrate the presence of gross chromosome deletions or duplications in tissues such as solid tumours which cannot be subjected to conventional chromosome analysis. Essentially CGH involves the competitive hybridization of a mixture of test DNA from the diagnostic sample and normal DNA from a control sample with normal human metaphase chromosomes. The test DNA is labelled with a green paint and the control

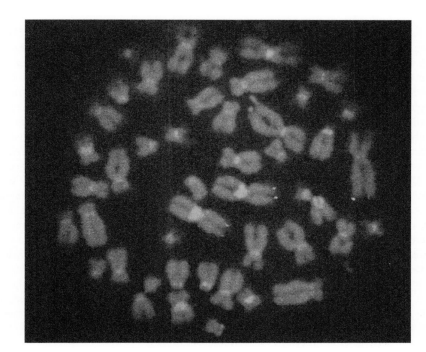

Fig. 2.5 FISH demonstration using telomeric probes for the short arm of chromosome 1 (red) and the long arm of chromosome 2 (green) showing monosomy for the terminal portion of chromosome 1p and trisomy for the terminal portion of chromosome 2q. Courtesy of the Department of Cytogenetics, City Hospital, Nottingham.

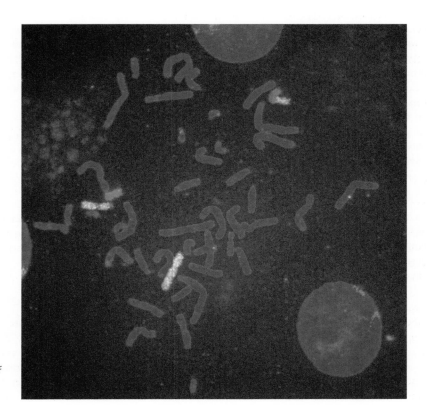

Fig. 2.6 Demonstration of a translocation between chromosomes 7 and 20 by chromosome painting. Courtesy of the Department of Cytogenetics, City Hospital, Nottingham.

BOX 2.2 CASE HISTORY: A CRYPTIC TRANSLOCATION

A young couple were referred for genetic consultation after the birth of their second child, who was noted to have a number of congenital abnormalities. Their first child, a boy, was entirely healthy, and initially the couple were not aware of any history of malformed or handicapped children on either side of the extended family. However, by the time they attended the genetics clinic, the father had learned from his mother that before his birth she had lost two babies in early infancy. Although information about these infants was limited, it was known that they had both been weak babies who were slow to gain weight and that one had also been a 'blue baby'. (This is a term used to describe cyanosis seen in babies with severe congenital heart defects.)

Examination of the couple's second child confirmed the presence of several abnormalities including a loud heart murmur and a cleft palate. Chromosome studies undertaken using lymphocytes from peripheral blood were reported as normal using conventional analysis. However, in view of the history of abnormalities in a previous generation, additional multisubtelomeric FISH studies were undertaken. These revealed that the baby had an abnormal karyotype in which the telomere at the end of the short arm of one number 1 chromosome was replaced by the telomere from the long arm of one number 2 chromosome (Fig. 2.5).

When the parents were seen again it was explained to them that their new baby had an unbalanced chromosome complement due to loss of material from the short arm of one number 1 chromosome with gain of material from the long arm of one number 2 chromosome. Both parents readily agreed to give blood for chromosome analysis. Two weeks later the cytogenetics laboratory reported that the father had been found to carry a balanced cryptic reciprocal translocation involving the telomeric regions of the short arm of one number 1 chromosome and the long arm of one number 2 chromosome. Subsequent studies revealed that this was also carried by the baby's paternal grandmother.

Cryptic chromosome rearrangements have been found to be an important cause of unexplained mental retardation in children, particularly when they are associated with multiple physical abnormalities and/or a positive family history. Their recognition enables all potential carriers in a family to be identified and to be counselled about possible risks of having abnormal children in future pregnancies. In this family the parents accepted the offer of prenatal testing in their next pregnancy and were keen that their healthy first child should also be tested. As this child showed no evidence of any developmental problems, and was therefore extremely unlikely to have an unbalanced karyotype, the parents agreed that testing should be delayed until he reached the age of consent when he would be able to make his own informed decision. It was recognized that with the benefit of hindsight it was very likely that the two babies lost in a previous generation had also had an unbalanced cryptic translocation and, with the family's full agreement, chromosome analysis was offered to all members deemed to be at risk of being carriers.

Reference

Knight SJL, Regan R, Nicod A *et al.* (1999) Subtle chromosome rearrangements in children with unexplained mental retardation. *Lancet*, **354**, 1676–1681.

DNA with a red paint. If the test sample contains more or less DNA from a particular chromosome region than the control DNA, this is revealed by an increase or decrease in the green to red fluorescence ratio as revealed by computer-based image processing (Fig. 2.8).

Key Point

Standard chromosome analysis is undertaken on cultured lymphocytes and requires a minimum of 3 days. Both FISH and CGH can be used to study chromosomes in cells which are not dividing.

The technological complexity of this technique, coupled with its expense, means that its use tends to be limited to the study of chromosome abnormalities in tumours.

The cell cycle and cell division

Most cells undergo regular division to sustain a total number of approximately 75×10^{12} cells in the human body. A knowledge of the cell cycle and the process of cell division is essential for an understanding of how chromosome abnormalities arise and of the genesis of many forms of human cancer.

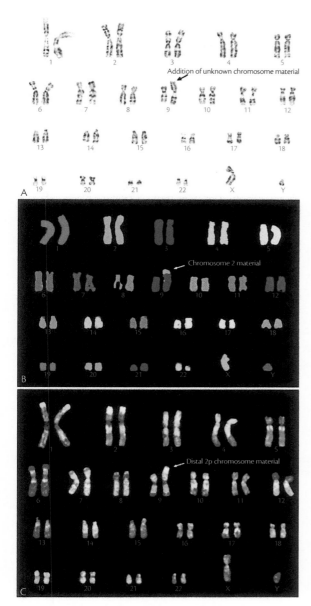

Addition of unknown chromosome material

Chromosome 2 material

Distal 2p chromosome material

Fig. 2.7 M-FISH showing identification of unknown chromosome material on chromosome 9. The lowest karyotype shows a special form of M-FISH, known as Rx-FISH, which can be used to identify specific chromosome regions and bands. Adapted with permission from Lee C, Murray MF, Marsden D, Irons M, Wilkins-Haug LE (2001). Multicolour karyotyping. *Lancet*, **357**, 1240. Kindly provided by Applied Imaging Corp and Dr Charles Lee, Harvard Medical School, Boston.

Interphase

Interphase is the period in the cell cycle between successive mitotic divisions. It consists of three main phas-

es: GAP1 (G_1), DNA synthesis (S), and GAP2 (G_2) (Fig. 2.9). G_1 usually lasts for around 10–12 hours, during which various *checkpoint* and *gatekeeper* genes, such as the retinoblastoma gene *RB1* (p. 191), regulate and monitor progress into the next phase of the cell cycle. Some cells, notably neurons and red blood cells, cease dividing at this point and enter an arrested stage known as G_0. One of the major challenges facing medicine is to try to reactivate these cells for the treatment of conditions such as transection of the spinal cord. Other cells which are deemed to be damaged irreversibly undergo cell death, a process known as **apoptosis**. Programmed apoptosis is an important component of several normal developmental processes such as the loss of normal embryonic webbing between the developing digits and the canalization of solid structures such as the urethra. Failure of apoptosis is a key factor in the genesis of cancer (p. 196).

After G_1 the cells move into the S phase, during which the extended DNA molecule in each chromosome replicates itself to yield two sister chromatids joined at the centromere. DNA synthesis commences at several hundred different sites known as *origins of replication*. On average the S phase lasts approximately 6–8 hours.

The next phase, G_2, is relatively short with an average duration of 2–4 hours. During G_2, DNA repair genes encode enzymes that correct errors which have occurred during DNA synthesis in the S phase. Mutations in the genes that encode these DNA repair enzymes result in a number of conditions associated with chromosome breakage (p. 66) and in an inherited form of colon cancer known as hereditary non-polyposis colon cancer (p. 200).

The typical cell cycle lasts approximately 24 hours, although in some cells with a rapid turnover, such as those lining the gastric mucosa, it is much more rapid than in cells such as myocytes or osteocytes. The rate of cell turnover, as indicated by the mitotic index, is related to the likelihood of an error occurring in DNA synthesis/repair and correspondingly to the risk of cancer in a relevant organ or tissue.

Mitosis

Mitosis can be defined as the process of cell division which results in the production of two genetically identical cells from a single parent cell (Fig. 2.10). Before mitosis, in the S phase of interphase, each chromosome has undergone DNA replication to generate two identical sister chromatids. During mitosis each pair of sister chromatids separates to segregate into independent

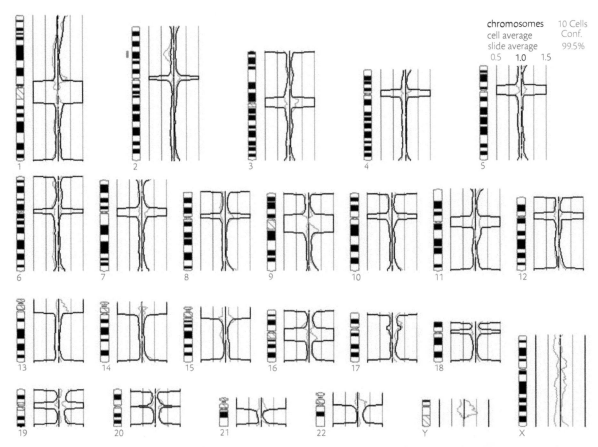

chromosomes   10 Cells
cell average   Conf.
slide average   99.5%
0.5   1.0   1.5

Fig. 2.8 Comparative genomic hybridization demonstrating a small deletion in the short arm of chromosome number 2. Courtesy of Dr T Gerdes, Department of Clinical Genetics, Juliane Marie Center, Rigshospitalet, Copenhagen and Applied Imaging Corp.

daughter cells. The process occurs over a period of 1–2 hours and is continuous, but for descriptive purposes four separate stages are recognized:

+ *Prophase.* The chromosomes start to condense, becoming more visible under light microscopy, and the nuclear membrane disappears. The spindle fibres start to radiate out from two centrioles, located at opposite poles of the cell, to become attached to each chromosome at the centromere.

+ *Metaphase.* The chromosomes are now aligned along the equator of the cell where they are attached by the spindle to each centriole. They are also maximally condensed and thus at their most visible. This is why chromosome analysis is usually undertaken at this stage.

+ *Anaphase.* The sister chromatids separate at the centromere to become independent daughter chromo-

somes and are pulled by the spindle fibres to opposite ends of the cell.

+ *Telophase.* New nuclear membranes form around each daughter set of 46 chromosomes. Cytoplasmic cleavage occurs, a process known as cytokinesis, leading to the formation of two new cells each of which has its own nucleus containing 46 chromosomes.

Meiosis

Meiosis is the name given to the process whereby diploid cells in the ovaries and testes, known as primary oocytes and spermatocytes respectively, divide to form haploid gametes. Two cell divisions are involved: meiosis I, which is also known as the reduction division, and meiosis II, which is a simple mitotic division (Fig. 2.11). The process of meiosis occurs as the final

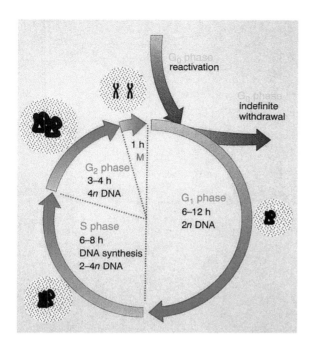

Fig. 2.9 The cell cycle. M = mitosis. 2n and 4n represent the quantity of DNA in each cell before and after DNA synthesis. Reproduced with permission from Lewin B (2000) *Genes VII*. Oxford University Press, New York.

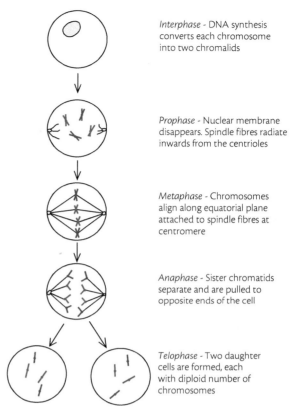

Interphase - DNA synthesis converts each chromosome into two chromalids

Prophase - Nuclear membrane disappears. Spindle fibres radiate inwards from the centrioles

Metaphase - Chromosomes align along equatorial plane attached to spindle fibres at centromere

Anaphase - Sister chromatids separate and are pulled to opposite ends of the cell

Telophase - Two daughter cells are formed, each with diploid number of chromosomes

Fig. 2.10 Diagrammatic representation of mitosis showing division of a cell with two pairs of chromosomes.

stage of gamete formation and takes approximately 60 days in men and up to 50 years in women.

Meiosis I

Conventionally this is divided into the stages of prophase I, metaphase I, anaphase I, and telophase I, although, as with mitosis, the process is essentially continuous.

Prophase I

As in mitosis, the chromosomes enter this phase having already undergone DNA replication to form sister chromatids. Homologous chromosomes pair by a process known as **synapsis** to form **bivalents**, which are held together by **synaptonemal complexes**. Material is exchanged between non-sister chromatids by the process known as **crossing-over** or **recombination**. Non-sister chromatids are chromatids from opposing homologues. Approximately two recombination events occur per bivalent. The chromatids then gradually separate but remain attached to points where recombination has occurred. These points are known as **chiasmata**. This relatively lengthy phase of meiosis is usually considered as consisting of five stages.

- leptotene—the chromosomes condense
- zygotene—synapsis occurs
- pachytene—crossing-over occurs
- diplotene—separation commences
- diakinesis—further condensation occurs.

Metaphase I, anaphase I, telophase I

These processes are very similar to those that occur in mitosis. However, each daughter cell, known as a secondary oocyte or spermatocyte, contains only 23 chromosomes, each of which consists of two joined sister chromatids.

Meiosis II

This is essentially a mitotic division during which each chromosome divides, with the chromatids, now independent daughter chromosomes, moving to opposite poles to form mature gametes known as ova or spermatids (which mature, without a division, into spermatozoa). In the female, each complete meiosis leads to the formation of only one mature ovum, with the other three much smaller daughter cells consisting largely of nuclei and being known as *polar bodies*.

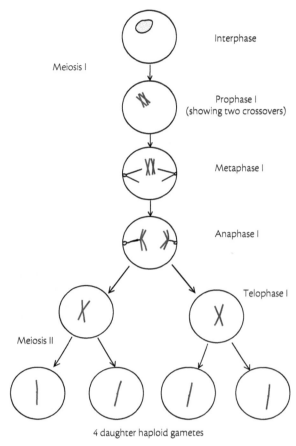

Meiosis I

Interphase

Prophase I
(showing two crossovers)

Metaphase I

Anaphase I

Telophase I

Meiosis II

4 daughter haploid gametes

Fig. 2.11 Diagrammatic representation of meiosis showing behaviour of a single pair of homologous chromosomes.

TABLE 2.5	Comparison of mitosis and meiosis	
	Mitosis	**Meiosis**
Number of cell divisions	1	2
Duration	1–2 hours	60 days in males 10–50 years in females
Location	All dividing somatic cells and gametogenesis up to formation of primary oocytes and spermatocytes	Final two divisions in gametogenesis
Crossing-over between non-sister chromatids	Very uncommon	Approximately 2 crossovers per bivalent
Outcome	2 identical daughter cells, each with 46 (n = 2) chromosomes	4 spermatids or 1 ovum and 3 polar bodies, all with 23 (n = 1) chromosomes

fundamental differences between the processes of mitosis and meiosis are summarized in Table 2.5.

Gametogenesis

An appreciation of how mature ova and spermatozoa are formed provides insight into the origin of many genetic and chromosome abnormalities. Although they share some similarities, the processes of gamete formation in males and females show several fundamental differences (Fig. 2.12).

Oogenesis

Most of this process occurs before birth in the developing ovaries. During embryonic and fetal life the primordial germ cells give rise to ova progenitors known as **oogonia**. These undergo 20–30 mitotic divisions before maturing into primary oocytes. In late fetal life these enter the first meiotic division. During prophase I they enter a stage of maturation arrest, known as *dictyotene*, in which they remain until ovulation occurs up to 50 years later. This culminates in completion of meiosis I with formation of a single secondary oocyte and a polar body. Meiosis II is completed after fertilization in the Fallopian tube.

The very long interval between the onset and completion of meiosis, which can vary from 10 to over 50 years, is probably relevant to the well-documented strong association between advancing maternal age and increased risk of non-disjunction. There is only a

After meiosis, the diploid chromosome complement is halved so that each mature gamete receives only one member of each homologous pair (i.e. a haploid complement). The diploid number is restored at fertilization, with each parent making an equal contribution of chromosomes. In addition, the diversity of the genetic constitution in each gamete is assured by the independent segregation of the chromosomes to opposite poles and hence to daughter cells. The probability that any two gametes will inherit exactly the same set of chromosomes is $(1/2)^{23}$ or approximately 1 in 8 000 000. Genetic diversity is further increased by the process of recombination. The

Key Point

In mitosis each daughter cell receives the diploid chromosome complement. In meiosis each gamete receives a haploid chromosome complement.

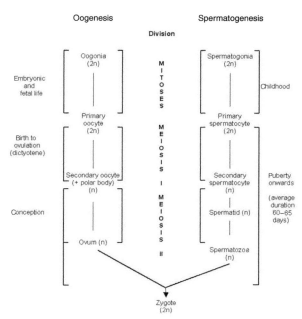

Fig. 2.12 Diagrammatic representation of gametogenesis showing the differences between oogenesis and spermatogenesis.

weak and clinically insignificant association between advancing paternal age and the risk of non-disjunction.

Spermatogenesis

Spermatogonia enjoy a relatively inert existence in the male testis, undergoing up to 30 mitotic divisions by the onset of puberty. Thereafter rapid division leads to the formation of primary spermatocytes, which enter meiosis I and emerge 60–65 days later as mature haploid spermatids. These then mature into spermatozoa.

It has already been mentioned that paternal age bears little if any relationship to the incidence of non-disjunction as gauged by offspring with trisomy. However, children of older fathers show a significant increase in the incidence of new dominant mutations, almost certainly as a consequence of copy errors occurring during DNA replication in the many mitotic divisions involved in the manufacture of mature spermatozoa. The numerical risk of non-disjunction in offspring of older mothers exceeds that of new mutations in the children of older fathers, although the long time interval that would have to elapse before all new muta-

> **Key Point**
>
> Most trisomy syndromes arise due to an error in oogenesis. Most new point mutations arise due to an error in spermatogenesis.

tions in offspring would be recognized makes it very difficult to determine the overall risk.

Chromosome abnormalities

Chromosome abnormalities are an important cause of morbidity and mortality in humans. They can be classified on the basis of whether they are numerical, involving for example the presence of an additional chromosome, or structural, involving a change or changes in the integrity of one or more chromosomes. A third group, classified as mixoploid, consists of the presence in an individual of two or more cell lines with different chromosome constitutions.

Clinically, the outcome of a chromosome abnormality depends on whether there is any loss or gain of chromosome material, and on this basis chromosome abnormalities can be classified as balanced or unbalanced. Unbalanced chromosome abnormalities almost always have an adverse effect on growth and development. In contrast, balanced rearrangements usually have no effect upon the individual in whom they are present, unless the function of an important gene has been disturbed by damage at one of the chromosome breakpoints. However, the importance of recognizing balanced rearrangements lies in their ability to cause problems in future generations if a child inherits an unbalanced chromosome complement from a balanced carrier parent.

Numerical chromosome abnormalities

Any alteration in the normal diploid number of 46 chromosomes is referred to as **heteroploidy**. Two forms of heteroploidy can occur: aneuploidy and polyploidy (Table 2.6). **Aneuploidy** refers to loss or gain of one or more chromosomes. Examples of aneuploidy include loss or gain of a single chromosome, referred to as **monosomy** or **trisomy** respectively, gain of two copies of a single chromosome, known as **tetrasomy**, and gain of two different chromosomes, known as **double trisomy**.

Polyploidy refers to the presence of one or more additional sets of the haploid (n = 23) chromosome complement. Two types of polyploidy have been identified in human miscarriages: **triploidy** (3n) with 69 chromosomes and **tetraploidy** (4n) with 92 chromosomes.

Trisomy

Gain of a single autosome is usually incompatible with survival beyond early pregnancy, as illustrated by the observation that trisomy constitutes approximately half of the chromosome abnormalities identified in spontaneous first-trimester miscarriages. However, trisomy 13 (Patau syndrome), trisomy 18 (Edwards

TABLE 2.6 Numerical chromosome abnormalities

Abnormality	Example	Syndrome
Aneuploidy		
Trisomy	47,XX,+13	Patau syndrome
	46,XY,+18	Edwards syndrome
	47,XX,+21	Down syndrome
	47,XXY	Klinefelter syndrome
	47,XYY	XYY syndrome
	47,XXX	Triple X syndrome
Tetrasomy	48,XXXX	Tetrasomy X
Double trisomy	48,XXY,+18	
Monosomy	45,X	Turner syndrome
Polyploidy		
Triploidy	69,XXX	
Tetraploidy	92,XXXX	
Mixoploidy		
Mosiaicism	47,XY,+21/46,XY	
Chimaerism	46,XX/46,XY	

syndrome), and trisomy 21 (Down syndrome) represent recognized clinical syndromes and are discussed further in the next chapter. Of these, trisomy 21 (Fig. 2.13) is by far the most common. The presence of an additional sex chromosome, X or Y, has little effect on embryonic survival and all of the sex chromosome trisomy syndromes (47,XXY; 47,XYY; 47,XXX) are compatible with relatively normal development, as reviewed in the next chapter.

The mechanism underlying most cases of trisomy is **non-disjunction**. This involves failure of a pair of chromosomes to separate normally during one of the meiotic divisions. In approximately 90% of cases of Down syndrome the non-disjunctional event occurs in anaphase of maternal meiosis I, so the fetus inherits both copies of its mother's number 21 chromosomes, plus one from its father, resulting in trisomy. If non-disjunction occurs in maternal meiosis II, then the fetus inherits two copies of one of its mother's number 21 chromosomes. Either way, the clinical outcome is that of the typical picture of Down syndrome.

The underlying cause of non-disjunction is not clear. The fact that Down syndrome is more common in the offspring of older mothers, and that the non-disjunctional event has usually occurred in the mother, indicates that the lengthy period during which meiosis occurs in women, as compared to men, is relevant. Significant alterations in patterns of recombination involving chromosome 21 have been noted in children with Down syndrome, and in particular it appears that recombination close to the centromere makes the chromosomes more susceptible to non-disjunction. Another possibility is that the chromosomes never joined normally as a bivalent, so that their destiny at the end of

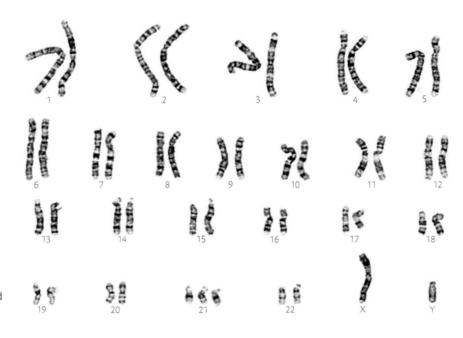

Fig. 2.13 Karyotype showing trisomy 21. Courtesy of Applied Imaging Corp.

meiosis I is random. In this situation the term non-disjunction is inappropriate as the chromosomes have never actually joined before becoming 'disjoined'.

Monosomy

Monosomy for any of the autosomes is incompatible with survival. Monosomy for an X chromosome (45,X) with loss of the X or Y chromosome from the other gamete, results in Turner syndrome (p. 000). Monosomy represents the outcome complementary to trisomy from non-disjunction. It can also arise due to the failure of a chromosome to reach the daughter cell nucleus due to damage to part of the spindle. This is referred to as **anaphase lag**.

Polyploidy

Both triploidy (Fig. 2.14) and tetraploidy are incompatible with long-term survival and are relatively common findings in spontaneous miscarriages. Triploidy can be paternal in origin due to the fertilization of a single ovum by two sperm, or maternal in origin due to the retention of a polar body in the secondary oocyte or ovum. These have slightly different consequences, as discussed in the next chapter.

Structural chromosome abnormalities

These result from damage to the structural integrity of one or more individual chromosomes, usually as a result of breakage with reconstitution in a new form (Fig. 2.15). Balanced rearrangements are usually harmless, whereas unbalanced rearrangements, particularly those involving one or more of the autosomes (numbers 1–22), generally cause severe developmental abnormalities (Table 2.7).

Translocations

A **translocation** can be defined as the transfer of a segment of chromatin from one chromosome to another. In a **reciprocal translocation** segments are exchanged between two or more chromosomes. When both 'derivative' chromosomes are present, the rearrangement is

TABLE 2.7 Structural chromosome abnormalities	
Balanced	**Unbalanced**
Deletion and insertion	Deletion or insertion
Inversions—paracentric and pericentric	Inversions—if recombination has occurred in loop in meiosis
Reciprocal translocations with both rearranged chromosomes	Reciprocal translocations with only one rearranged chromosome
Robertsonian translocations (total count = 45)	Robertsonian translocations (total count = 46) Isochromosomes Ring chromosomes

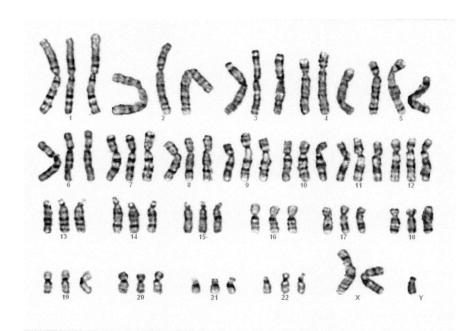

Fig. 2.14 Karyotype showing triploidy. Courtesy of Applied Imaging Corp.

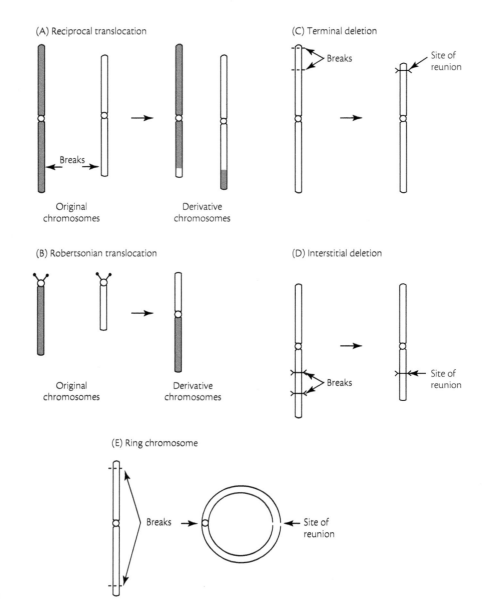

Fig. 2.15 Diagrammatic representation of different types of structural chromosome abnormalities.

balanced. However, transmission of only one of the rearranged chromosomes will result in an unbalanced karyotype, as discussed in the next section.

A **Robertsonian translocation** involves fusion of two acrocentric chromosomes (numbers 13, 14, 15, 21, 22, and Y) at or close to their centromeres with loss of the short arms. An individual with a balanced Robertsonian translocation will have a total count of 45 chromosomes, but the karyotype is balanced because the correct amount of chromosome material is present, as loss of the short arms is of no clinical significance. However, as with carriers of reciprocal translocations, carriers of balanced Robertsonian translocations are at risk of having children with unbalanced chromosome complements. Surveys of normal newborn infants have shown that balanced Robertsonian translocations are approximately 10 times more common than balanced reciprocal translocations, with incidence figures of 1 in 1000 and 1 in 10 000 respectively.

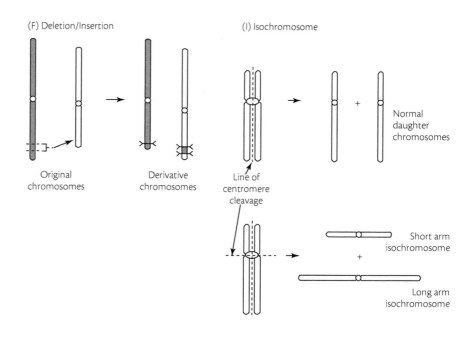

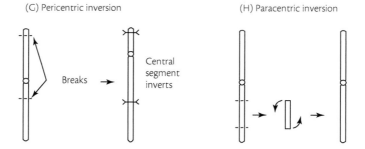

Fig. 2.15 *Cont'd.*

Deletions

A **deletion** involves loss of chromosome material. It can be terminal or interstitial. If the deleted material is inserted elsewhere in the chromosome complement, then the combined deletion–insertion is balanced. When present in isolation the deletion results in an unbalanced karyotype. Deletions which are smaller than 5–10 Mb cannot be detected using conventional light microscopy. These are referred to as **micro-deletions**.

Insertions

An **insertion** results from the movement of a segment of one chromosome into another. As indicated above, a deletion–insertion is balanced, whereas an isolated deletion or insertion is unbalanced. Carriers of a balanced deletion–insertion are at high risk of having an abnormal child.

Rings

A **ring** chromosome forms when breaks occur at both ends of a chromosome with loss of the terminal segments and union of the sticky ends. Ring chromosomes are unbalanced as a result of loss of the terminal segments.

Inversions

An **inversion** results when two breaks occur in a chromosome and the intervening segment turns upside down before being re-inserted. An inversion is described as **pericentric** if it involves the centromere (peri = round or about). A **paracentric** inversion involves only one arm of a chromosome (para = beside

or beyond). Inversions are balanced rearrangements, but carriers of both peri- and paracentric inversions are at risk of producing gametes with unbalanced chromosome complements.

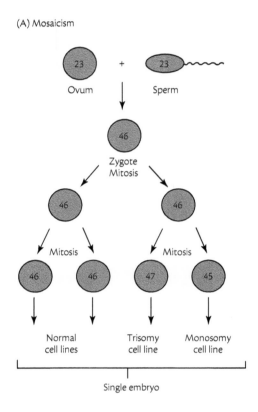

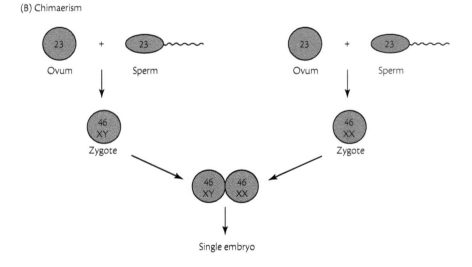

Fig. 2.16 Demonstration of the difference between (A) mosaicism and (B) chimaerism. In mosaicism the two (or more) cell lines arise from a single zygote. In chimaerism the two cell lines arise from different zygotes.

Isochromosomes

An **isochromosome** consists of either two short arms or two long arms (iso = same). It results from abnormal separation of the centromere in meiosis. Isochromosomes are unbalanced as they lead to monosomy for the arm which is missing and trisomy for the arm which is duplicated.

Mixoploidy

The presence of two or more cell lines with different chromosome constitutions in an individual can be the result of either mosaicism or chimaerism (Fig. 2.16). In **mosaicism** the different cell lines are derived from a single zygote and usually arise due to non-disjunction occurring after conception in a mitotic division in the rapidly dividing zygote or embryo. In **chimaerism**, the cell lines originate from different zygotes. Natural chimaerism is encountered only rarely, usually as a result of the mixing of stem cells through a shared placental circulation in non-identical twins. Individuals who have undergone bone marrow transplantation sometimes show acquired chimaerism in their blood if their own bone marrow stem cells have not been totally ablated.

Consequences of chromosome abnormalities

Specific syndromes due to chromosome abnormalities are discussed in the next chapter. As a general rule, any degree of chromosome imbalance involving one or more of the autosomes will have very serious adverse effects. Most of the autosomal trisomies and all of the autosomal monosomies are incompatible with survival beyond early pregnancy. Similarly, large degrees of imbalance, resulting for example from a large insertion or deletion, are rarely seen in surviving newborn infants. Careful study of the relationship between chromosome abnormalities and outcomes in pregnancy has shown that there is a correlation between the amount of additional or missing chromosome material and outcome, although some chromosomes are much more gene rich than others (Table 2.8). In general it is very unusual for a baby to survive with trisomy for more than 4% of total autosome material or monosomy for more than 2%. All observations indicate that the effects of monosomy are more severe than those of trisomy.

Infants who do survive with autosomal imbalance almost invariably have severe physical and mental developmental problems. Often they have multiple congenital malformations such as congenital heart defects and cleft lip or palate, together with unusual facial features, a finding referred to as facial dysmorphism. Many children

TABLE 2.8 Numbers of gene loci on the human chromosomes

Chromosome	Total number of gene loci
1	869
2	566
3	490
4	348
5	435
6	564
7	419
8	324
9	326
10	307
11	579
12	476
13	158
14	277
15	263
16	344
17	530
18	137
19	599
20	215
21	119
22	228
X	537
Y	46

Reproduced with permission from NCBI GenBank® OMIM Statistics. November 2004.

with severe chromosome abnormalities die before reaching adult life as a result of severe internal malformations and increased susceptibility to infection.

Abnormalities of the sex chromosomes tend to result in much less serious problems. All of the sex chromosome trisomy syndromes are compatible with relatively normal mental and physical development. Monosomy for a single X chromosome (45,X) usually results in loss of the fetus in early pregnancy. Survivors have the condition known as Turner syndrome. The much milder effects of sex chromosome imbalance can be readily explained by the very small number of genes on the Y chromosome and the process of X chromosome inactivation which renders most of all but one of the X chromosomes inactive (p. 78).

Chromosome abnormalities and meiosis

Adults with structural or numerical autosomal imbalance usually have severe developmental problems and are rarely able to reproduce. However, women with Down syndrome are able to become pregnant (sometimes involuntarily, because of their vulnerability to abuse). The incidence of Down syndrome in their offspring is approximately 1 in 3, an observation explained by random segregation of the additional number 21 chromosome at meiosis followed by selective spontaneous loss of Down syndrome conceptions during pregnancy. Men with Down syndrome are rarely able to reproduce.

Carriers of balanced rearrangements require very careful evaluation to determine the likelihood of imbalance in a child. This applies particularly for translocations and inversions as discussed below. A carrier of a balanced deletion–insertion faces a risk of 1 in 2 for producing an unbalanced gamete with either the insertion or the deletion.

Reciprocal translocations and meiosis

Normally homologous chromosomes pair at meiosis I to form bivalents. However, chromosomes involved in a reciprocal translocation cannot pair in this way and instead form a structure known as a **pachytene quadrivalent** (Fig. 2.17). This enables all segments of homologous chromosome material to align opposite each other. As the spindle starts to pull the chromosomes apart in anaphase of meiosis I it is impossible to predict which chromosomes will separate (segregate) to each pole of the cell. Segregation can involve two chromosomes going in each direction—2 : 2 segregation— or three going in one direction and only one in the other—3 : 1 segregation. If alternate chromosomes segregate together then the gamete acquires either a normal or a balanced chromosome constitution, and fertilization will result in a normal infant. However, either form of adjacent segregation results in a gamete with an unbalanced chromosome complement, which, if fertilized, will lead either to spontaneous pregnancy loss or to the birth of an abnormal infant. Essentially, both forms of adjacent segregation result in gametes receiving only one of the derivative chromosomes, and

the outcome is that the balanced parental rearrangement has generated an unbalanced gamete. All forms of 3 : 1 segregation will also generate unbalanced gametes. In practice the likelihood of abnormal segregation and associated risk of viable imbalance should be assessed for each translocation on an individual basis. Some translocations will result in major imbalance, which will almost invariably result in spontaneous pregnancy loss. Other translocations, particularly those with breakpoints near the ends of chromosome arms, can be associated with high risks for the birth of an abnormal baby (see Box 2.2).

Robertsonian translocations and meiosis

When considering how a Robertsonian translocation behaves at meiosis, three chromosomes have to be taken into account, although one of these essentially consists of two acrocentric chromosomes fused at their centromeres. In meiosis I these form a trivalent. This can result in the production of gametes with unbalanced chromosome complements (Fig. 2.18). Robertsonian translocations occur relatively frequently and are an important cause of Down syndrome. For example, the empiric risk that a female carrier of a balanced 14;21 Robertsonian translocation, as shown in Fig. 2.18, will have a child with Down syndrome is approximately 10–15%. For a male carrier this risk is around 1–2%. Risks for Down syndrome are considered at greater length in the next chapter.

Inversions and meiosis

Inversions also present the chromosomes with a challenge if complete homologous pairing is to occur. If an inversion is very large, then the chromosomes will probably pair upside down, with the small terminal portions remaining unpaired. Smaller inversions achieve homologous pairing by forming an inversion loop (Fig. 2.19). When the chromosomes separate each gamete will inherit a balanced rearrangement as long as a crossover has not occurred in the inversion loop. If a crossover does occur, this will result in chromosome imbalance with duplication of one end of the chromosome and loss of the other end. In the case of paracentric inversions, unbalanced chromosomes will have either two centromeres (a dicentric chromosome) or no centromere (an acentric fragment). Both dicentric chromosomes and acentric fragments are inherently unstable in mitosis, so an abnormal pregnancy resulting from a paracentric inversion usually ends in early spontaneous miscarriage. However, a pericentric inversion can result in an abnormal and viable pregnancy as unbalanced derivative chromosomes will have only a

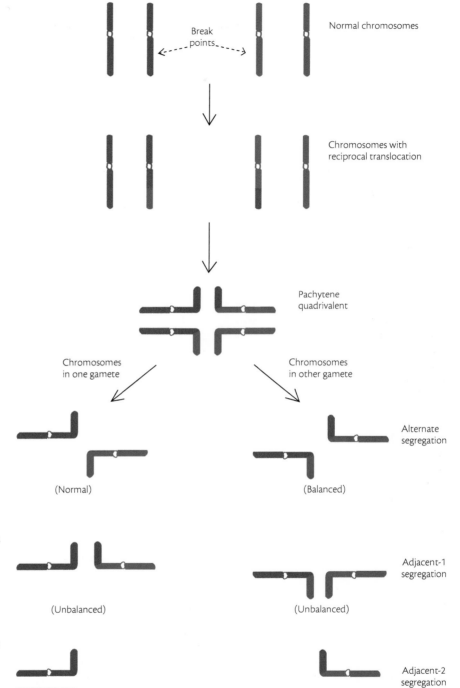

Fig. 2.17 Segregation of a reciprocal translocation in meiosis. The four chromosomes form a quadrivalent in pachytene in meiosis I. Alternate segregation results in all gametes having either a normal or a balanced chromosome complement. Both adjacent-1 (non-homologous centromeres segregate together) and adjacent-2 (homologous centromeres segregate together) segregation result in all gametes having an unbalanced chromosome complement.

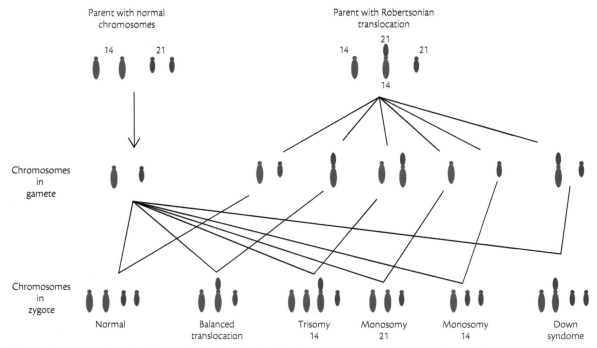

Fig. 2.18 How segregation of a Robertsonian translocation involving chromosomes 14 and 21 can cause Down syndrome.

(A) Pericentric inversion

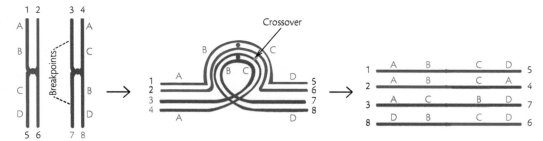

(B) Paracentric inversion

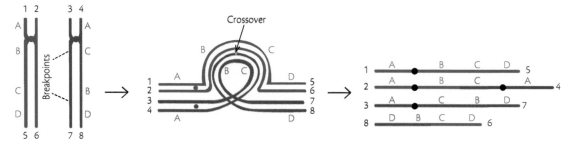

Fig. 2.19 Shows how segregation in meiosis of (A) a pericentric and (B) a paracentric inversion can generate unbalanced chromosome complements if a crossover occurs in the loop which forms to facilitate homologous pairing.

single centromere and therefore undergo stable division in mitosis. This possibility of viable imbalance resulting in an abnormal liveborn infant explains why balanced pericentric inversions are a greater cause of concern than balanced paracentric inversions.

> ### Key Point
> A carrier of a balanced rearrangement such as an inversion or a translocation can produce gametes with an unbalanced chromosome complement.

Chromosomes and cancer

It is now well recognized that chromosome abnormalities show a strong causal association with cancer. This association is mainly with acquired somatic abnormalities which have been observed in specific forms of leukaemia and in many solid tumours (p. 190). Their identification has been greatly facilitated by the development of techniques such as CGH, which enables chromosome analysis without a prior need for cell culture. These acquired chromosome abnormalities and rearrangements result in changes in the expression patterns of genes, such as oncogenes and tumour suppressor genes, which are involved in the regulation of cell division and proliferation (Chapter 10).

A small number of constitutional chromosome abnormalities convey a slightly increased risk for developing specific forms of leukaemia (e.g. Down syndrome—see p. 55) or embryonal tumours such as nephroblastoma (e.g. the WAGR microdeletion syndrome—see p. 64). However, in general the incidence of cancer is not usually increased in children with constitutional chromosomal abnormalities. The chromosome breakage syndromes (p. 66) represent important exceptions to this generalization.

Summary

A chromosome consists of a long strand of DNA, tightly coiled around structural proteins known as histones. Each chromosome has a primary constriction, known as a centromere, and two telomeres at the end of the short (p) and long (q) arms. The normal human chromosome complement consists of 44 autosomes plus 2 sex chromosomes, XX in the female and XY in the male.

Routine chromosome analysis is normally undertaken using cultured lymphocytes. Specialized diagnostic techniques include fluorescence *in situ* hybridization (FISH) and comparative genomic hybridization (CGH). FISH can be used to screen before birth for the common trisomy syndromes and to identify deletions of less than 10 Mb (microdeletions). CGH is the preferred method for the identification of chromosome abnormalities in tumours.

Mitosis is the process of cell division in which each chromosome, consisting of two identical sister chromatids, becomes aligned along the equator of the cell before separation of the chromatids to form daughter chromosomes. In meiosis, homologous chromosomes become aligned as pairs before undergoing recombination. They then separate to form haploid gametes as part of the processes of oogenesis and spermatogenesis.

Chromosome abnormalities can be numerical or structural. Numerical abnormalities include loss or gain of a single chromosome, monosomy or trisomy, or gain of a complete set of haploid chromosomes, triploidy. Structural abnormalities include translocations, deletions, insertions, rings, inversions and isochromosomes. In a balanced rearrangement there is no overall loss or gain of chromosome material, so that carriers of balanced rearrangements are usually normal. However, a carrier of a balanced rearrangement can produce gametes with an unbalanced complement, which can result in either miscarriage or the birth of an abnormal infant.

Further reading

Gardner RJM, Sutherland GR (2004) *Chromosome abnormalities and genetic counselling*, 3rd edn. Oxford University Press, New York.

Gersen SL, Keagle MB (ed.) (1999) *The principles of clinical cytogenetics*. Human Press, Totowa, NJ.

Rooney DE (ed.) (2001) *Human cytogenetic constitutional analysis*, 3rd edn. Oxford University Press, Oxford.

Multiple choice questions

1 Routine chromosome analysis

   (a) is carried out using red blood cells

   (b) requires a minimum of 3 days

   (c) detects all known chromosome abnormalities

   (d) involves analysis of cells in metaphase

   (e) is based on study of a single cell

2 Characteristics of meiosis are that:

   (a) each daughter cell has 23 chromosomes

   (b) all of the daughter cells have identical chromosomes

   (c) recombination occurs in prophase II

   (d) the duration is longer in males than in females

   (e) a meiotic error resulting in a male with a 47,XYY karyotype must have occurred in paternal meiosis II

3 The following are examples of aneuploidy

   (a) 46,XY

   (b) 69,XXX

   (c) 47,XXY

   (d) 45,X

   (e) 92,XXXX

4 The following chromosome complements or rearrangements are balanced:

   (a) 47,XY,+21

   (b) 45,XY,rob(14;21)(q10;q10)

   (c) a ring

   (d) a deletion

   (e) a deletion and insertion

5 Chromosome studies are undertaken on a couple who have lost a baby soon after birth because of multiple abnormalities. The father is found to have a balanced reciprocal translocation involving chromosomes 4 and 11. Therefore the parents should be told that

   (a) all of their future babies will be abnormal

   (b) the abnormalities in their baby were probably caused by triploidy

   (c) it is very unlikely that this translocation will cause ill health in the father but there is a risk that he and his partner will have another abnormal baby

   (d) balanced reciprocal translocations are usually harmless so it is very unlikely that this finding is relevant to their baby's abnormalities

   (e) the rearrangement in the father occurred as a result of non-disjunction

Answers

1 (a) false—chromosome analysis can only be carried out using cells which have a nucleus and are dividing, e.g. white blood cells

(b) true—for routine analysis the white cells have to be cultured for at least 3 days

(c) false—routine analysis will not detect micro-deletions or the chromosome breakage syndromes (see Chapter 4)

(d) true—this is when the chromosomes are maximally condensed

(e) false—usually at least 10 cells are analysed

2 (a) true—the main purpose of meiosis is to generate haploid (n=23) gametes

(b) false—chromosome segregation and recombination mean most of the gametes will have different chromosomes

(c) false—recombination occurs in prophase I

(d) false—meiosis lasts several years in females compared to around 60 days in males

(e) true—the only meiotic error which could generate two Y chromosomes in a gamete is non-disjunction in male meiosis II

3 (a) false—this is a normal karyotype

(b) false—this is an example of triploidy, a form of polyploidy

(c) true—aneuploidy refers to loss or gain of one or more chromosomes. This karyotype has an additional X chromosome

(d) true- this is an example of monosomy

(e) false—this is an example of tetraploidy, a form of polyploidy

4 (a) false—this child has an additional number 21 chromosome

(b) true—this is a balanced Robertsonian translocation

(c) false—a ring is unbalanced as both ends of the chromosome have been lost

(d) false—a deletion involves loss of part of a chromosome

(e) true—if the deleted material has been inserted elsewhere in the karyotype then there is no overall loss or gain of chromosome material

5 (a) false—alternate segregation will result in a balanced gamete and a normal infant

(b) false—the abnormalities were probably caused by chromosome imbalance resulting from the paternal translocation

(c) true—there is a significant risk that a future baby could inherit an unbalanced gamete

(d) false—balanced translocations are usually harmless but they can generate unbalanced rearrangements in offspring which are almost always harmful

(e) false—the translocation in the father could have arisen *de novo* or it could have been inherited from a carrier parent due to alternate segregation in meiosis I

Common chromosome disorders

Any disturbance in the correct amount of chromosome material has very serious consequences for human development. Chromosome imbalance is relatively common at conception, as judged by the presence of chromosome abnormalities in around 50% of all spontaneous first-trimester miscarriages (Table 3.1). In

TABLE 3.1 Incidence of chromosome abnormalities

	Incidence (%)
First-trimester miscarriages	
Trisomy (16 is most common)	25
Monosomy (X is most common)	10
Triploidy	8
Tetraploidy	2
Other	5
TOTAL	50
Stillborn infants	5
Liveborn infants	
Balanced rearrangements	0.3
Unbalanced rearrangements	
Autosomal trisomy	0.2
Sex chromosome trisomy	0.3
Sex chromosome monosomy	0.01
Other	0.1
TOTAL	0.9
Children with mental retardation	35–40
Children with congenital malformations	5–10
Couples with recurrent miscarriages	3

liveborn infants the incidence of unbalanced rearrangements has fallen to around 1 in 200, although the figure in stillborn infants is much higher.

Not all chromosome abnormalities are obvious at birth. Some may present in childhood as the cause of mild learning disability and others may not become apparent until adult life when investigations are undertaken because of infertility or a history of recurrent pregnancy loss. In addition to these congenital chromosome abnormalities, which are described as constitutional because they are present from conception, chromosome abnormalities also arise due to errors in mitosis throughout life and are a major factor in the genesis of cancer, as discussed in Chapter 10.

Well over 1000 constitutional chromosome abnormalities have been reported. Most of them are extremely rare and in some instances unique. For example, a child with mental retardation may be found to have a small interstitial deletion which has never been reported previously. Information about such children is sometimes published in relevant medical journals and subsequently collated in commercially available computerized databases. A support group, known appropriately as Unique (www.rarechromo.org), offers parents and professionals access to information about the effects and consequences of many of the rarely encountered chromosome abnormalities.

Clearly it is not possible to be familiar with all of the known chromosome abnormality syndromes. However, a small number of these disorders are relatively common and knowledge of these conditions is expected. This applies particularly to disorders such as Down syndrome and those sex chromosome aneuploidy syndromes that are associated with specific long-term medical complications.

Autosomal aneuploidy syndromes

As discussed in Chapter 2, aneuploidy refers to the loss or gain of one or more chromosomes. For many years it was suspected that autosomal aneuploidy would not be compatible with survival beyond early pregnancy, and it is now recognized that in most instances autosomal trisomy or monosomy results in either failure of embryogenesis to become established or early spontaneous miscarriage. For example, trisomy 16 constitutes approximately 15% of all chromosome abnormalities seen in spontaneous abortions but has never been reported in a liveborn infant. Similarly, none of the autosomal monosomy states is associated with a recognized syndrome in newborn infants.

However, with the advent of a reliable technique for analysing human chromosomes in 1956, it soon became apparent that a few of the autosomal trisomy states were compatible with survival to term. These comprise full trisomy for chromosomes 13, 18, and 21, together with mosaic trisomy for chromosome 8. Trisomy 21 is by far the most important of these conditions. In all of the full trisomy syndromes the additional chromosome usually originates from a non-disjunction event in one of the maternal meiotic divisions. All three of the recognized autosomal trisomy syndromes (trisomy 13, 18, and 21) show an association of increasing incidence with advancing maternal age.

Trisomy 21 (Down syndrome)

Trisomy 21 is the most common and widely recognized malformation syndrome. Its eponymous title comes from the description by Dr Langdon-Down in 1866 in the Clinical Lecture reports of the London Hospital. The presence of an additional number 21 chromosome as the probable cause of Down syndrome was first reported by Lejeune and colleagues in 1959 (Box 3.1). Shortly afterwards it was recognized that Down syndrome could also be caused by an unbalanced Robertsonian translocation and by mosaicism for trisomy 21.

Prevalence

Before the introduction of prenatal diagnostic and screening procedures the incidence of Down syndrome at birth was around 1 in 650 to 1 in 800 infants in almost all populations studied, with no major ethnic differences. In countries where prenatal diagnostic programmes with selective termination of pregnancy have been established, 25–50% of all cases undergo termination and liveborn incidence figures have therefore fallen. Several surveys have established that Down syndrome accounts for approximately 30% of all children with mental retardation and around 15% of all adults living in institutions because of learning disability.

Natural history

Approximately 70% of all Down syndrome pregnancies miscarry spontaneously. On the basis of the different prevalence figures noted at various stages of pregnancy, it has been estimated that approximately 40% and 25% of Down syndrome pregnancies are lost spontaneously after 10 and 16 weeks gestation respectively.

The average life expectancy for children with Down syndrome has increased from less than 10 years in 1930 to over 50 years at present; 96% of Down syndrome babies without heart defects survive to the age of at

BOX 3.1 LANDMARK PUBLICATION: THE CAUSE OF DOWN SYNDROME

Although Down syndrome was first reported in 1866, the underlying cause remained unknown for almost 100 years. A chromosome abnormality was suspected, and it was also recognized from studies in *Drosophila* that an association of increased incidence with advancing maternal age, as had been observed for Down syndrome, could be related to non-disjunction. However, the technical problems associated with chromosome analysis had made it impossible to undertake reliable chromosome studies in children.

All this changed with the development of a reliable technique for chromosome analysis in 1956 (see Box 2.1). In 1959 Dr Jérôme Lejeune and his colleagues published a short paper—only $1^1/_2$ pages—in which they reported their results of chromosome analysis in 9 'enfants mongoliens', in all of whom they identified an additional small 'telocentric' chromosome. (Telocentric would now be classified as acrocentric.) The authors were not able to determine whether the additional chromosome represented an intact normal chromosome or a 'fragment resulting from another type of aberration'. Subsequent studies confirmed that the additional chromosome in Down syndrome is an intact number 21 chromosome. In fact it later emerged that the chromosome designated as number 21 is actually smaller than the number 22 chromosome, but to avoid confusion it was agreed that the existing nomenclature should remain unchanged, so by convention it is the number 21 chromosome that is trisomic in Down syndrome.

Soon afterwards the chromosomal causes of the other common aneuploidy syndromes were established and it was also recognized that Robertsonian translocations accounted for most of the rare reports of Down syndrome occurring in more than one member of a family.

Dr Lejeune went on to become the first Professor of Fundamental Genetics in Paris where he also worked as a clinician at the Hospital for Sick Children. He was a deeply religious man who held very strong views that human life begins at conception and that termination of a Down syndrome pregnancy is morally wrong. Consequently the realization that his discovery had led indirectly to the development of reliable diagnostic techniques which could be used to facilitate termination of a Down syndrome pregnancy caused him great anguish. He was later to be appointed by Pope John Paul II to the presidency of the Pontifical Academy of Life, but in other circles his views generated great controversy and were not always well received. Further details of Jerome Lejeune's life and of the challenges which he faced can be found in the biography written by his daughter Clara.

References

Lejeune C (2000) *Life is a blessing.* Ignatius Press, San Francisco.
Lejeune J, Gautier M, Turpin R (1959) Étude des chromosomes somatique de neuf enfants mongoliens. *Comp Rend Acad Sci*, **248**, 1721-1722.

least 1 year. The comparable figure for Down syndrome infants with heart defects is 80%, but survival amongst these children is increasing in parallel with improvements in cardiac surgery.

Intellectual skills

Surveys of children with Down syndrome indicate that approximately 10% show profound intellectual disability, with IQ scores of less than 20; 70% have severe mental retardation, with IQ scores of between 20 and 50; and the remaining 20% are either mildly affected, with IQ scores between 50 and 70, or fall within the normal range. The mean IQ for children and young adults with Down syndrome is approximately 40–45. It can be argued that formal IQ assessment is an unsatisfactory method of assessing the skills and potential of a child with Down syndrome, and many parents feel that it is unfair to use a system which ignores the more positive aspects of a Down syndrome child's personality. An alternative indication of intellectual development, which is sometimes used when counselling prospective parents of a baby detected prenatally, is that the average mental age of young adults with Down syndrome is roughly comparable to that of a normal 5–6-year-old child. Approximately 50% of children with Down syndrome learn to read, but most have very limited arithmetical skills. Almost all adults with Down syndrome require a degree of supervision, particularly in later life.

Clinical phenotype

Children and adults with Down syndrome have a characteristic facial appearance (Fig. 3.1) with upward sloping palpebral fissures, prominent epicanthic folds, small ears, and protruding tongue. Over 50% have bilateral single palmar creases and there is usually a wide

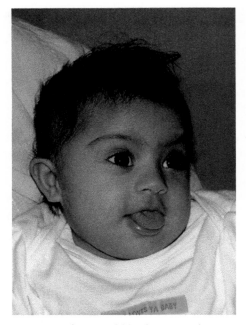

Fig. 3.1 Facial view of a young child with Down syndrome (courtesy of Dr Rachel Harrison).

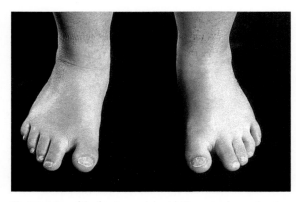

Fig. 3.2 View of the feet in an adult with Down syndrome showing wide 'sandal' gap between the first and second toes.

TABLE 3.2 Clinical findings in Down syndrome	
Congenital abnormalities	**Frequency (%)**
Cardiac	
Atrioventricular canal defect	15
Ventricular septal defect	10
Atrial septal defect	5–10
Patent ductus arteriosus	3–5
Other	5
Total	40–45
Gastrointestinal	
Duodenal atresia	5
Hirschsprung disease	3
Oesophageal atresia	2
Total	10
Other	5
Total	60

behavioural disturbance such as hyperactivity or attention deficit disorder. Most adults with Down syndrome begin to develop Alzheimer disease in mid life and almost all are severely affected by the age of 70 years.

Congenital abnormalities

Approximately 60% of babies with Down syndrome have one or more congenital abnormalities (Table 3.2). Cardiac defects are present in 40–45% and gastrointestinal malformations in 10%. Atrioventricular canal defects are roughly 400 times more common in Down syndrome than in the general population. Similarly, duodenal atresia is about 300 times more common in Down syndrome than in unaffected children. Consequently the discovery of either of these findings on prenatal ultrasonography raises suspicion that the unborn baby could have Down syndrome.

Many other abnormalities show a marginally increased incidence in Down syndrome. These include undescended testes (also known as cryptorchidism) in 10–20%, urinary tract malformations in 5%, congenital cataracts in 1%, and cleft lip or palate in 0.5%. The high incidence of malformations in Down syndrome means that all affected babies should undergo careful evaluation, including echocardiography, in the neonatal period.

Medical complications

In addition to a high incidence of congenital malformations, children and adults with Down syndrome show

'sandal' gap between the first and second toes (Fig. 3.2). Average adult height is 141 cm in women and 151 cm in men. Babies with Down syndrome are usually hypotonic and sleepy, with poor primitive reflexes and hyperextensible joints. The presence of excess skin at the back of the neck is a useful diagnostic feature. This reflects the increased nuchal swelling which forms the basis of early screening programmes in pregnancy.

Most children with Down syndrome have an engaging personality, with social skills more advanced than other intellectual abilities. Although many are happy and affectionate, approximately 1 in 6 manifests a an increased incidence of acquired disorders. A transient form of megakaryoblastic leukaemia occurs in 10% of Down syndrome infants. This resolves spontaneously, but approximately 1–2% subsequently develop acute megakaryocytic leukaemia by the age of 4 years. Hearing loss occurs in up to 50% of children with Down syndrome and can be conductive (middle ear), sensorineural (inner ear), or mixed in origin. Hypothyroidism occurs in approximately 3% of children and 10% of adults. Strabismus and nystagmus are common eye findings. Adult onset epilepsy occurs in 10–15%. The prevalence of Alzheimer disease rises from around 10% between the ages of 40 and 49 years to almost 100% by 70 years. Other conditions with an increased incidence include coeliac disease, arthritis, diabetes mellitus, and obstructive sleep apnoea.

This long list of potential medical complications emphasizes the importance of regular surveillance throughout childhood and adult life with special attention being paid to visual, hearing, and thyroid function assessment.

Chromosome findings

Full trisomy 21 accounts for 95% of all cases (Table 3.3). Usually the additional chromosome 21 has originated from the mother, most often as a result of non-disjunction in meiosis I. In almost half of these cases there has been a complete absence of recombination between the maternal number 21 chromosomes, implying that normal pairing has not occurred during prophase in meiosis I. Recombination in women normally occurs in the fetal ovary before birth, so that in some instances the original error predisposing to the birth of a child with Down syndrome may actually have occurred in the child's own mother before she herself was born.

Unbalanced Robertsonian translocations account for approximately 4% of all cases, roughly 75% of which have arisen as new (*de novo*) events. The clinical phenotype in these cases is identical to that seen in trisomy 21. Parental chromosome studies are essen-

tial when a Robertsonian translocation is identified, to ensure that correct information can be given about recurrence risks to the parents and that other relevant family members can be offered chromosome analysis to establish if they are also carriers.

Finally, roughly 1% of children with Down syndrome are found to have a mosaic karyotype, usually for a normal cell line and a cell line with trisomy 21. These children are usually less severely affected than those with full trisomy 21.

Molecular correlations

The recent sequencing of the human genome has led to the identification of a large number of genes on chromosome 21 (120 at the last count). These include 16 genes which have a role in mitochondrial energy generation and 10 which are involved in brain development and/or function. However, to date it has not proved possible to correlate the role of these genes with the clinical phenotype observed in Down syndrome, although it has been recognized that children who have trisomy for just the distal portion of the long arm of chromosome 21 usually have the typical facial features of Down syndrome. This is referred to as the Down syndrome *critical region*.

Two specific genes have been implicated as the probable cause of recognized complications of Down syndrome. The amyloid precursor protein gene (*APP*) at 21q21 is thought to play a major role in conveying susceptibility to Alzheimer disease through a direct gene dosage effect, as the deposition of amyloid plaques in the brain is one of the characteristic neuropathological findings seen in Alzheimer disease. Another gene, *GATA1*, located on the X chromosome, is involved in the growth and maturation of red blood cells and megakaryocytes. For reasons which are not understood, mutations in this gene are a major factor in the development of acute megakaryocytic leukaemia in children with Down syndrome.

Genetic counselling

For parents who have had a child with trisomy 21 the recurrence risk in a future pregnancy has been shown to be approximately 1% in addition to the pre-existing risk associated with the mother's age (Table 3.4). A similar risk applies to parents of a child with a *de novo* unbalanced Robertsonian translocation.

Specific risks for carriers of balanced parental Robertsonian translocations have been derived. These are around 10–15% and 1–2% for female and male carriers respectively. However, male and female carriers of a 21q;21q translocation have much greater risks as all of

TABLE 3.3   Chromosome findings in Down syndrome

Chromosome abnormality	Frequency (%)
Full trisomy 21	95
Origin of extra chromosome	
Maternal meiosis	90 ($^3/_4$ meiosis I, $^1/_4$ meiosis II)
Paternal meiosis	5 ($^3/_4$ meiosis I, $^1/_4$ meiosis II)
Robertsonian translocation	4–5 ($^3/_4$ de novo, $^1/_4$ inherited)
Mosaic trisomy 21	1

TABLE 3.4	Down syndrome and maternal age		
Maternal age (years)	**Incidence of Down syndrome**	**Maternal age (years)**	**Incidence of Down syndrome**
15–19	1 in 1500	37	1 in 250
20–24	1 in 1500	38	1 in 200
25–29	1 in 1200	39	1 in 150
30	1 in 900	40	1 in 100
31	1 in 800	41	1 in 85
32	1 in 700	42	1 in 65
33	1 in 600	43	1 in 50
34	1 in 500	44	1 in 40
35	1 in 400	45	1 in 30
36	1 in 300	46+	1 in 20

Figures relate to maternal age at delivery and incidence of Down syndrome in liveborn infants assuming no prenatal diagnosis and termination of pregnancy. Incidence figures have been rounded off and taken from Cuckle HS, Wald NJ, Thompson SG (1987) Estimating a woman's risk of having a pregnancy associated with Down syndrome using her age and serum alpha-protein level. *British Journal of Obstetrics and Gynaecology*, **94**, 387–402.

their offspring will have either only one number 21 chromosome, leading to early pregnancy loss, or effectively have three number 21 chromosomes, resulting in Down syndrome. Thus the likelihood of Down syndrome in their liveborn infants is 100%

Trisomy 13 (Patau syndrome)

Trisomy 13, also known as Patau syndrome after the lead author of the paper first describing the condition in 1960, is a severe multiple malformation syndrome which is almost invariably associated with an extreme-

> **Key Point**
>
> Trisomy 21 accounts for 95% of all cases of Down syndrome. Usually this has arisen as a result of non-disjunction in maternal meiosis I. Trisomy 21 conveys a low recurrence risk of around 1% for siblings. An unbalanced Robertsonian translocation is found in 4–5% of cases and in roughly one third of these one of the parents (usually the mother) carries the translocation in a balanced form. In these situations there is a much higher recurrence risk for siblings.

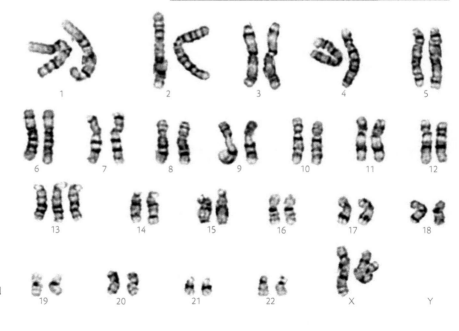

Fig. 3.3 Karyotype showing trisomy 13. Courtesy of Applied Imaging Corp.

ly poor outcome. In 90% of cases it is caused by the presence of an additional number 13 chromosome (Fig. 3.3) which is usually of maternal origin. The remaining 10% of cases result from unbalanced Robertsonian translocations. The incidence in liveborn infants is around 1 in 10 000. The diagnosis can be suspected in over 50% of cases through fetal anomaly scanning in the second trimester of pregnancy.

Clinical features

Affected infants show severe facial dysmorphism with either deep-set small eyes and cleft lip or palate, or holoprosencephaly in which there is an underlying abnormality of the forebrain in association with failure of development of the central part of the upper lip (p. 174). Other characteristic features include occipital scalp defects, postaxial polydactyly (i.e. arising from the ulnar side of the hand), and genital anomalies. Cardiac malformations are present in 90% of affected babies.

Fig. 3.4 A rare example of long-term survival in trisomy 13. Clare at age 16 years (see Box 3.2).

BOX 3.2 CASE CÉLÈBRE: LONG-TERM SURVIVAL IN TRISOMY 13

The following is a true account of the life of Clare, who is the longest known survivor with trisomy 13. Clare's parents have kindly granted permission for her details to be included here.

Clare weighed 2.8 kg when she was born in 1964 at full term. Shortly after birth she was noted to have extra digits on both hands and her eyes were found to be small, with cataracts. During early childhood it became apparent that Clare's development was severely delayed but she had a happy and mischievous personality, which made her a very popular and much loved member of her family. Clare began to crawl at the age of $2^1/_2$ years and by late childhood she was able to bear her own weight and take a few steps with assistance. However, she never achieved satisfactory independent locomotion. Nor was she ever able to use meaningful speech, although she had no difficulty in conveying her wishes by means of gesture and the use of recognizable sounds.

At age 8 years Clare required treatment for a lacrimal cyst and from 11 years onwards she had convulsions and recurrent urinary tract infections in association with renal calculi. In adult life she developed a severe kyphoscoliosis. She also had recurrent chest infections and skin abscesses. Despite frequent ill health she retained her happy personality and particularly enjoyed human company, laughter, and music, until she died at the age of 37 years as a result of severe pneumonia.

Chromosome studies were carried out on samples of blood from Clare on four occasions, on each of which all cells analysed showed a 47,XX,+13 karyotype. No studies were ever undertaken using other tissue samples such as cultured skin fibroblasts or bone marrow cells, so mosaicism for trisomy 13 cannot be ruled out. However, review of Clare's medical records and family photographs (Fig. 3.4) suggests that Clare almost certainly did have full trisomy 13. Her remarkable survival was probably due to a combination of factors including the absence of a serious cardiac malformation, a loving family environment, and an innate robust constitution (Clare came from a family of long-living ancestors).

In most cases of both trisomy 13 and trisomy 18 the diagnosis is now suspected in pregnancy and many prospective parents opt for termination of pregnancy. Those babies who do survive to term usually perish in early infancy. When viewed against this background, Clare's survival to the age of 37 years is quite remarkable.

References

Wyllie JP, Wright MJ, Burn J, Hunter S (1994) Natural history of trisomy 13. *Archives of Disease in Childhood*, **71**, 343–345.

Root S, Carey JC (1994) Survival in trisomy 18. *American Journal of Medical Genetics*, **49**, 170–174.

Other common internal anomalies include polycystic kidneys, hydronephrosis, and bicornuate uterus.

Approximately 50% of affected infants die within the first week of life as a result of the combination of severe internal malformations and poor respiratory effort. Most of the remaining 50% die by the age of 6 months. Survival into adult life has been reported in exceptional cases (Fig. 3.4 and Box 3.2). Children who survive long term usually show severe neurodevelopmental delay.

Trisomy 18 (Edwards syndrome)

This condition shares many features with trisomy 13. The underlying chromosome abnormality was first described by Edwards *et al.* in 1960. Almost all cases are caused by the presence of an additional number 18 chromosome (Fig. 3.5), which is maternal in origin in 90% of cases. The incidence in liveborn infants is approximately 1 in 5000 and, as with trisomy 13, over 50% of cases can be suspected on the basis of ultrasound scanning in mid pregnancy.

Clinical features

Babies with trisomy 18 are small and have a characteristic appearance with short palpebral fissures, prominent occiput, small jaw, and overlapping clenched fingers. The feet may have a 'rocker bottom' appearance or show severe talipes (Fig. 3.6). Congenital heart

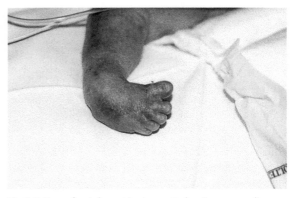

Fig. 3.6 Foot of an infant with trisomy 18 showing severe talipes equinovarus. Courtesy of Dr William Reardon, National Centre for Medical Genetics, Dublin.

abnormalities are present in 90% of cases. Other common internal abnormalities include oesophageal atresia with a tracheo-oesophageal fistula, diaphragmatic hernia, a horseshoe kidney or cystic kidneys, and absence of the corpus callosum. Long-term survival is very poor. As with trisomy 13, approximately 50% of affected babies die in the first week of life and survival beyond 1 year is very unusual (Fig. 3.7). Those children with full trisomy 18 who survive usually show severe neurodevelopmental retardation. Children with mosaic trisomy 18 are more mildly affected.

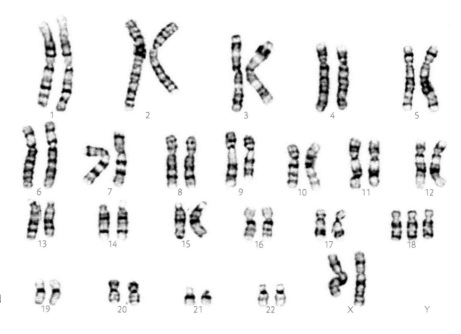

Fig. 3.5 Karyotype showing trisomy 18. Courtesy of Applied Imaging Corp.

Fig. 3.7 A rare example of long-term survival in trisomy 18. This child was aged 8 years at the time of this photograph. Courtesy of the Richard Lemon Studio.

Sex chromosome aneuploidy syndromes

With the exception of monosomy Y (45,Y), which is probably not viable after fertilization, all of the sex chromosome aneuploidy states involving monosomy or trisomy are compatible with relatively normal growth and development. This can be explained by inactivation of all but one of the X chromosomes in each cell in the case of X chromosome aneuploidy and by the small number of genes located on the Y chromosome in boys with two or more Y chromosomes. The main features of the sex chromosome aneuploidy syndromes are summarized in Table 3.5.

Turner syndrome (45,X)

This condition was first described in 1938 by Dr H. Turner, an American endocrinologist, and the underlying chromosomal cause was established in 1959. Turner syndrome affects approximately 1 in 5000 newborn female infants, although the incidence at conception is much greater, with around 97% of all Turner syndrome pregnancies resulting in spontaneous loss. Although it is a rare condition, awareness of Turner syndrome is important because of its potential complications and the possibility of treatment.

TABLE 3.5	Features of the common sex chromosome aneuploidy syndromes		
Syndrome	**Karyotype**	**Liveborn incidence**	**Clinical features**
Turner	45,X (50%)	1 in 5000 females	Intrauterine oedema
	45,X/46,XX (20%)		Neck webbing
	46,X,i(Xq) (15%)		Short stature
	46,X,r(X) (5%)		Infertility
	46,X,del(Xp) (5%)		
	Other (5%)		
Klinefelter	47,XXY	1 in 1000 males	Tall stature
			Gynaecomastia
			Infertility
			Mild learning difficulties
Triple X	47,XXX	1 in 1000 females	No physical abnormalities
			Normal fertility
			Mild learning difficulties
XYY	47,XYY	1 in 1000 males	Tall stature
			Normal fertility
			Mild learning difficulties
			Behavioural problems—usually minor

Clinical features

Most Turner syndrome conceptions develop severe intrauterine oedema, also known as *hydrops fetalis*, which can be detected in mid pregnancy by ultrasonography. This is thought to be due to delayed maturation of the lymphatic drainage system. At birth, affected infants, who almost always have a normal female phenotype, often manifest a residue of this intrauterine oedema in the form of neck webbing and puffy extremities. Short stature usually becomes apparent in early childhood and, if untreated, average adult height is 140–145 cm. Treatment with growth hormone has been reported to add between 5 and 10 cm to final adult height.

Infertility is almost invariable in women with full Turner syndrome. This results from failure of normal development of the ovaries (*ovarian dysgenesis*) from mid pregnancy onwards, leading to the presence of small streak gonads. Oestrogen replacement therapy should be introduced from the age of 12 years onwards to promote pubertal development and prevent the onset of osteoporosis. Pregnancy has been achieved in a small number of women with Turner syndrome by means of embryo transplantation following *in vitro* fertilization using their partner's sperm and a donor egg.

Women with Turner syndrome are of normal intelligence, although subtle defects in visuospatial perception and fine motor skills are reported. They show an increased incidence of internal anomalies, notably coarctation of the aorta in 15% and renal malformations in 40%. In adults the incidence of hypothyroidism is 20%, and recent studies suggest an increased incidence of aortic dilatation with an associated risk of dissection. The small number of women with Turner syndrome due to mosaicism for a 45,X/46,XY karyotype have an increased risk of developing a gonadoblastoma, so removal of the gonads is recommended in these individuals. Mosaicism for a 45,X/46,XY karyotype usually results in a normal male phenotype but can occasionally lead to sexual ambiguity or a Turner syndrome phenotype in a female.

Chromosome findings

Approximately 50% of women with Turner syndrome have a 45,X karyotype and in 80% of these the meiotic error will have occurred in a paternal meiotic division. Most of the remaining 50% have a karyotype consisting of either 45,X/46,XX mosaicism, or 46 chromosomes with one normal X and the other X being abnormal in the form of either a ring, a long arm isochromosome, or a partially deleted X chromosome (see Table 3.5).

Molecular correlations

It is not clear why the presence of a single X chromosome results in the Turner syndrome phenotype. Most of the second X chromosome is inactivated in normal women who obviously do not show Turner syndrome features. Amongst the genes present in the pseudoautosomal regions, which are not inactivated, attention has focused on *SHOX* (short stature homeobox-containing gene). Mutations in, or deletions of, *SHOX* result in a rare skeletal disorder known as dyschondrosteosis, in which affected males and females have mild short stature. Women with dyschondrosteosis also have an abnormality of the wrist, known as Madelung deformity, caused by partial subluxation of the distal ends of the radius and ulna. Madelung deformity also occurs in a small proportion of women with Turner syndrome. Haploinsufficiency for *SHOX* is believed to account for some but not all of the short stature seen in Turner syndrome.

Klinefelter syndrome (47,XXY)

The clinical syndrome now associated with the presence of a 47,XXY karyotype was first described by Klinefelter *et al.* in 1942, 17 years before the underlying chromosome abnormality was identified. The incidence is approximately 1 in 1000 newborn boys. Approximately 80% of males with Klinefelter syndrome have a 47,XXY karyotype with the additional X chromosome being derived equally from meiotic errors in each parent. Other karyotypes associated with this diagnosis include 47,XXY/46XY mosaicism and more severe X chromosome aneuploidy such as 48,XXXY and 49,XXXXY.

Clinical features

Newborn boys with Klinefelter syndrome are clinically normal. The diagnosis is usually first suspected in mid childhood because of mild learning difficulties or in adult life because of infertility (Box 3.3). Affected boys show difficulty in acquiring verbal skills and have intelligence scores which are on average 10–20 points lower than those of their unaffected siblings. They tend to be relatively passive and to lack self-confidence. Physical abnormalities are usually limited to a mild increase in stature due to relatively long limbs. Adults with Klinefelter syndrome have small testes and are almost always infertile. Approximately 50% show gynaecomastia and adult males have an increased risk of breast cancer. Fertility has been achieved using haploid spermatocytes obtained by testicular biopsy. Sexual orientation is usually normal and many men with Klinefelter syndrome are happily married.

BOX 3.3 CASE HISTORY: KLINEFELTER SYNDROME

John first came to medical attention at the age of 7 years when it became apparent that he was struggling to keep up with his peers at school, particularly in reading and spelling. His teachers had noticed that he was a rather shy and nervous boy who lacked self-confidence and found it difficult to make friends. John's parents recognized that he was different from his older brother and particularly recalled that he had been a 'good' baby who did not cry as much as his siblings or seem to need as much attention.

At school it was arranged for John to receive extra help in subjects involving verbal skills, and with the support of his parents and sympathetic teachers he became a popular member of his class. However, problems recurred when he started secondary school where his placid nature made him an easy prey for bullying and the relatively poor development of his genitalia led to teasing and embarrassment. He was referred for investigations which showed high levels of gonadotrophins with a low level of testosterone, findings consistent with delayed onset of puberty due to primary testicular failure. John's karyotype was reported as 47,XXY.

The diagnosis of Klinefelter syndrome was explained and John was started on small but increasing doses of testosterone given by regular monthly injection. These resulted in the onset of puberty with enlarge-

ment of John's phallus and a general improvement in his level of self-confidence. Regular monitoring of John's hormone levels confirmed that he had very low levels of endogenous testosterone production and so it was decided to continue regular testosterone injections in order to maintain John's general sense of well-being and sexual development as well as prevent the onset of osteoporosis in later life.

On leaving school at the age of 16 years John successfully applied for a job in an office where he was able to work at his own pace in a supportive environment. Several years later he married his long-term girlfriend who was aware that he was likely to be infertile. John's most recent visit to hospital was to accompany his wife to an infertility clinic, where options such as the use of donor sperm (AID = artificial insemination by donor sperm) and possible sperm retrieval by testicular biopsy were discussed.

This case history is typical of many men with Klinefelter syndrome who generally cope well in supportive surroundings but who sometimes struggle to make friends and develop relationships with the opposite sex. Early treatment with testosterone, when endogenous levels have been shown to be low, often has beneficial effects, not only on physical development but also on self-confidence, energy levels, and motivation.

The presence of more than two X chromosomes (i.e. 48,XXXY and 49,XXXXY) is associated with much more severe learning difficulties and marked hypogonadism with very small testes and a small phallus.

Trisomy X syndrome (47,XXX)

Trisomy X is present in approximately 1 in 1000 women, most of whom never come to medical attention. In 95% of cases the additional X chromosome originates from an error in maternal meiosis. Women with a 47,XXX karyotype show normal physical and sexual development, although reduced fertility and early menopause have been reported in a few cases. Surveys of newborn infants with a 47,XXX karyotype, who are ascertained in an unbiased fashion and followed up through childhood, have shown that most girls with this karyotype show mild delay in acquiring both expressive and receptive language skills with lower full-scale intelligence scores than their unaffected siblings. However, severe learning difficulties are unusual and most women with a 47,XXX karyotype

lead normal lives with the diagnosis never being suspected.

47,XYY syndrome

Newborn surveys indicate that this condition affects approximately 1 in 1000 males, and as with trisomy X the diagnosis often never comes to light. Physical abnormalities are limited to mild proportionate tall stature with enlarged teeth and occasionally an increased susceptibility to develop acne. As in Klinefelter syndrome, verbal skills are delayed and some affected boys show muscle weakness with poor coordination. Intelligence scores are mildly reduced by on average 10–20 points.

The 47,XYY karyotype gained undue notoriety in the early 1960s when it was suggested that the presence of an additional Y chromosome could predispose to serious violent criminal behaviour. Subsequent research showed that men with a 47,XYY karyotype show an increased tendency to minor criminal behaviour, often associated with impulsiveness and poor self-control.

However, severe violent or psychopathic behaviour is very unusual and many males with this karyotype live normal lives with the diagnosis never being suspected. Fertility in men with a 47,XYY karyotype is normal.

> ### Key Point
>
> Sex chromosome aneuploidy usually has much less severe effects than autosomal aneuploidy. Children with a 47,XXX, 47,XXY, or 47,XYY karyotype often show mild delay in acquiring verbal skills. Individuals with a 45,X or 47,XXY karyotype are usually infertile.

Genetic counselling and the sex chromosome aneuploidy syndromes

All of the four common sex chromosome aneuploidy syndromes (45,X; 47,XXY; 47,XXX; 47,XYY) usually occur sporadically within a family and there is very little evidence to suggest that the recurrence risk is increased for siblings. However, on the basis that a few families may harbour a poorly understood tendency for recurrent non-disjunction, most parents who have had a child with one of these conditions would be offered prenatal diagnosis in future pregnancies.

Fertility in the 45,X and 47,XXY syndromes is severely impaired, so risks to offspring are usually not relevant. Fertility in the 47,XXX and 47,XYY syndromes is normal and the incidence of sex chromosome abnormalities in the offspring of affected individuals is very low. This is probably because of selection against gametes that have an additional sex chromosome.

The chance detection of a sex chromosome abnormality during pregnancy, when a test has been carried out looking primarily for Down syndrome, creates a very difficult ethical counselling situation. Experience indicates that when the prospective parents are given full details of the relevant condition in as unbiased a manner as possible so that they can make their own informed decisions, approximately 50% choose to continue the pregnancy.

Polyploidy

Tetraploidy (n = 92) accounts for approximately 5% of all chromosomally abnormal spontaneous abortions, but tetraploid conceptions rarely survive beyond the first trimester. Triploidy (69,XXX; 69,XXY; 69,XYY) is approximately three times more common than tetraploidy amongst spontaneous abortions and, unlike tetraploidy, some triploid conceptions survive beyond

the second trimester and even occasionally to term. Babies with triploidy show severe intrauterine growth retardation with relative preservation of head size in association with a very small trunk. Abnormalities such as syndactyly (webbing) between the third and fourth fingers or second and third toes, neural tube defects, and urogenital malformations are common. Survival beyond early infancy is very unusual.

In triploid conceptions with two maternal haploid contributions originating from retention of a polar body in a meiotic division, the placenta is small and the fetus is usually reasonably well developed. In contrast, in triploid conceptions with two paternal haploid contributions, resulting from fertilization of a single ovum by two sperm, the placenta is large and often shows partial hydatidiform changes (Fig. 3.8), while the fetus is small and poorly developed. These observations are consistent with an imprinting effect (p. 82), whereby the paternally derived genome is essential for placental and membrane development, whereas the maternally

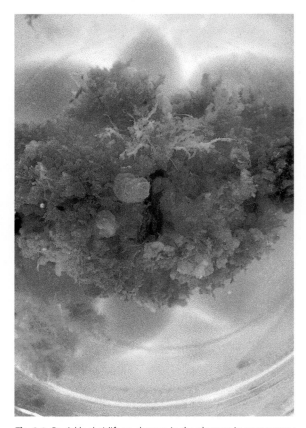

Fig. 3.8 Partial hydatidiform changes in the placenta in a pregnancy with triploidy. Courtesy of Dr Helen Porter, Leicester Royal Infirmary, Leicester.

derived genome is particularly important for fetal development. Conceptions which are exclusively paternal in origin, as can arise when two sperm fertilize an empty ovum, develop into a mass of cystic proliferating tissue known as a *hydatidiform mole*. This shows malignant potential and often develops into an invasive choriocarcinoma which has to be treated vigorously with chemotherapy.

> **Key Point**
>
> Complete hydatidiform moles have a diploid karyotype, which is exclusively paternal in origin. These show a high incidence of malignant change. Partial hydatidiform changes, also known as a partial mole, are seen in triploidy if the additional set of haploid chromosomes is derived from the father.

Common deletion syndromes

Most of the terminal and interstitial deletion syndromes which have been identified are extremely rare. However, two specific deletion syndromes, involving the short arms of chromosomes 4 and 5, occur with sufficient frequency to be classified as well-recognized chromosome disorders.

Wolf–Hirschhorn syndrome [del(4p)]

Babies with the Wolf–Hirschhorn syndrome have a deletion of variable size involving the short arm of chromosome 4 but always including the band 4p16. They are small with microcephaly, a characteristic facial appearance (Fig. 3.9), and a high incidence of congenital heart defects and other malformations. Over 70% survive beyond early childhood. Usually they show

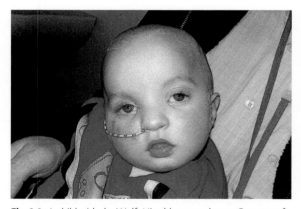

Fig. 3.9 A child with the Wolf–Hirschhorn syndrome. Courtesy of Dr William Reardon, National Centre for Medical Genetics, Dublin.

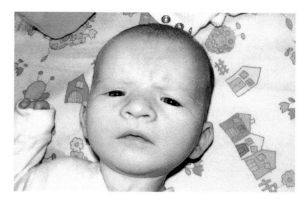

Fig. 3.10 Facial view of an infant with the cri-du-chat syndrome. Courtesy of Dr William Reardon, National Centre for Medical Genetics, Dublin.

severe psychomotor and growth retardation. Around 80% of all cases arise as *de novo* deletions, with the remaining 20% occurring as the result of a parentally transmitted unbalanced translocation. The incidence at birth is approximately 1 in 90 000.

Cri-du-chat syndrome [del(5p)]

This condition is so named because babies with a deletion involving the terminal region of the short arm of chromosome 5 often present with a high-pitched cat-like cry. Other features include a round face with microcephaly (Fig. 3.10), and up to 50% have a cardiac abnormality. The rate of survival is high, but most affected children show severe psychomotor retardation, often in association with difficult behaviour. The incidence at birth is approximately 1 in 50 000.

Microdeletion syndromes

The limit of resolution using conventional light microscopy is approximately 10 Mb. Deletions which are smaller than this, and which therefore require special molecular cytogenetic techniques for their detection, are referred to as microdeletions. Using FISH it has been established that a large number of previously well-recognized but unexplained syndromes are caused by chromosome microdeletions (see Table 3.6). These are also referred to as **contiguous** gene deletion syndromes as they are caused by loss of a relatively small number of genes.

Williams syndrome

This is caused by a 1.5-Mb interstitial microdeletion involving chromosome 7q11.2. Affected children have a

TABLE 3.6 Features of the common microdeletion syndromes

Syndrome	MIM number	Deletion	Clinical features
Williams	194059	7q11.2	Supravalvular aortic stenosis, outgoing sociable personality, mild–moderate learning difficulties
WAGR	194072	11p13	Wilms tumour, aniridia, genitourinary abnormalities, mental retardation
Angelman	105830	15q11–13	Ataxia, convulsions, severe mental retardation
Prader–Willi	176270	15q11–q13	Neonatal hypotonia, obesity, hypogonadism, mental retardation
DiGeorge/Shprintzen	188400	22q11	Cardiac defects, cleft palate, mild learning difficulties

characteristic facial appearance with sagging cheeks and wide mouth in association with an engaging 'cocktail party' personality which conceals mild learning disability. Over 50% have either supravalvular aortic stenosis or pulmonary stenosis. At least 17 genes have been identified in the deletion region. These include *ELN*, the gene which encodes for elastin and which in turn regulates cell proliferation in arterial smooth muscle. Haploinsufficiency for *ELN* is thought to explain the proliferation of arterial smooth muscle cells which leads to vascular stenosis.

WAGR syndrome

The association of <u>W</u>ilms tumour, <u>a</u>niridia, <u>g</u>enitourinary abnormalities, and mental <u>r</u>etardation (hence WAGR) is caused by a deletion involving chromosome 11p13. The region contains several genes including *WT1*, mutations in which cause Wilms tumour in young children, and *PAX6*, loss of which causes aniridia.

Angelman and Prader–Willi syndromes

These conditions are considered in greater length in the section on genomic imprinting and uniparental disomy in Chapter 4. In approximately 70% of cases, Angelman syndrome and Prader-Willi syndrome result from *de novo* microdeletions involving chromosome 15q11–13. In Angelman syndrome these are maternally derived, whereas in the Prader-Willi syndrome they are paternally derived. Alternatively, these conditions can be caused by paternal or maternal uniparental

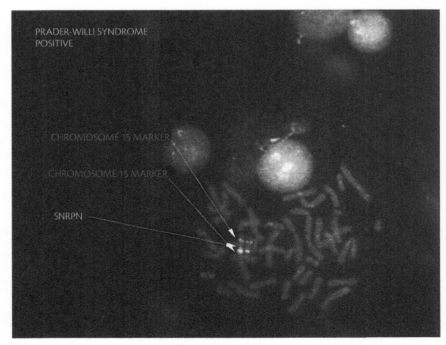

Fig. 3.11 FISH demonstration of a microdeletion in chromosome 15 causing the Prader–Willi syndrome. Courtesy of the Department of Cytogenetics, City Hospital, Nottingham.

disomy respectively (see p. 83). Essentially, absence of the chromosomal region 15q11–13 in the maternally derived chromosome causes Angelman syndrome; absence of the same region in the paternally derived chromosome causes the Prader–Willi syndrome.

Although the precise pathogenesis of these conditions has not been fully elucidated, much of the severe psychomotor retardation in Angelman syndrome is attributed to abnormal expression of *UBE3A*. This encodes a protein ligase thought to play an important role in the localization of proteins in the brain. *UBE3A* is expressed chiefly in the brain and only from the maternally derived allele. The main gene implicated in the Prader–Willi syndrome is known as *SNRPN* (small nucleoribonucleoprotein N). This encodes a protein involved in pre-mRNA splicing and processing in brain and muscle. Locus-specific probes for *UBE3A* and *SNRPN* are used in FISH to diagnosed microdeletions in the Angelman and Prader–Willi syndromes respectively (Fig. 3.11).

DiGeorge/Shprintzen syndrome

This condition has proved to be a relatively common cause of congenital heart defects such as interrupted aortic arch and ventricular septal defect. It is caused by a microdeletion involving chromosome 22q11, which in 90% of cases is 3 Mb in size and encompasses an estimated 30 genes. Affected children usually show mild

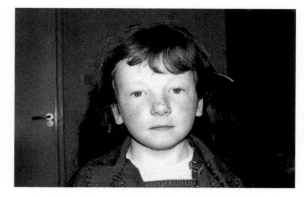

Fig. 3.12 Facial view of a child with the DiGeorge/Shprintzen syndrome. Courtesy of Dr William Reardon, National Centre for Medical Genetics, Dublin.

developmental delay as well as a cardiac abnormality, cleft palate, and a characteristic facial appearance (Fig. 3.12). Adults with the deletion show an increased incidence of severe psychiatric illness, including schizophrenia (p. 130).

Despite extensive research, no clear genotype/phenotype correlations have emerged for the common findings in this condition. However, in mice it has been shown that haploinsufficiency for *Tbx1*, the human homologue of which (*TBX1*) is located in the 22q11 deletion region, causes cardiac abnormalities. Mutations in

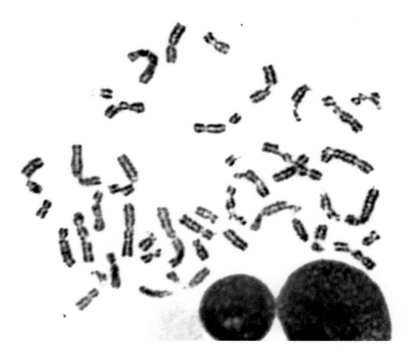

Fig. 3.13 Chromosome breakage in Fanconi syndrome. Courtesy of Applied Imaging Corp.

TBX1 have been identified in a few children who show clinical features of the DiGeorge/Shprintzen syndrome but do not have any obvious deletion in the critical region. Consequently haploinsufficiency for *TBX1* is thought to account for many of the cardiac defects seen in children with the DiGeorge/Shprintzen syndrome.

Chromosome instability syndromes

The complexity of the human genome is such that during cell division, when 3.2×10^9 DNA bases have to be faithfully replicated, it is almost inevitable that errors will occur. An elaborate system of DNA repair enzymes normally monitors the synthesis of DNA in the cell cycle and acts either by preventing replication of an error by arresting cell division before the S phase to allow repair, or by initiating repair after DNA synthesis. The repair system consists of two main pathways, one dedicated to the repair of single base errors, known as base excision repair, and the other responsible for the replacement of a tract of DNA up to 38 nucleotides in size.

Inherited abnormalities of these DNA repair mechanisms lead to a number of rare cancer-predisposing conditions. Some of these can be detected by the presence of increased chromosome breakage when chromosomes are exposed to agents, such as radiation, which can cause chromosome damage (Fig. 3.13). Agents which can induce chromosome breakage are known as **clastogens**.

The chromosome breakage syndromes listed in Table 3.7 are all rare and all show autosomal recessive inheritance. Ataxia-telangiectasia is characterized, as the name implies, by cerebellar ataxia, manifesting as an unsteady gait with poor coordination, and the development of small dilated blood vessels known as telangiectasia. In Bloom syndrome and xeroderma pigmentosum there is marked photosensitivity. Children with Fanconi syndrome develop anaemia in mid childhood and often have congenital malformations involving the radius and thumb. All of these conditions are associated with an increased risk of developing malignancies such as lymphomas in ataxia telangiectasia, lymphoreticular tumours in Bloom syndrome, leukaemia in Fanconi syndrome, and skin tumours in xeroderma pigmentosum. When children with these disorders are being investigated, it is important that special care is taken to limit the use of diagnostic radiation as this can increase the underlying risk of tumour development.

Summary

Chromosome abnormalities are found in 50% of spontaneous miscarriages and in approximately 1 in 200 liveborn infants. The most common aneuploidy condition is Down syndrome. This is caused by the presence of an additional number 21 chromosome (trisomy 21) in 95% of cases. Sixty per cent of babies with Down syndrome have one or more congenital abnormalities. The average mental age of young adults with Down syndrome is around 5–6 years. In the absence of a severe cardiac malformation, life expectancy in Down syndrome is good. The other autosomal trisomy syndromes (trisomy 13—Patau syndrome, trisomy 18—Edwards' syndrome) are associated with a high incidence of severe life-threatening malformations and poor survival beyond infancy.

There are four common sex chromosome aneuploidy syndromes: 45,X (Turner syndrome), 47,XXY (Klinefelter syndrome), 47,XXX (triple X syndrome), and 47,XYY. Physical abnormalities in these conditions are mild and

TABLE 3.7　Chromosome breakage syndromes

Disorder	MIM number	Basic defect	Chromosome breakage demonstrated by
Ataxia telangiectasia	208900	Failure to arrest cell division to allow normal DNA repair	Radiation
Bloom syndrome	210900	Helicase which unwinds DNA prior to replication and repair	Increased sister chromatid exchanges in mitosis[a]
Fanconi anaemia	227650	Repair of abnormal interstrand crosslinks and double-strand breaks	Mitomycin C and diepoxybutane
Xeroderma pigmentosum	278700	Nucleotide excision repair pathway	UV light and increased sister chromatid exchange[a]

[a] Normally up to 10 exchanges occur between sister chromatids during mitosis in a dividing cell. This number is greatly increased in Bloom syndrome and xeroderma pigmentosum. Sister chromatid exchanges can be demonstrated using a special culture technique.

intelligence is either normal or only mildly reduced. Individuals with Turner syndrome and Klinefelter syndrome are usually infertile.

Large numbers of deletion and microdeletion syndromes have been identified. Microdeletions, as identified by FISH, have been shown to be the cause of several previously unexplained conditions including Williams syndrome, Angelman syndrome, Prader–Willi syndrome, and DiGeorge/Shprintzen syndrome. An understanding of the basic molecular pathogenesis of some of these conditions is emerging as genes within the deletion regions are identified.

A small number of rare conditions, including ataxia telangiectasia, Bloom syndrome, Fanconi anaemia, and xeroderma pigmentosum, are associated with an increase in laboratory-induced chromosome breakage. These conditions all show autosomal recessive inheritance and convey increased susceptibility to malignancy.

Further reading

Gorlin RJ, Cohen MM, Hennekam RCM (2001) *Syndromes of the head and neck*, 4th edn. Oxford University Press, Oxford.

Jones KL (1997) *Smith's recognizable patterns of human malformation*, 5th edn. WB Saunders, Philadelphia.

Schinzel A (2001) Catalogue of unbalanced chromosome aberrations in man, 2nd edn. De Gruyter, Berlin.

Multiple choice questions

1 Characteristics of Down syndrome are that

  (a) karyotype analysis always shows a total of 47 chromosomes

  (b) the additional chromosome has usually originated in oogenesis

  (c) the karyotype usually shows triploidy

  (d) children with mosaicism are the most severely affected

  (e) all affected children have at least one serious congenital malformation

2 An individual with a 47,XXY karyotype will

  (a) have female external genitalia

  (b) be infertile

  (c) be at increased risk of psychopathic behaviour

  (d) probably have affected siblings

  (e) look normal at birth

3 Severe mental retardation is a feature of

  (a) Turner syndrome

  (b) Klinefelter syndrome

  (c) Wolf–Hirschhorn syndrome

  (d) Williams syndrome

  (e) DiGeorge/Shprintzen syndrome

4 Routine chromosome analysis would be expected to show an abnormality in

  (a) all children with Down syndrome

  (b) the mothers of most children with Down syndrome

  (c) approximately 50% of spontaneous first-trimester miscarriages

  (d) all children with Angelman syndrome

  (e) children with Fanconi syndrome

5 A newborn child with Down syndrome is found to have an unbalanced Robertsonian 14q;21q translocation and you are asked to counsel the parents. It would be correct to tell them that

  (a) Down syndrome is less severe when caused by a translocation as compared with full trisomy

  (b) both parents should be offered chromosome analysis as one of them could carry the translocation in a balanced form

  (c) if the mother is found to be a carrier then the recurrence risk is 50%

  (d) their baby should be referred for a cardiac evaluation

  (e) average life expectancy is less than 10 years

Answers

1 (a) false—most children with Down syndrome have 47 chromosomes but those with a translocation have a total chromosome count of 46

 (b) true—non-disjunction has usually occurred in maternal meiosis I

 (c) false—the karyotype usually shows trisomy, not triploidy

 (d) false—children with mosaicism are usually the most mildly affected

 (e) false—approximately 60% have a serious malformation

2 (a) false—a 47,XXY karyotype causes Klinefelter syndrome in which the phenotype is male. In general the presence of a Y chromosome always results in maleness, regardless of the number of X chromosomes present

 (b) true—most men with Klinefelter syndrome are infertile

 (c) false—men with Klinefelter syndrome tend to be passive rather than aggressive. Men with a 47,XYY karyotype have been noted to show psychopathic behaviour but this is uncommon

 (d) false—all of the sex chromosomal aneuploidy syndromes usually occur as sporadic events in a family

 (e) true—they show no external abnormalities at birth

3 (a) false—women with Turner syndrome are usually of normal intelligence

 (b) false—males with Klinefelter syndrome may have mild learning difficulties but these are rarely severe

 (c) true—this condition almost always results in severe learning disability

 (d) false—individuals with Williams syndrome usually have only mild learning disability

 (e) false—children with this condition often have a degree of learning disability but this is rarely severe

4 (a) true—either trisomy, an unbalanced translocation or mosaicism

 (b) false—in most cases both parents have normal chromosomes

 (c) true—it is often difficult to carry out chromosome analysis on abortus material but if successful this will show an abnormality in around 50% of cases

 (d) false—around 70% of cases have a microdeletion but this is usually only detectable by FISH

 (e) false—special techniques are usually needed to demonstrate increased breakage in Fanconi syndrome and the other breakage syndromes

5 (a) false—the clinical effects of trisomy 21 and an unbalanced Robertsonian translocation are identical

 (b) true—in around 25% of translocation cases one of the parents is a carrier

 (c) false—the recurrence risk is 10%–15% when the mother is a carrier

 (d) true—all babies with Down syndrome should be referred for cardiac evaluation

 (e) false—average life expectancy is over 50 years

Single-gene (Mendelian) inheritance

Human characteristics or disorders that have an underlying genetic basis are generally classified under the headings of chromosomal, single-gene, and polygenic (multifactorial). Chromosome disorders, as outlined in Chapters 2 and 3, result from loss or gain of a significant amount of chromosome material. Polygenic disorders (Chapter 6) are believed to represent the combined effects of several genes, each making a small contribution, possibly interacting with environmental factors (hence 'multifactorial').

In contrast, single-gene disorders are caused, as the name suggests, by an error in a single gene. Several patterns of inheritance can result, depending upon the chromosome on which the relevant gene is located and whether a disorder is caused by a mutation in only one of the relevant genes or in both copies. Over 12 000 human characteristics or disorders are now known to be caused by mutations in a single gene. Up-to-date information on all of these can be accessed at Online Mendelian Inheritance in Man (OMIM®) at *www3.ncbi.nlm.nih.gov/entrez/query.fcgi?db=OMIM* (p. 262). Note that the term 'Mendelian' is often applied in the context of single-gene disorders and their associated patterns of inheritance in recognition of the enormous contribution made by Gregor Mendel (Box 4.1).

Before considering individual patterns of single-gene inheritance it is necessary to explain a few important terms. An **autosomal** gene is one which is located on an autosome (i.e. one of the chromosomes numbered 1 to 22). An **X-linked** or **Y-linked** gene is located on the X or Y chromosome respectively. A **locus** (plural **loci**) is the position or site on a chromosome where a gene is

BOX 4.1 LANDMARK PUBLICATION: GREGOR MENDEL'S PLANT BREEDING EXPERIMENTS

Fig. 4.1 Etching of Gregor Mendel by August Potucek. Reproduced with permission of the Osterreichische Nationalbibliothek, Vienna.

Gregor Mendel (Fig. 4.1) was born in 1822 in the small town of Heinzendorf (now Hyncice in Moravia, part of the Czech Republic.) With the possible exception of the elucidation of the structure of DNA, his observations based on pea-breeding experiments constitute the greatest discovery in the history of genetics.

Mendel entered the Augustinian monastery in Brunn, now Brno, in 1843 and was ordained as a priest 4 years later. It seems that he was not particularly well suited to the life of a parish priest, preferring instead a vocation as a teacher. However, success in gaining teaching qualifications eluded him, and his intellectual energy and curiosity were invested instead in botany and specifically in a meticulous 8-year study of the patterns of inheritance of various observable characteristics of peas. In total he cross-pollinated 34 different kinds of pea, each with carefully selected features, raising over 10 000 plants in the process. His strategy was to use plants which were pure-bred (homozygous) for a characteristic such as height (tall or short) or seed pattern (smooth or wrinkled). When tall and short plants were cross-bred all of the first hybrid (F1) generation were tall. Similarly, when smooth and wrinkled plants were cross-bred, all of the F1 progeny were wrinkled. However, when plants from the F1 generation were cross-bred, then tall and short plants re-emerged in a ratio of three to one. Similarly with

the seed patterns, wrinkled and smooth plants were obtained in a ratio of 3 to 1. This is precisely what would be expected if tallness and wrinkled seeds are dominant characteristics, with smallness and smooth seeds being recessive.

The novelty of Mendel's observations lay in their contradiction of the prevailing dogma, propounded by Francis Galton, a cousin of Charles Darwin, that inherited characteristics blended. All of the plants should have been of intermediate height and all of the seeds should have been mildly wrinkled. Mendel's results were published in 1866 in the *Proceedings of the Natural Association* in Brunn. Sadly, his death from the combined effects of cardiac and renal failure in 1884 occurred long before his work began to receive due scientific recognition with the republication of his original paper in English in 1901 (see below for details). Based on the results of Mendel's experiments two fundamental laws of inheritance became established. These are generally referred to as Mendel's laws. The *law of segregation* states that each individual possesses two units of inheritance (genes) for each characteristic and that only one of these is transmitted to each offspring. The *law of independent assortment* states that genes at different loci segregate independently. Whilst this is generally true, it does not apply if two loci are linked (p. 88), an occurrence which Mendel could not have anticipated as the chromosomal basis of inheritance was not recognized until long after his death in 1903.

In the annals of genetics, Gregor Mendel , through his fundamental and painstaking observations, has achieved iconic status. His name lives on in OMIM (Online Mendelian Inheritance in Man) and in recognition of his contribution the term 'Mendelian' is used to refer to single gene inheritance. Although he was aware that his observations were of great potential interest and importance, it is a cruel irony that he died without ever receiving any public recognition of the enormity of his achievement.

Reference

Mendel G (1901) Experiments in plant hybridisation. *Journal of the Royal Horticultural Society*, **XXVI**, 1–32.

located. **Alleles** are alternative forms of the same gene. A **heterozygote** (adjective **heterozygous**) has different alleles at the same locus. A **homozygote** (adjective homozygous) has identical alleles at the same locus. **Genotype** is an individual's genetic constitution. **Phenotype** is the observed effect of the action of a gene or genes. **Heterogeneity** refers to diversity. **Locus heterogeneity** implies that a condition can be caused by mutations at different loci. **Allelic heterogeneity** indicates that a disease can be caused by several different mutations at the same locus. Finally, **dominant** implies that a single copy of an abnormal gene causes problems, whereas **recessive** indicates that both copies of the gene have to be abnormal for ill-effects to occur. It is useful to remember that dominant and recessive refer to the effects of the gene rather than to the gene itself. Also note that it is common practice to state that a particular disorder is 'recessive', when more correctly it should be stated that the disorder shows recessive inheritance.

Almost every consultation at a genetic clinic will involve the construction of a family tree, more formally referred to as a **pedigree**. This is one of the basic skills expected of a medical student. An approach to pedigree construction is provided in Chapter 13 (p. 244). To facilitate the process a standardized system of pedigree notation has been devised. The most commonly used symbols are shown in Fig. 4.2.

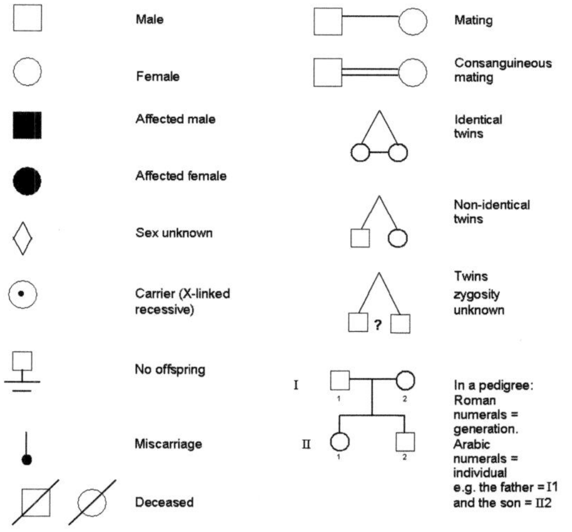

Autosomal dominant inheritance

A disorder which shows autosomal dominant inheritance is caused by an error in a single copy of a gene located on one of the autosomes. Thus an individual will be affected if he or she has one normal or **wild-type** allele, and one abnormal or **mutant** allele (Fig. 4.3). When this individual has children there is one chance in two that each child will inherit the mutant allele and therefore be affected. Thus, on average, half of the children will be affected and half will be unaffected.

The main characteristics of autosomal dominant inheritance are as follows:

♦ Both males and females can be affected and both can transmit the disorder to sons and daughters. This contrasts with X-linked disorders (p. 77).

Fig. 4.2 Examples of pedigree symbols.

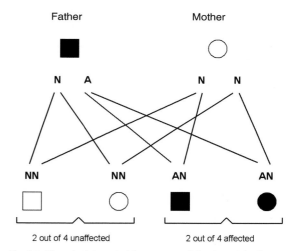

Fig. 4.3 The principle underlying autosomal dominant inheritance. N = normal (wild type) allele; A = abnormal (mutant) allele.

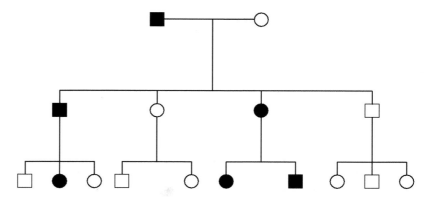

Fig. 4.4 A typical autosomal dominant pedigree showing male to male transmission.

◆ When an affected individual has children there is one chance in two that each child will be affected. This assumes that the other parent is not affected.

◆ In a family the disorder passes down through the generations, showing what is sometimes referred to as vertical transmission (Fig. 4.4).

Some of the more common disorders which show autosomal dominant inheritance, together with those which are used to illustrate various important principles throughout this book, are listed in Table 4.1.

Variable expression/expressivity

If a disorder shows variable **expression** this indicates that the disease manifestations or degree of severity vary from individual to individual within a family. This is illustrated by neurofibromatosis type 1 (p. 198), which can cause a wide range of abnormalities including mild learning disability, scoliosis, and malignancy. Fortunately most affected individuals show only multiple areas of increased skin pigmentation known as café-au-lait patches in childhood, with subsequent development of small soft benign skin tumours, known as neurofibromata, in adult life.

Reduced penetrance

Rarely, a disorder showing autosomal dominant inheritance will appear to skip a generation, passing from grandparent to grandchild with the intervening relative showing absolutely no manifestations. This is re-

ferred to as an example of non-penetrance. **Penetrance** is expressed mathematically as a proportion of 1 or as a percentage. Thus if only 90 out of 100 heterozygotes show evidence of the effects of a mutation, the penetrance is 0.9 or 90%. Well-known examples of disorders which show reduced penetrance are retinoblastoma and the rare autosomal dominant forms of breast cancer (p. 200). In the case of retinoblastoma (p. 191), approximately 90% of children who inherit a mutation develop one or more tumours. The 10% who escape tumour development probably do so because of failure of a 'second hit' (p. 193) to occur in a retinal cell at a critical time in early retinal development. Penetrance in the rare autosomal dominant forms of breast cancer is age dependent, and once again this probably reflects the accumulation of other mutations with the passage of time. Note that penetrance is an all or nothing phenomenon. Either there are manifestations, or there are none. As an aide-memoire, remember that penetrance is a population (p for p) statistic, whereas expression refers to the variable effects (e for e) in an individual.

Homozygosity

Most individuals affected with an autosomal dominant disorder are heterozygotes. Rarely homozygosity can occur, most commonly because of a mating between two heterozygotes, i.e. both parents are affected (Fig. 4.5). In most instances homozygosity results in a much more severe phenotype, as illustrated by death before or soon after birth in a baby homozygous for achondroplasia (p. 173), or early coronary artery disease in someone who is homozygous for familial hypercholesterolaemia (p. 213). Among disorders showing autosomal dominant inheritance, Huntington disease (p. 105) is almost unique in that homozygotes are no more severely affected than heterozygotes, probably

TABLE 4.1 Disorders which show autosomal dominant inheritance

Achondroplasia
Adult polycystic kidney disease
Familial adenomatous polyposis coli
Familial hypercholesterolaemia
Hereditary motor and sensory neuropathy
Hereditary non-polyposis colon cancer
Huntington disease
Marfan syndrome
Multiple endocrine adenomatosis (types 1 and 2)
Myotonic dystrophy
Neurofibromatosis (types 1 and 2)
Osteogenesis imperfecta
Retinoblastoma
Von Hippel–Lindau disease

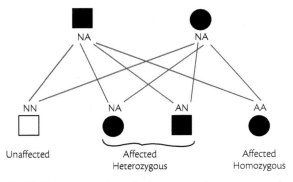

Fig. 4.5 The outcome of a mating between two heterozygotes each affected with an autosomal dominant disorder.

because the mutation has a gain-of-function effect (p. 13), which results in a non-dosage-dependent disruption of cell function. This very rare example of homozygosity being no more severe than heterozygosity is in accordance with the classical definition of dominance as occurring when the phenotype of a homozygote closely resembles that of a heterozygote for the same gene. In practice this does not usually apply in medical genetics.

> ### Key Point
>
> An autosomal dominant disorder is expressed in a heterozygote, who by definition has a single copy of the mutant allele. Both males and females are affected and both can transmit the disorder to children of either sex. Some autosomal dominant disorders show variable expression and reduced penetrance. Homozygotes are usually more severely affected than heterozygotes.

Codominance

When two dominant alleles are expressed equally when present in the heterozygous state, they are said to be **codominant**. Examples include genetic systems such as RFLPs (p. 142) and the ABO blood groups (p. 145). A child who inherits blood group A from its father and blood group B from its mother has an AB blood group as the A and B blood groups are codominant.

Autosomal recessive inheritance

A disorder which shows autosomal recessive inheritance is manifest only in the homozygous state, with the affected individual having two mutant alleles. With very rare exceptions, one of these mutant alleles will have been inherited from each heterozygous (carrier) parent (Fig. 4.6). When two parents who are both carriers have children, then on average half of their children will be carriers, one quarter will be homozygous unaffected and the remaining quarter will be homozygous affected. Thus the main characteristics of autosomal recessive inheritance are as follows:

- Both males and females are affected.

- Generally only members of a single sibship are affected (Fig. 4.7). In contrast to the vertical pattern of inheritance seen in autosomal dominant inheritance, this is sometimes referred to as horizontal inheritance.

- The probability that the sibling (brother or sister) of an affected individual will also be affected is 1 in 4 (25%).

- The probability that a *future* sibling of an affected individual will be a carrier is 1 in 2 (50%). However, the probability that an *existing unaffected* sibling is a carrier is 2 out of 3 (67%). This is because the fact that the sibling concerned is unaffected means that the denominator (i.e. 4 in 2/4) must now exclude the possibility of being affected.

Over 1000 disorders or traits showing autosomal recessive inheritance have been identified. These include most of the known inborn errors of metabolism (Chapter 11) and the common haemoglobinopathies (Chapter 8). The most common autosomal recessive disorders, and those used to illustrate important principles elsewhere in this book, are listed in Table 4.2.

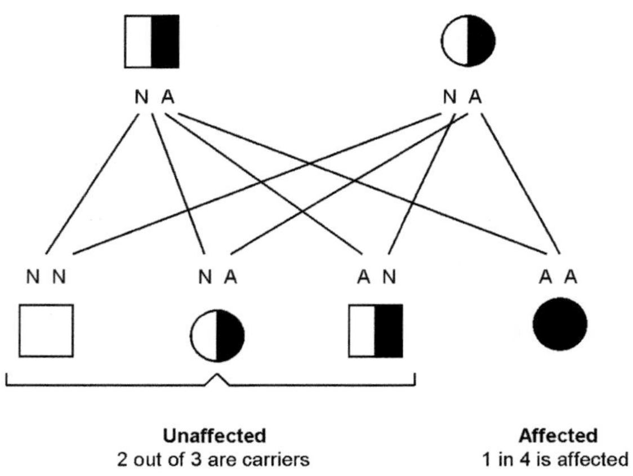

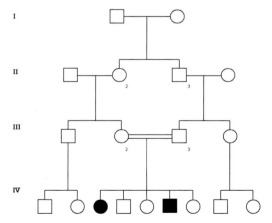

Fig. 4.6 The principle underlying autosomal recessive inheritance. N = normal (wild type) allele; A = abnormal (mutant) allele.

Fig. 4.7 A typical autosomal recessive pedigree showing affected brother and sister whose parents are first cousins.

TABLE 4.2 Disorders which show autosomal recessive inheritance

Alkaptonuria

Chromosome breakage syndromes (see Chapter 3)

Congenital adrenal hyperplasia

Cystic fibrosis

Galactosaemia

Haemochromatosis

Oculocutaneous albinism

Phenylketonuria

Pseudocholinesterase deficiency

Sickle-cell disease

Tay–Sachs disease

α- and β-Thalassaemia

Also most inborn errors of metabolism (see Chapter 11)

Autosomal recessive inheritance and consanguinity

Two individuals are **consanguineous** if they are related by descent from a common ancestor. Consanguineous relationships show an association with an increased risk for autosomal recessive disorders in offspring because related parents are more likely to carry the same rare mutant allele than unrelated members of the general population. In general, the rarer an autosomal recessive disorder, then the greater the frequency of consanguinity in the parents of affected children. Toulouse-Lautrec, the famous French impressionist painter, almost certainly had a rare bone disease known as pyknodysostosis as a result of his parents being first cousins (Box 4.2).

Generally it is assumed that most humans carry at least one deleterious autosomal recessive mutation. An estimate of the probability that related individuals will have a child with an autosomal recessive disorder can be made by simply calculating the probability that their child will inherit two copies of each common ancestor's mutant allele. Thus for the offspring of first cousins, there is 1 chance in 64 that a child will be homozygous for its great-grandfather's mutant allele and 1 chance in 64 that it will be homozygous for its great-grandmother's mutant allele (Fig. 4.9). This gives a total risk of 1 in 32 (3%) for a child of first cousins having an autosomal recessive disorder as a result of the parental relationship. Studies of the incidence of autosomal recessive disorders in the offspring of consanguineous matings are in agreement with this degree of

risk, and it is largely on the basis of the results of these studies that the assumption of one mutant allele per individual has been made.

Note that relationships between individuals who are closer that first cousins are associated with much greater genetic risks for offspring. For individuals who are second-degree relatives (see Table 6.1) the risk is 5–10%. For first-degree relationships, which are defined as incestuous and are almost universally illegal, the overall risk of abnormality in offspring is approximately 50%. Much of this risk is for mental retardation.

Pseudodominant inheritance

This term is applied when an autosomal recessive disorder appears to show autosomal dominant inheritance in a pedigree. This situation arises when there is a high incidence of carriers within the extended family so that many of the matings resulting in affected children are between heterozygotes and homozygotes (Fig. 4.10). Such a pedigree is sometimes encountered in a small inbred community or when there is assortative mating (p. 140) for a condition such as autosomal recessive hearing loss.

Compound and double heterozygosity

An individual with an autosomal recessive disorder who has different mutations in the two alleles which are causing his or her disease is referred to as a **compound heterozygote**. Compound heterozygosity is common in disorders such as cystic fibrosis (p. 107) and β-thalassaemia (p. 161), which both show marked allelic heterogeneity. In contrast, an individual who is a carrier of two different autosomal recessive disorders is referred to as a **double heterozygote**. He or she carries two conditions but is not affected with either.

The distinction between compound and double heterozygosity is illustrated in Fig. 4.11. This shows the outcome for children of parents with different forms of haemoglobinopathy. β-Thalassaemia and sickle-cell disease are both caused by mutations involving the β-globin locus on chromosome 11 (p. 154). Thus each child of the parents shown in Fig. 4.11A inherits a β-globin mutation for β-thalassaemia from its father and a β-globin mutation for sickle-cell disease from its mother. Each child will be a compound heterozygote, and clinically these children present with a picture similar to sickle-cell disease. (This is because the β-thalassaemia mutant globin chain is not synthesized, as explained in

BOX 4.2 CASE CÉLÈBRE: HENRI DE TOULOUSE-LAUTREC

Fig. 4.8 Henri de Toulouse-Lautrec

Henri de Toulouse-Lautrec was born in 1864 into a wealthy family of French aristocrats in which there was a strong tradition of artistic talent and a fondness for alcohol. Both of these familial tendencies were to characterize the life of this remarkable man, widely acknowledged to have been one of the greatest artists of his age. Even those who have little knowledge of, or interest in, art cannot fail to be impressed by the colour and vivacity depicted in his work, as illustrated by his renowned lithograph of the Moulin Rouge.

Unfortunately for Henri, he also inherited a form of short stature that blighted his childhood and made it impossible for him to live up to his austere father's high expectations. In early childhood he was noted to be short and to have a large head with an open anterior fontanelle. (This is the 'soft spot' in a baby's skull which normally closes at around 15 months.) His bones were abnormally fragile and he sustained fractures to both femora, which left him with an awkward gait. As an adult his unusual appearance made it difficult for him to develop normal social relationships. Increasingly he combined business with pleasure by painting, and overindulging in, the less salubrious aspects of Paris night-life, where he sought refuge in alcohol-induced oblivion and the company of prostitutes. Sadly, Henri's unhealthy lifestyle eventually took its inevitable toll, culminating in his premature death in 1901 from the combined ravages of alcoholism and syphilis.

Review of Henri's medical details has prompted the suggestion that he may have had a rare skeletal disorder known as pyknodystosis (p. 108). Henri's parents were first cousins, through his grandmothers who were sisters. Pyknodystosis shows autosomal recessive inheritance, which would be consistent with this history of parental consanguinity. The typical features include a large head, short stature, and a tendency to pathological fractures. The disorder is caused by mutations in *cathepsin K*, which encodes one of the lysosomal enzymes responsible for normal bone resorption in osteoclasts. To date it has not proved possible to prove this diagnosis through analysis of his remains or the surviving descendants of his close relatives. He himself had no (known) offspring. Nevertheless, on clinical grounds alone the diagnosis appears to be secure.

Those who wish to know of Henri's medical history can consult the references below. Alternatively, look out for the original version of *Moulin Rouge* directed by John Huston and released in 1952. The film starred José Ferrer, who walked on his knees in his portrayal of Henri.

References

Maroteaux P, Lamy M (1965) The malady of Toulouse-Lautrec. *Journal of the American Medical Association*, **191**, 715–717.
Frey J (1995) What dwarfed Toulouse-Lautrec? *Nature Genetics*, **10**, 128–130.

Chapter 8.) In Fig. 4.11B the father has α-thalassaemia due to mutations (usually deletions) involving the α-globin loci on chromosome 16 (p. 159), while, as in Fig. 4.11A, the mother has sickle-cell disease. Each child inherits an α-globin thalassaemia mutation from its father and a β-globin sickle-cell mutation from its mother. Thus each child is a double heterozygote. As these genes are at different loci the child is simply a carrier of both disorders and is not affected with either.

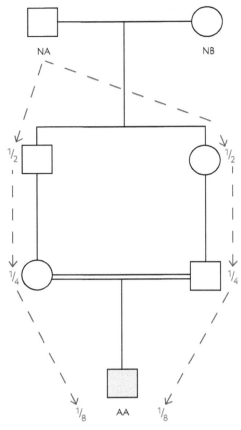

Fig. 4.9 The probability that a child born to first cousins will be homozygous for their common grandfather's mutant recessive allele (A). There is a probability of 1/8 × 1/8 = 1/64 that the child will inherit two copies of this allele.

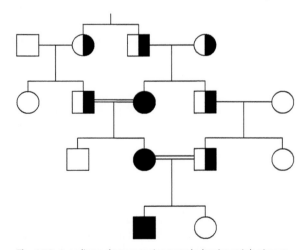

Fig. 4.10 A pedigree demonstrating pseudodominant inheritance. Heterozygotes are indicated as half-shaded.

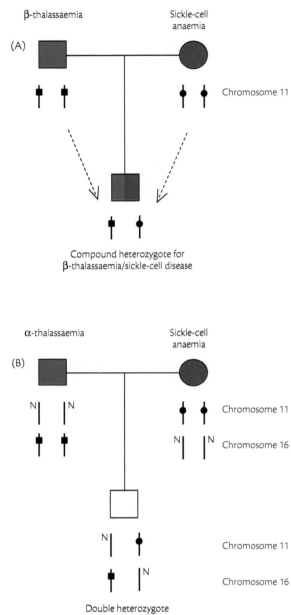

Compound heterozygote for β-thalassaemia/sickle-cell disease

Double heterozygote (unaffected)

Fig. 4.11 Diagrammatic representation of (A) compound heterozygosity and (B) double heterozygosity. N = normal allele.

A note on nomenclature—dominant and recessive

Autosomal dominant disorders are expressed in the heterozygote, autosomal recessive disorders in the homozygote. So how should we classify disorders in which heterozygotes have minimal or mild manifestations, while homozygotes are more seriously affected? For example, heterozygotes for familial hypercholesterolaemia (p. 213) have moderately elevated levels of serum cholesterol and develop coronary artery disease in middle age. Homozygotes have grossly elevated serum cholesterol levels leading to coronary artery disease in early adult life. Heterozygotes for sickle-cell disease can be detected by a simple blood test and may undergo red blood cell sickling at very low levels of oxygen tension, although in practice they are usually entirely healthy. Homozygotes have chronic anaemia and are subject to recurrent painful crises (p. 159).

It could be argued that both of these conditions show autosomal dominant inheritance, as they are both manifest in the heterozygote. Alternatively, they could both be said to show to show autosomal recessive inheritance as homozygotes, particularly in the case of sickle-cell disease, are much more likely to encounter serious ill health than heterozygotes. For practical purposes it is accepted that if a single copy of a mutant allele causes an adverse phenotypic effect, then the disorder is said to show dominant inheritance, whereas if two copies are needed to cause significant ill health then the disorder is said to show recessive inheritance. Thus familial hypercholesterolaemia is classified as an autosomal dominant condition, whereas sickle-cell disease is classified as autosomal recessive because only the homozygote (or compound heterozygote) becomes seriously ill.

Key Point

An autosomal recessive disorder is expressed only in homozygotes, who by definition have two copies of the mutant allele. Carriers (heterozygotes) are unaffected. When two carriers have children there is 1 chance in 4 that each child will be affected. The probability that an unaffected sibling is a carrier is 2/3. Rare autosomal recessive disorders are more common in the offspring of consanguineous relationships, as marriage between close relatives provides an opportunity for a rare recessive mutant allele to encounter itself.

X-linked recessive inheritance

A disorder or characteristic which shows X-linked recessive inheritance is caused by a mutation in a gene located on the X chromosome and is usually only manifest in males. Females are not affected as they normally have two X chromosomes and the normal allele compensates for the effects of the mutant allele. However, in

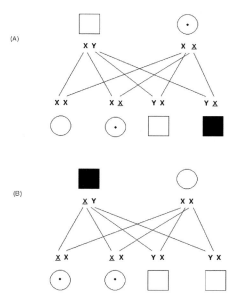

Fig. 4.12 The principle underlying X-linked recessive inheritance. $\underline{X}$ represents the mutant allele.

males, who normally have only a single X chromosome and are therefore described as being **hemizygous**, the X and Y chromosomes show little homology, so the condition resulting from the mutant allele is expressed.

The mechanism underlying X-linked recessive inheritance is illustrated in Fig. 4.12. A heterozygous female, who is often referred to as a carrier, transmits the mutant allele on average to half of her daughters, who will be carriers, and to half of her sons, who will be affected (Fig. 4.12A). An affected male transmits the mutant allele to all of his daughters, who are therefore *obligate* carriers, and to none of his sons (Fig. 4.12B). Thus the typical characteristics of X-linked recessive inheritance are as follows:

◆ Only males directly related through the female line are affected (Fig. 4.13).

◆ There is a 1 in 2 chance that each son born to a carrier female will be affected and a 1 in 2 chance that each daughter will be a carrier.

◆ All of the daughters of an affected male will be carriers, unless by chance they also inherit a mutant allele from their mother, in which case they will be affected. As a man transmits his Y chromosome to all of his sons, they will all be unaffected. Thus, in contrast to autosomal dominant inheritance, male to male transmission does not occur.

Over 200 disorders or characteristics showing X-linked recessive inheritance have been described. Some of the more common, together with those used

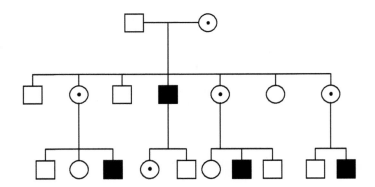

Fig. 4.13 A typical X-linked recessive pedigree.

TABLE 4.3	Disorders which show X-linked recessive inheritance

Androgen insensitivity

Bruton agammaglobulinaemia

Duchenne and Becker muscular dystrophy

Glucose-6-phosphate dehydrogenase deficiency

Haemophilia (A and B)

Hunter syndrome

Lesch–Nyhan syndrome

Menkes disease

Red–green colour blindness

to illustrate important basic principles elsewhere in this book, are listed in Table 4.3.

X-chromosome inactivation

Women have two X chromosomes, men have only one. Therefore in theory women would be expected to have twice the level of X-chromosome-encoded proteins as men. However, in practice women and men have roughly the same levels of X chromosome gene products in all body tissues.

This is explained by the fact that most of one X chromosome is inactivated in every cell in a developing female embryo at 1–2 weeks after conception. This is achieved by the spread of an inactivation signal from a gene known as *XIST* (X inactivation specific transcript) located at the X inactivation centre at Xq13.3. *XIST* is expressed on only one X chromosome in each cell, unless there are more than two X chromosomes, in which case it is expressed on one less than the total number of X chromosomes. Thus in a female with three X chromosomes (47,XXX—p. 61)

two X chromosomes are inactivated, and in a male with two X chromosomes (47,XXY) one is inactivated.

The inactivation signal extends to most of the X chromosome, but excludes a small portion at the tip of the short arm known as the **pseudoautosomal region**. The inactivation signal is maintained by methylation of CpG dinucleotides in the 5′ region of genes which would otherwise be expressed. In each cell it is random whether the paternally derived or the maternally derived X chromosome is inactivated, so that on average roughly half the cells in a female embryo express the paternal X and the other half express the maternal X. However, once an X chromosome has been inactivated in a cell, then all cells descended from that cell show the same inactivation pattern. Thus unless the process of X inactivation is non-random, as discussed later, roughly half the cells in an adult female will be expressing one X chromosome with the remaining half expressing the other X chromosome.

This process, which is sometimes referred to as **Lyonization**, in honour of Dr Mary Lyon who first proposed it on the basis of her observations in mice, explains how 'dosage compensation' occurs so that men have the same level of X chromosome activity as women. It also explains why women have a **Barr body** in each cell, a finding not present in normal men. The Barr body is a small mass of sex chromatin, which can be seen in interphase in the nucleus of cells from normal females. This represents the inactive X chromosome which replicates more slowly than the active X chromosome. Women with a 45,X karyotype (Turner syndrome, p. 59) have no Barr body whereas women with a 47,XXX karyotype have two. Before standard chromosome analysis became reliable, analysis of Barr bodies in a sample of cells from the buccal mucosa (hence the 'buccal smear') was used as a very unreliable test for confirming sex at sporting events.

A female with an X-linked recessive disorder

Rarely, a female is encountered with an X-linked recessive disorder. This can occur for one of several reasons:

◆ *Homozygosity.* If a female is homozygous for an X-linked recessive mutation then she will be affected in the same way as a male who is hemizygous. Homozygosity can arise in a female if her mother is a carrier and her father is affected, or if only one of her parents harbours the mutant allele and a new mutation occurs on the other X chromosome which she inherits. In practice it is very unusual to encounter a female who is homozygous for a serious X-linked recessive disorder, but homozygosity for less serious X-linked characteristics such as red–green colour blindness is quite common. This harmless type of colour blindness affects approximately 1 in 12 males and 1 in 144 females (i.e. 1/12 × 1/12)

◆ *Turner syndrome.* The commonest cause of Turner's syndrome is absence of the second sex chromosome (45,X—p. 59). From the point of view of an X-linked recessive mutation, a female with a 45,X karyotype is in the same position as a male with a 46,XY karyotype as she does not have a normal homologous allele.

◆ *Androgen insensitivity.* This is a very rare disorder in which insensitivity to testosterone ('androgen insensitivity') results in a chromosomally male embryo developing into a phenotypic female (p. 179). An X-linked recessive mutation in such a 'female' will have the same effect as in a normal chromosomal male.

◆ *Skewed X chromosome inactivation.* The process of X chromosome inactivation is random in that it is a matter of chance whether the paternally or maternally derived X chromosome is inactivated in each cell. In most women the paternally and maternally derived X chromosomes are each active in approximately 50% of cells. Just occasionally by chance this 50 : 50 ratio is skewed so that if the majority of cells have an active X chromosome which bears a recessive mutation the woman may show features of its effects. If one of the X chromosomes in a female is structurally abnormal then the inactivation process may not be random. For example, an X chromosome can take the form of a ring because of loss of material from both arms with reunion of the 'sticky' ends (p. 41). Cells in which this ring X chromosome is active will lack the products of those genes that are missing from the ring and will therefore be disadvantaged. Such cells will probably not survive in competition with normal cells, or the inactivation process may intrinsically recognize that one of the

X chromosomes is abnormal, resulting in its preferential inactivation. Either way the net effect is that the normal X chromosome is active in most cells. Therefore any recessive mutation on the normal chromosome will result in clinical problems.

> ### Key Point
>
> An X-linked recessive disorder is caused by a mutation on an X chromosome which is expressed in males but not (usually) in females. There is 1 chance in 2 that each son born to a carrier female will be affected and that each of her daughters will be a carrier. All of the daughters of an affected male will be carriers. None of his sons will be affected. Male to male transmission does not occur.

Other patterns of single-gene inheritance

The three most commonly encountered patterns of single-gene inheritance are autosomal dominant, autosomal recessive, and X-linked recessive. Rarer forms, which can provide a trap for the unwary in examinations, are considered in this section.

X-linked dominant inheritance

A disorder which shows X-linked dominant inheritance is caused by a mutation in a gene on the X chromosome which is manifest in a heterozygous female. Thus both

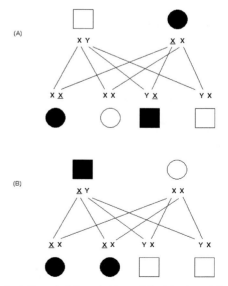

Fig. 4.14 The principle underlying X-linked dominant inheritance. X̲ represents the mutant allele.

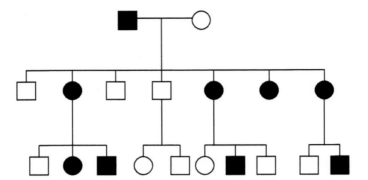

Fig. 4.15 A typical X-linked dominant pedigree. All the daughters and none of the sons of an affected male are affected.

males and females are affected. The underlying mechanism is illustrated in Fig. 4.14. The typical features of X-linked dominant inheritance are as follows:

◆ Both males and females are affected, although females are usually less severely affected than males.

◆ There is 1 chance in 2 that each son and daughter born to an affected female will be affected.

◆ All of the daughters and none of the sons of an affected male will be affected (Fig. 4.15). Male to male transmission cannot occur.

Only a small number of disorders showing X-linked dominant inheritance have been identified (Table 4.4). In several of these, affected females show patchy, mosaic involvement in tissues such as skin. This reflects the clonal origin of cells in the relevant tissue, with the areas of normal skin representing expression of the normal allele and the abnormal areas of skin reflecting survival of cell lines in which the mutant allele is active. This is particularly well illustrated by the condition known as incontinentia pigmenti (MIM 308310—Fig. 4.16), which is caused by mutations in *NEMO*. This gene normally activates a regulator of cell death, the process known as apoptosis (p. 168), so that

TABLE 4.4 Disorders which show X-linked dominant inheritance

Both males and females affected
Fragile X syndrome
Ornithine transcarbamylase deficiency
Vitamin D resistant rickets

Lethal in males
Aicardi syndrome
Incontinentia pigmenti
Rett syndrome

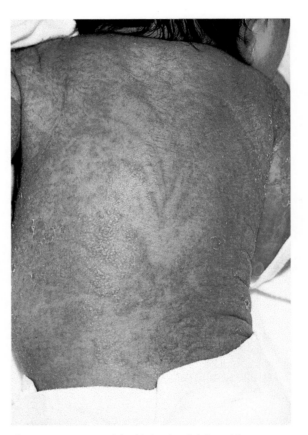

Fig. 4.16 Appearance of the skin in a female infant with incontinentia pigmenti showing mosaic involvement following the developmental lines of Blaschko.

in individuals with incontinentia pigmenti cells are very susceptible to apoptosis. This probably accounts for the fact that incontinentia pigmenti is usually lethal in affected males who only rarely survive to term. Thus affected females have surviving children with a female to male ratio of 2 : 1, consisting of affected females, unaffected females, and unaffected males in equal numbers.

In a few very rare X-linked dominant disorders women are so severely affected that they are unable to reproduce, so that all affected women represent single or 'isolated' cases. An example of such a condition is Rett syndrome (MIM 312750) in which affected girls show severe developmental delay with characteristic hand-wringing movements. Rett syndrome has been shown to be caused by mutations in *MECP2*, which normally suppresses gene transcription. The severe retardation seen in Rett syndrome probably results from uncontrolled transcription of genes expressed in early brain development.

> ### Key Point
>
> An X-linked dominant disorder differs from an X-linked recessive disorder in that it is expressed in heterozygous females who are usually less severely affected than hemizygous males. Some X-linked dominant disorders are lethal in males.

Y-linked inheritance

Only a small number of genes have been identified on the Y chromosome, most of which are involved in the determination of maleness and the maintenance of spermatogenesis. These include *SRY* (sex determining region of the Y chromosome—p. 178) and *DAZ* (deleted in azoospermia). A disorder which is Y-linked is said to show **holandric** inheritance and is transmitted by a man to all of his sons and to none of his daughters. The recognition of Y-linked genes necessary for spermatogenesis has had implications for procedures such as IVF with sperm retrieval for couples where the male partner has unexplained infertility. If this happens to be due to a mutation in a Y-linked gene then it will automatically be transmitted to all males in the next generation.

Pseudoautosomal inheritance

A gene is said to show **pseudoautosomal** inheritance, also known as partial sex-linkage, if it is located on the pseudoautosomal region of the X and Y chromosomes. These small regions at the tip of the short arm of each of these chromosomes are homologous and pair during meiosis I (see Fig. 9.14). A pseudoautosomal gene can be transferred between these chromosomes by a crossover event so that any associated disorder can appear to be X-linked in some parts of the pedigree and Y-linked in other parts. An example of such a condition is dyschondrosteosis (MIM 127300), also known as the Leri–Weill syndrome, which is caused by mutations in, or large deletions of, *SHOX*, which is located in the pseudoautosomal region. Individuals with dyschondrosteosis show short stature and subtle skeletal abnormalities. Loss of one copy of *SHOX* accounts for some of the short stature seen in girls with Turner's syndrome (p. 60).

Atypical Mendelian inheritance

Progress in molecular genetics has revealed that Mendelian inheritance is not always as straightforward as Mendel would have predicted. Several unexpected new genetic mechanisms have emerged.

Anticipation

A small number of disorders showing single-gene inheritance manifest at an earlier stage or with increasing severity in succeeding generations. This is referred to as **anticipation**. The underlying molecular mechanism involves expansion of a triplet repeat mutation (p. 12). Disorders caused by triplet repeat mutations can be divided into two groups, the first consisting of neurological conditions, such as Huntington disease (p. 105), which are caused by small CAG (polyglutamine) expansions. The second group, including myotonic dystrophy (p. 13) and fragile X syndrome (p. 112), are caused by much larger expansions. The size of the expansion is relevant to the observation of anticipation. Small expansions which can be carried by spermatozoa are associated with anticipation in Huntington disease when transmitted by men, whereas larger expansions, as seen in myotonic dystrophy and the fragile X syndrome, result in anticipation only when transmitted by women. This is almost certainly because large expansions can only be accommodated by ova.

New mutations and germline mosaicism

Many individuals with a single-gene disorder represent isolated cases within a family. This does not mean that their condition can be attributed to a non-genetic cause. In a few, very rare situations the condition may have resulted from environmental factors leading to a phenotype closely resembling that of the relevant genetic condition. This would be referred to as a **phenocopy**. For example, microcephaly, which can be defined as a head circumference falling three or more standard deviations below the mean, can be caused by mutations in several single genes or by environmental factors such as exposure to radiation or congenital infection (p. 182).

However, in most situations the disorder in question is clearly genetic in origin and will have arisen as the result of a new mutation. For example, all children with achondroplasia (p. 173) have a mutation in *FGFR3* (p. 171) and there is no other possible explanation. When an affected child reaches adult life, there is a probability of 1 in 2 that each child conceived by that individual will be affected. Although it is reasonably straightforward to counsel the affected individual, it is much more difficult to offer accurate information about recurrence risks to his or her unaffected parents. This is because it is impossible to know at what stage in gametogenesis the mutation first arose. If it occurred in a late mitotic division just before meiosis or in meiosis itself, then the recurrence risk for a future sibling will be essentially negligible. If, however, it arose in an early mitotic division in a germline stem cell then the relevant parent will almost certainly be producing other mutant-bearing gametes. This is referred to as **germline** or **gonadal mosaicism**. Although generally uncommon, and largely unexpected until recent years, germline mosaicism is encountered in a few conditions such as severe osteogenesis imperfecta (brittle bone disease—p. 15), in which the sibling recurrence risk, after the birth of an affected child to unaffected parents, has been shown to be approximately 5%

The issue of counselling the parents of an isolated case of an X-linked disorder such as Duchenne muscular dystrophy (p. 109) is also difficult. In this situation there are three possibilities. The mother of the affected boy could be a carrier with a recurrence risk of 1 in 2 for sons, or she could show germline mosaicism with an indeterminate recurrence risk of up to 1 in 2, or the mutation could have arisen in a meiotic division before conception, in which case the recurrence risk is negligible. Research has shown that the mother is a carrier in approximately two thirds of cases, that she shows germline mosaicism in 5–10% of cases, and that the remaining cases result from late mitotic or meiotic new mutations conveying a negligible recurrence risk for future siblings. In practice it can be extremely difficult to distinguish between these different possibilities.

> **Key Point**
>
> The possibility of parental germline mosaicism should be taken into account when counselling the parents of a child in whom a new mutation has occurred.

Digenic and triallelic inheritance

The notion that genetic disorders are either monogenic (single-gene) or polygenic has proved to be an oversimplification. A small but steadily increasing number of conditions have been shown to be caused by the additive effects of mutations in genes at two different loci, a concept referred to as *digenic* inheritance. This is not entirely surprising, as studies in mice have revealed that neural tube defects can be caused by the combined effects of mutations in two different genes, e.g. *Pax3*, a member of the Pax gene family, and *ct* (curly tail).

Different patterns of digenic inheritance have been recognized. One form of retinitis pigmentosa (MIM 180721), a progressive disorder which leads to severe visual loss, is caused by double heterozygosity for mutations in two unlinked genes, *ROM1* and *peripherin*, which both encode proteins present in photoreceptors. Individuals with only one of these mutations are not affected. A different mechanism has been described in another condition in which retinal changes occur, known as Bardet–Biedl syndrome (MIM 209901). Traditionally this was thought to show autosomal recessive inheritance, but in some families the disorder occurs only if an individual has both a dominantly inherited susceptibility mutation at one locus and recessive allelic mutations at another locus. This requirement for three mutations has been referred to as **triallelic** inheritance.

The observation that Hirschsprung disease, a disorder of intestinal neuronal development, can be caused by mutations at a relatively small number of loci has generated interest in the concepts of **oligogenic inheritance** and **synergistic heterozygosity**. This is considered further in Chapter 6.

Genomic imprinting

According to the principles of classical Mendelian inheritance, as outlined earlier in this chapter, identical alleles on homologous chromosomes are expressed equally. Although this holds true for the vast majority of the human genome, it has emerged that a small number of genes, located mainly on chromosomes 6, 7, 11, 14, and 15, show different expression dependent upon the parent of origin. This is referred to as genomic **imprinting**. Thus the genes are tagged (imprinted) with a distinguishing molecular message as they pass through meiosis in either a sperm or an ovum. The underlying mechanism whereby the imprint is established is thought to involve methylation leading to inactivation of transcriptional regulatory regions. This alteration in the expression of a gene without alteration of its structure is described as an **epigenetic** phenomenon. Once established, the parental imprint persists through DNA replication and cell division throughout life but is removed during gametogenesis and then re-established according to the sex of the transmitting parent.

An early example of imprinting noted in animal research was that a zygote generated from two sperm (*androgenetic*) developed a reasonable placental system but with very poor embryonic development, whereas the opposite effect was observed if the zygote was generated from two ova (*gynogenetic*). A parallel observation in humans is that an androgenetic zygote leads to absence of an embryo at the expense of a potentially malignant growth of the placenta known as a hydatidiform mole (see Fig. 3.8), whereas a gynogenetic zygote results in the development of an ovarian

teratoma. This phenomenon of genomic imprinting emphasizes the importance of an equal parental contribution to the zygote.

Uniparental disomy

Mendel's law of segregation (see Box 4.1) stipulates that genes, and by implication chromosomes also, segregate independently so that each member of a pair of homologous chromosomes has a different parental origin. In practice this is usually true, but a small number of individuals have been described in whom both homologous chromosomes have originated from the same parent. This is referred to as **uniparental disomy** (UPD). The most plausible explanation for UPD is that non-disjunction has occurred in one of the parents, resulting in the formation of a disomic gamete. This is then fertilized by a normal monosomic gamete, resulting in

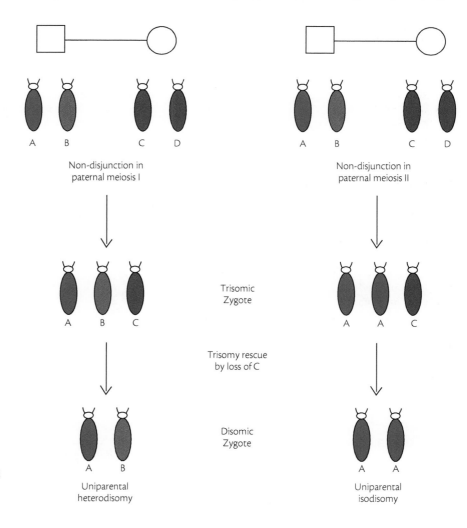

Fig. 4.17 The generation of uniparental heterodisomy and isodisomy through trisomy rescue.

a trisomic zygote (Fig. 4.17). Subsequent jettisoning of one of the three relevant chromosomes, a process referred to as *trisomy rescue*, restores the normal disomic constitution so that the zygote survives and development continues. If the jettisoned chromosome originated in the normal monosomic gamete then the conceptus will have UPD for the relevant chromosome pair. Less likely alternative explanations are that the disomic gamete is fertilized by a monosomic gamete (*gamete complementation*) or that there is duplication of a chromosome after fertilization to correct a monsomic gamete (*monosomy rescue*). If the UPD involves the presence of both copies of one parent's chromosomes, implying non-disjunction in meiosis I,

this is referred to as *uniparental heterodisomy*. *Uniparental isodisomy* indicates that two copies of one parental chromosome are present, consistent with non-disjunction in meiosis II or monosomy rescue.

UPD may often go unnoticed and is not detectable by routine chromosome analysis. Problems can arise if it involves a chromosome carrying a recessive mutation which becomes 'doubled-up' in the child; this results in the very rare scenario of parents, only one of whom is a carrier, having a child with an autosomal recessive disorder, or even more rarely a daughter with an X-linked recessive disorder. This has been reported for a small number of children with cystic fibrosis and other much rarer autosomal recessive disorders. UPD can also be unmasked if it involves an imprinted chromosome as the child in whom it has occurred will not have inherited a correctly imprinted chromosome from each parent.

Angelman and Prader–Willi syndromes

Confirmation that imprinting and UPD can be important in humans first came to light through the study of these two conditions. Children with Angelman syndrome (AS) (MIM 105830) show severe developmental delay in association with seizures, very limited speech, ataxia, and a distinctive and inappropriately happy personality (Fig. 4.18). Children with Prader–Willi syndrome (PWS) (MIM 176270) are floppy in infancy and show mild developmental delay in association with

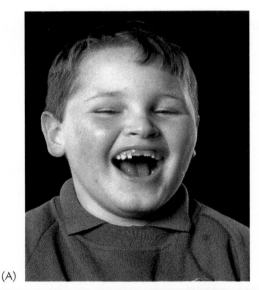

Fig. 4.18 The facial appearance of two boys with Angelman syndrome. The boy in (A) has paternal uniparental disomy. The boy in (B) has a mutation in *UBE3A*. (B) courtesy of Dr William Reardon, National Centre for Medical Genetics, Dublin.

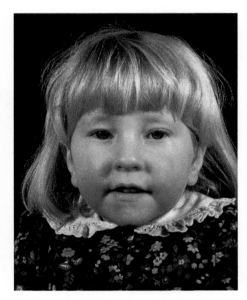

Fig. 4.19 The facial appearance of a young girl with Prader–Willi syndrome.

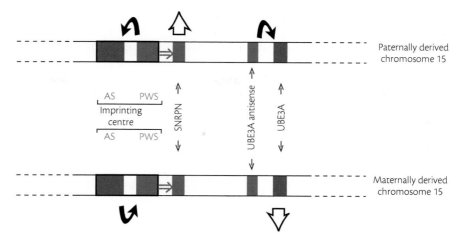

Fig. 4.20 Schematic representation of Prader–Willi syndrome/Angelman syndrome critical region on chromosome 15. Genes shown in blue are expressed from the paternal chromosome. Genes shown in red are expressed from the maternal chromosome. Arrows in black indicate suppression. Arrows in white indicate gene expression.

(A) Normal inheritance

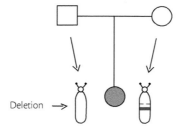

(B) Microdeletion

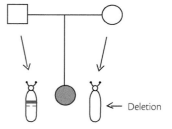

(C) Uniparental disomy

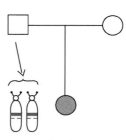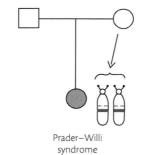

Fig. 4.21 Common causes of the Prader–Willi and Angelman syndromes.

Prader–Willi syndrome

Angelman syndrome

an insatiable appetite, obesity, hypogonadism, and challenging behaviour (Fig. 4.19).

Both of these conditions are caused by disturbed function of a group of genes located at the so-called PWS/AS critical region on chromosome 15q11–13 (Fig. 4.20). Key genes in this region include *SNRPN* and *UBE3A*, which are actively expressed only from the paternal and maternal alleles respectively, i.e. they are *imprinted*. Closely adjacent to the *SNRPN* locus there is a bipartite imprinting centre which shows differential expression depending upon the parent of origin. On the maternally derived chromosome the AS imprinting centre is active and is responsible for silencing *SNRPN* through methylation of its promoter. Conversely, on the paternally derived chromosome the PWS imprinting centre is active, leading to activation of several closely adjacent genes. These include the *UBE3A antisense* gene which produces a transcript preventing expression of *UBE3A*.

Anything that disturbs the normal function of the paternally or maternally derived PWS/AS critical region results in either Prader–Willi syndrome or Angelman syndrome respectively (an aide-memoire—paternal for PWS). Approximately 70% of cases of Prader–Willi syndrome result from 4-Mb microdeletions involving the paternally derived number 15 chromosome, as demonstrated by FISH (p. 64). Approximately 25% of cases result from maternal UPD for chromosome 15. The remaining cases are due to a mutation or tiny deletion in the PWS imprinting centre or *SNRPN* promoter. Similarly, Angelman syndrome is caused by microdeletions involving the maternally derived number 15 chromosome in approximately 70% of cases and by paternal UPD for chromosome 15 in 2–3% of cases (Fig. 4.21). Most of the remainder are caused by mutations in *UBE3A* or another gene involved in the imprinting process.

> ### Key Point
>
> A small proportion of the human genome is imprinted. Imprinted genes are expressed differently depending upon the parent of origin. A group of genes at chromosome 15q11–13 are imprinted. Loss of the paternal contribution results in Prader–Willi syndrome. Loss of the maternal contribution causes Angelman syndrome.

Mitochondrial inheritance

Most DNA in humans is located in the cell nucleus. However, mitochondria, which are present in the cytoplasm, contain a small circular chromosome, approximately 16.5 kb in length, consisting of 37 genes encoding ribosomal RNA, transfer RNA, and 13 subunits of the oxidative phosphorylation system. The human embryo receives its mitochondria exclusively from the mother. Each ovum contains approximately 100 000 mitochondria, in contrast to the sperm which contains less than 100. These are selectively eliminated soon after fertilization. This exclusive maternal transmission of mitochondria has been exploited by anthropologists interested in the origin of different populations such as the Polynesians, in whom the female population has been shown by molecular characterization of mitochondrial DNA to be largely descended from Papua New Guinea and surrounding

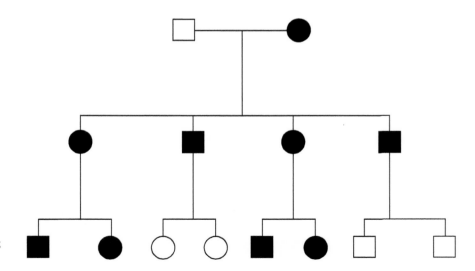

Fig. 4.22 A pedigree showing mitochondrial inheritance.

areas. Curiously, similar studies of the Y chromosome, which is of course exclusively transmitted through the male line, have revealed that approximately only two thirds of Polynesian males are of South-east Asian origin with the remainder having a Y chromosome of European origin.

The exclusive transmission of mitochondria by the female means that any disorder resulting from a mutation in mitochondrial DNA can only be maternally transmitted, resulting in a unique matrilineal pattern of transmission (Fig. 4.22) referred to as *cytoplasmic* or *mitochondrial inheritance*. Diseases caused by mutations in mitochondrial DNA mainly involve organs or tissues with high energy requirements such as muscle, heart, and brain (p. 216). Examples include a disorder known as MELAS (mitochondrial encephalopathy, lactic acidosis, and stroke-like episodes—MIM 540000) and Leber hereditary optic neuropathy (MIM 535000), which is characterized by sudden onset of severe visual loss in young adults (Box 4.3).

BOX 4.3 CASE HISTORY: LEBER HEREDITARY OPTIC ATROPHY

A healthy 26-year-old man, Mr C, was driving home from work when he noticed that the vision in his left eye was becoming blurred. On waking next morning he could not see clearly out of either eye. His local doctor could not find anything abnormal on examination but referred him for an urgent expert opinion. Later that day he was seen by an experienced ophthalmologist who noted very subtle changes in the appearance of the optic discs. At this point Mr C remembered that a great-uncle, a brother of his maternal grandmother, had lost his vision quite suddenly while fighting in the trenches in the First World War. At that time he was accused of being a malingerer and he was eventually discharged from the army with a diagnosis of 'shell-shock'. The fact that this great-uncle's vision had shown a slow partial recovery afterwards had simply served to confirm suspicion that his sudden blindness had been psychological in origin.

Fortunately Mr C was not referred for psychiatric assessment and instead a number of investigations were initiated, including mitochondrial DNA analysis for the common mutations which cause Leber hereditary optic atrophy (LHON). It subsequently emerged that he showed homoplasmy for the mitochondrial T14484C mutation and on the basis of this result it was possible to offer him cautious encouragement that his visual loss, which was now severe, might show a degree of spontaneous recovery. It was not lost on the family that their great-uncle had almost certainly also been affected and that posthumously he was now exonerated from any suggestion of 'cowardice' or 'psychological weakness'.

LHON is the most common mitochondrial disease with an incidence of approximately 1 in 25 000. It is characterized by acute onset of blurring or clouding of vision in one or both eyes, usually presenting in young adults and progressing to severe and often irreversible visual loss. Ninety-five per cent of cases are caused by one of the mutations, G3460A, G11778A, and T14484C. These all involve genes which encode complex 1 subunits of the respiratory chain. For reasons that are not understood, the T14484C mutation is associated with partial spontaneous recovery in around 50% of cases. Clinically, affected individuals eventually show quite marked optic atrophy, but in the early stages of the disease the eyes may appear normal, so a diagnosis of 'hysterical blindness' is sometimes considered.

A particularly curious feature of LHON is that only 50% of males and 10% of females with a pathogenic mutation ever develop symptoms. Possible explanations for this example of incomplete penetrance include the proportion of mitochondria harbouring the mutation (i.e. homoplasmy or heteroplasmy), the need for another coexisting nuclear mutation to be present, or exposure to a precipitating environmental factor such as cigarette smoke or vitamin deficiency. Despite extensive research, the underlying explanation remains unknown.

Unfortunately there is no effective treatment or cure for LHON and most affected individuals experience severe life-long visual loss. Genetic counselling is difficult because of the reduced penetrance. Affected males can be reassured that the risk to offspring is negligible because of the exclusive maternal inheritance of mitochondria, but risks for the offspring of maternally related females are considerable. For example, the observed risk that a maternally related male first cousin of a man with the T14484C mutation will be affected is close to 1 in 5.

Reference

Man PYW, Turnbull DM, Chinnery PF (2002) Leber hereditary optic neuropathy. *Journal of Medical Genetics*, **39**, 162–169.

In practice it is often difficult to predict accurate risks for offspring when a mother has a disorder caused by a mutation in mitochondrial DNA. This is because of the very large number of mitochondria in each cell and the uncertainty as to whether some or all of them harbour the mutation. These states are referred to as **heteroplasmy** and **homoplasmy** respectively. Heteroplasmy probably accounts for the wide range of phenotypic variability seen in disorders caused by mutations in mitochondrial DNA.

> ### Key Point
>
> Mitochondria are inherited exclusively from the ovum. Thus a disorder caused by a mutation in mitochondrial DNA can be transmitted by a woman to some or all of her children but cannot be transmitted by a man.

Genetic linkage

If two loci are closely adjacent on the same chromosome and alleles at these loci are observed to be inherited together more often than would be expected by chance, then these loci are said to be **linked**. Essentially this means that the loci are so close together that it is unlikely that they will be separated by a crossover during prophase of meiosis I (p. 35). Linked alleles on the same chromosome are said to be in **coupling**; those on opposite chromosomes are in **repulsion**. This positioning of linked alleles, i.e. in coupling or in repulsion, is referred to as the **linkage phase**.

The underlying principle is illustrated in Fig. 4.23. This shows three loci—L1, L2, and L3—with alleles A and a, B and b, and C and c respectively. On average there are two to three crossovers per pair of homologous chromosomes during each meiosis. Thus for loci which are widely separated, such as L1 and L2 in Fig. 4.23, there is a high probability that at least one crossover will occur between them. If it is observed that gametes contain an equal proportion of chromosomes with AB, Ab, aB and ab genotypes, then this confirms that the L1 and L2 loci are not linked.

For loci which are close together, such as L2 and L3, it is unlikely that they will be separated by a crossover, so gametes will contain significantly more chromosomes

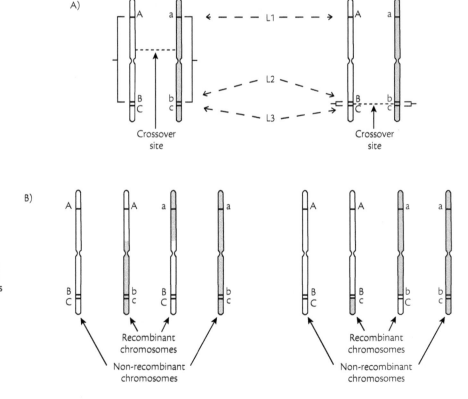

Fig. 4.23 Three loci L1, L2, L3, with alleles A and a, B and b, C and c, on a pair of homologous chromosomes. Loci L1 and L2 are widely separated and are not linked. Loci L2 and L3 are closely adjacent so that it is unlikely that a crossover will occur between them. Thus these loci are linked.

with either BC or bc than with Bc or bC. This demonstration would confirm that the loci are linked. Note that a group of alleles such as BC and bc, which are in coupling at closely linked loci, constitute what is known as a **haplotype**.

It is important to emphasize that linkage has nothing to do with cause. If two loci are linked, this simply indicates that they are closely adjacent on the same chromosome. Linkage does not in any way imply that an allele at one locus influences expression of an allele at the other locus.

> **Key Point**
>
> Two loci are linked if alleles at these loci do not segregate independently.

Recombination fraction

When the linkage phase at two linked loci is known in a parent, the study of offspring enables the frequency with which a crossover occurs between these two loci to be determined. This is known as the **recombination fraction** and is usually denoted as θ (Greek letter theta). In Fig. 4.24 the inheritance or **segregation** of alleles at

two linked loci is shown in a three-generation family. The linkage phase is known with certainty in II1 as he must have inherited an AB chromosome from his father (I1) and an ab chromosome from his mother (I2). Thus in II1 A and B are in coupling and A and b are in repulsion. Inspection of the genotypes in the 10 offspring in generation III reveals that AB or ab have been transmitted intact by the father to his first 9 children. However, his tenth child has inherited a recombinant chromosome with an Ab haplotype. This means that a rough estimate of 1 in 10 can be made for the recombination fraction. In practice many more families would be studied to obtain a more accurate indication of the correct figure.

If study of a large sample indicates that the recombination fraction is definitely less than 0.5, this being the value that is obtained by simple chance segregation of any two randomly selected loci, then this indicates that the loci are definitely linked. In clinical practice DNA markers at loci known to be closely linked to a disease locus can be used for preclinical diagnosis, prenatal diagnosis, and carrier detection. This is explained later in this chapter. Generally this approach is only used when specific mutation analysis is not possible and when the disease and marker loci are linked with a recombination fraction of 0.05 (1 in 20) or less.

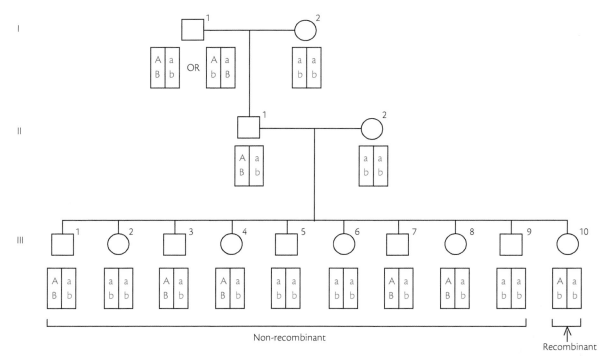

Fig. 4.24 Segregation of two linked markers in three generations. Analysis of the pattern of segregation in the third generation indicates that there has only been one crossover (in III10) indicating that the recombination fraction is 0.1.

> **Key Point**
>
> The recombination fraction, denoted by θ, indicates the probability that alleles in coupling at linked loci will be separated by a recombination event in meiosis 1.

Linkage and physical distance

The linkage distance, also known as the *genetic distance*, between two loci reflects the likelihood that a crossover, also known as a *recombination*, will occur between them. This is measured in **centiMorgans** (cM), which are also known as map units. If two loci are 1 cM apart, then a crossover will occur between them once in every 100 meioses. Thus, for two linked loci, if θ = 0.01, the distance between the loci is 1 cM. Similarly, if θ = 0.02, the distance between the two loci is 2 cM.

Note that genetic or linkage distance is not the same as physical distance, which is measured in base pairs. The genetic length of the human genome in males is approximately 3000 cM. The physical length is approximately 3.2×10^9 base-pairs. Thus 1 cM corresponds to approximately one million (1×10^6) base-pairs or 1 Mb. The relationship between genetic and physical length is not linear; some areas of the genome are more prone to recombination events than others, being referred to as recombination *hotspots*. For reasons which are not understood, but which probably relate to the fundamental differences in meiosis between the sexes, recombination occurs much less often in males than in females in whom the linkage length of the genome has been estimated to be approximately 4200 cM.

LOD scores

The method most commonly used to confirm or refute linkage between two loci involves a maximum likelihood estimate, which requires calculation of a LOD score (LOD stands for log of the odds). A LOD score, which is sometimes denoted by Z, represents the logarithm to the base 10 of the likelihood ratio of linkage for a particular value of θ to non-linkage. LOD scores are often referred to in scientific papers announcing the results of research aimed at identifying a disease locus. For practical purposes a LOD score of 3 or more is taken as mathematical proof of linkage, with a score of –2 or less taken as confirmation that two loci are not linked. Thus a statement in an abstract to the effect that 'Linkage between disease x and polymorphic marker locus D2S37 has been demonstrated with a maximum LOD score of 3.0 at a recombination fraction of 0.05' can be interpreted as indicating that the locus for disease x has been mapped to chromosome 2, where it is closely linked to marker locus D2S37 such that a crossover occurs between the two loci during on average 1 in 20 meioses. The mathematical probability that these two loci are linked as opposed to not linked is 1000 to 1 (if the LOD score is 3 then the odds for linkage are 10^3 to 1).

The principle underlying the use of LOD scores can be demonstrated using the pedigree shown in Fig. 4.24. If the loci are linked then the linkage phase is known with certainty in individual II2, in whom A and B are in coupling. For θ = 0.1, the probability that III1 would inherit both A and B from his father is 0.9. A similar probability applies to individuals III2 to III9. The probability that III10 would inherit A and b from his father is 0.1 (as there has been a crossover in the meiosis that generated his paternal gamete). Thus the probability of observing the genotypes in generation III if the loci are linked given a value of θ of 0.1, is $0.9^9 \times 0.1$. The probability of observing the genotypes in generation III if the loci are not linked is 0.5^{10}, as there is 1 chance in 2 (0.5) that alleles at any two unlinked loci will segregate together. Therefore the LOD score for a value of θ of 0.1 is $\log_{10} (0.9^9 \times 0.1)/0.5^{10}$, which equals $\log_{10} 44.08 = 1.644$. In practice LOD scores would be calculated for other values of θ to identify the value which gives the maximum LOD score and correspondingly the maximum likelihood value of θ.

Students should not be unduly concerned if they find this to be a difficult concept, as it would be very unusual for a student to be asked to calculate a LOD score. Simply remember that the calculation of LOD scores provides a method for proving or disproving linkage between two loci.

Linkage disequilibrium

Linkage disequilibrium can be defined as the association of specific alleles at linked loci more or less often than would be expected by chance. Consider two closely linked loci A and B, at each of which there are two possible alleles, A1 and A2, and B1 and B2. Each of these alleles has a frequency of 0.5. Therefore the frequency of each haplotype would be expected to be 0.25, i.e.

Haplotype	Frequency
A1 B1	$0.5 \times 0.5 = 0.25$
A1 B2	$0.5 \times 0.5 = 0.25$
A2 B1	$0.5 \times 0.5 = 0.25$
A2 B2	$0.5 \times 0.5 = 0.25$

If formal analysis of these haplotypes in a population shows that they all have a frequency of 0.25 then these loci would be said to be in linkage equilibrium. However, if significant deviation from these frequencies is observed, this would be an example of linkage disequilibrium

How does linkage disequilibrium arise? In fact it would be more appropriate to reflect on how linkage equilibrium arises. Consider once again the two-locus, two-allele A1A2 B1B2 system. At some point in history only a single allele existed at one of these loci, e.g. A1. Then a mutation occurred giving rise to A2 on a chromosome where there was a closely adjacent B2 allele. Thus for several generations in this population allele A2 was in coupling with B2, until crossovers began to lead to a small proportion of chromosomes having an A2B1 haplotype. Eventually, many generations later, equilibrium was achieved when the proportions of A2B2 and A2B1 haplotypes became equal. The number of generations required to reach equilibrium depends on the genetic distance between the two loci; the closer they are, the longer it will be before equilibrium is achieved.

Against this background linkage disequilibrium can be viewed as a snapshot phenomenon occurring during the evolution of a population. With the passage of time it diminishes, as a result of recombination between the relevant loci. Knowledge of disequilibrium between a disease locus and adjacent loci can be used to deduce the order of linked loci in positional cloning (p. 98) and to help identify susceptibility loci for multifactorial disorders (p. 123).

> **Key Point**
>
> Linkage disequilibrium refers to the presence of specific alleles in coupling at linked loci more often than would be expected by chance.

The use of linked markers in clinical genetics

If it is known that a disease locus is linked to another locus at which multiple alleles occur (i.e. a polymorphic locus), then this information can be used in a clinical setting to facilitate predictive testing, carrier detection, or prenatal diagnosis. This approach, which used to be referred to as *gene tracking*, is particularly useful when specific testing for the mutation is not possible, either because the disease locus has not been identified or because the disease gene is very large and the mutation cannot be identified in the relevant family.

Examples of how this technique can be applied in practice are given below. Note that a polymorphic marker, such as an SNP (p. 141) or a microsatellite

(p. 143), can be within the disease locus, in which case the recombination fraction can usually be assumed to be zero as it is very unlikely that recombination will occur between the mutation causing the disease and the closely adjacent polymorphism. Alternatively it can be tightly linked with a recombination fraction of 0.05 or less, so that the probability of a crossover between the disease and marker loci is less than 1 in 20. In the following examples and in most examples set in student examinations the possibility of recombination can be ignored.

Autosomal dominant inheritance

Figure 4.25 shows a three-generation family in which three individuals are affected with Marfan syndrome (p. 106). This disorder tends to be variable in its manifestations (i.e. it shows variable expression) and the parents in the second generation are concerned that one or more of their three apparently unaffected children could have inherited the disease mutation and therefore be at risk of developing complications in the future.

Linkage analysis using an intragenic polymorphism with alleles A and a shows that the disease mutation is linked with the A allele transmitted by I1 to his son II2. II3 has not inherited this A allele, so she can be confidently reassured that neither she nor her two children, III3 and III4, have inherited the condition. Note that both children have an A allele, but this has been transmitted from their father so it is a different A allele from that which is segregating with the disease allele in their uncle and cousin.

The parents of the other children, III1 and III2, can also be reassured that their son III1 has not inherited the condition as he has inherited his paternal grandmother's a allele rather than his paternal grandfather's A allele which is linked to the disease mutation.

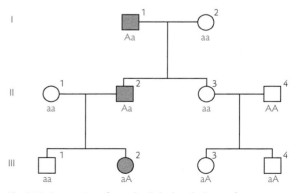

Fig. 4.25 Segregation of a marker linked to the locus of an autosomal dominant disorder in a three-generation pedigree.

Autosomal recessive inheritance

Figure 4.26 shows a nuclear pedigree in which the first child born to healthy parents has cystic fibrosis (p. 107). The parents request that prenatal diagnosis be carried out in their next pregnancy (II2). Mutation analysis is carried out in II1 but no mutations can be identified. However, an informative intragenic marker is found as shown in the pedigree. Inspection reveals that the disease allele is linked with allele A in both parents. Thus if the fetus inherits two A alleles it will be affected. If it inherits one A allele and one a allele it will be an unaffected carrier, although it will not be possible to determine which allele has been inherited from each parent. Finally, if it inherits two a alleles it will be unaffected as it will also have inherited its parents' two normal cystic fibrosis genes.

Figure 4.27 shows a similar situation, but this time the pedigree is not fully informative as it is not possible to establish the linkage phase in either parent. If the fetus, II2, inherits an AA or an aa genotype then the parents can be reassured that it will not be affected. If it inherits an Aa genotype then there is an equal chance that it will be affected (if it has inherited the same cystic fibrosis alleles as its brother II1) or unaffected (if it has inherited the opposite cystic fibrosis alleles to its brother). In this situation further analysis would have to be undertaken to try to identify a fully informative different polymorphic marker.

Note that the term 'informative' is used to indicate whether the results of the marker studies provide clear information about the linkage phase in the relevant transmitting parent(s). In Fig. 4.26 the results are fully informative. In Fig. 4.27 the results are only partially informative. If I1, I2, and II1 in Fig. 4.27 were all found to have an AA genotype, this marker would be classified as being totally uninformative.

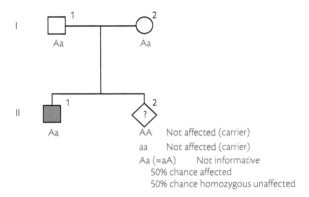

Not affected (carrier)
aa   Not affected (carrier)
Aa (=aA)   Not informative
    50% chance affected
    50% chance homozygous unaffected

Family 4.27. Segregation of a genetic marker linked to the locus of an autosomal recessive disorder in a two-generation family. In this pedigree the linked markers are not informative.

X-linked recessive inheritance

The pedigree shown in Fig. 4.28 is that of a family in which two boys, II3 and III2, are or have been affected with Duchenne muscular dystrophy (p. 109). The affected boy in the first generation (II3) died many years ago and none of his DNA is available for analysis. However, the fact that he was affected means that his sister, II2, who has had an affected son, must be a carrier. No specific mutation can be identified in this sister or in her affected son, III2. However, an informative intragenic linked marker has been identified as indicated, in that the mother is heterozygous for alleles A and a, having transmitted the a allele to her affected son. Furthermore it is known that she inherited an A allele from her unaffected father. Thus in this mother the disease mutation must be in coupling with marker allele a.

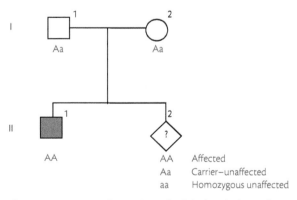

AA   Affected
Aa   Carrier–unaffected
aa   Homozygous unaffected

Fig. 4.26 Segregation of a genetic marker linked to the locus of an autosomal recessive disorder in a two-generation family.

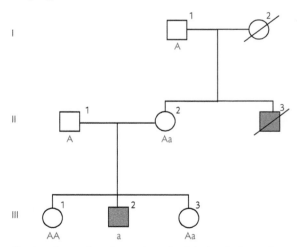

Fig. 4.28 Segregation of a genetic marker linked to the locus of an X-linked recessive disorder in a three-generation pedigree.

For most disorders it is assumed that the likelihood of recombination between an intragenic marker and a disease mutation is so low that it can be ignored, but because of the large size of the Duchenne muscular dystrophy gene it is generally assumed that there is a possibility of recombination between an intragenic marker and a Duchenne muscular dystrophy mutation, with a value of θ of 0.05. The two sisters of the affected boy wish to know if they are carriers. Both sisters must have inherited an A allele from their father as he will have transmitted his X chromosome to both of his daughters. The older sister, III1, has inherited an A allele from her mother whereas the younger sister, III3, has inherited her mother's a allele. This means that the older sister will only be a carrier if a recombination has occurred between the disease mutation and the marker locus in the maternal meiosis which preceded her conception, so she can be told that there is only a 5% (1 in 20) chance that she is a carrier. Unfortunately the younger sister, III3, has inherited the high-risk marker allele and therefore has to be informed that there is a 95% (19 out of 20) chance that she is a carrier.

> ### Key Point
>
> Polymorphic markers at a locus closely linked to a disease locus can be used to predict carrier and disease status in a family as long as the pattern of segregation is informative.

Summary

An autosomal dominant disorder is caused by a mutation in a single copy of a gene which can affect individuals in multiple generations. Both males and females are affected and male to male transmission can occur. The probability that an affected individual will transmit the disorder to each of his or her children is 1 in 2. An autosomal recessive disorder is caused by mutations in both copies of a gene and usually only affects members of a single sibship. Rare autosomal recessive disorders are more common in the offspring of consanguineous relationships. The probability that parents who are both carriers will have an affected child is 1 in 4.

An X-linked recessive disorder is caused by a mutation in an X chromosome which is recessive in females, who are carriers, but males are affected as they are hemizygous for the X chromosome. An affected male transmits his mutation to all of his daughters who are obligatory carriers. He transmits his Y chromosome to all of his sons, who are unaffected. When a carrier female has children there is 1 chance in 2 that each of her sons will be affected and 1 chance in 2 that each of her daughters will be a carrier. An X-linked dominant disorder is caused by a mutation which exerts a dominant effect in both males and females, so that both sexes are affected. Usually males are more severely affected than females and some X-linked dominant disorders are lethal in males. An affected male transmits the disorder to all of his daughters and to none of his sons. An affected female transmits the disorder on average to half of her sons and to half of her daughters.

Exceptions to Mendelian inheritance include anticipation, digenic inheritance, and uniparental disomy. Anticipation implies that a disorder becomes more severe or presents at an earlier age in successive generations. Disorders which show anticipation are caused by expansion of an unstable triplet repeat mutation. Disorders which show digenic inheritance are caused by the interaction of mutations at two different loci. Uniparental disomy occurs when an individual inherits both copies of a chromosome pair from one parent. This can cause a clinical problem if it results in the doubling up of a recessive mutation or if the chromosome is imprinted. A chromosome is imprinted if all or part of it is expressed differently depending upon the parent of origin.

Two loci are linked if alleles at these loci segregate together more often than would be expected by chance. The recombination fraction indicates the likelihood that alleles at two linked loci will be separated by a recombination event in meiosis I. A polymorphic marker locus linked to a disease locus can be used to predict disease or carrier status when specific disease mutation analysis is not possible.

Further reading

Engel E, Antonarakis SE (2002) *Genomic imprinting and uniparental disomy in medicine.* Wiley-Liss, New York.

Kingston H M (2001) *An ABC of clinical genetics*, 3rd edn. BMJ Books, London.

Ott J (1991) *Analysis of human genetic linkage.* Johns Hopkins University Press, Baltimore.

Vogel F, Motulsky AG (1996) *Human genetics*, 3rd edn. Springer-Verlag, Berlin.

Multiple choice questions

1 In achondroplasia all heterozygotes show marked short stature. Inheritance is autosomal dominant. Parents of normal stature have two affected children. The most likely explanation for this is

(a) variable expression

(b) reduced penetrance

(c) germline mosaicism

(d) pseudodominant inheritance

(e) pseudoautosomal inheritance

2 A woman is found to be affected with the common form of X-linked recessive colour blindness. This could be because

(a) she has Klinefelter syndrome

(b) she has Turner syndrome

(c) her mother is a carrier and her father is affected

(d) she has uniparental disomy for her father's X chromosome

(e) X-linked colour blindness is caused by a mutation in mitochondrial DNA

3 A boy is known to have Prader–Willi syndrome. This could have been caused by

(a) maternal uniparental disomy

(b) paternal uniparental disomy

(c) a deletion in the maternally inherited number 15 chromosome

(d) a deletion in the paternally inherited number 15 chromosome

(e) all of the above

4 If two loci show genetic linkage then

(a) they must be on the same chromosome

(b) mutations at both loci cause the same disease

(c) alleles at these loci interact to cause the same disease

(d) if the recombination fraction between these loci is 0.05 then this means that on average alleles in coupling at these loci will cosegregate in 95% of meioses

(e) alleles at these loci must show linkage disequilibrium

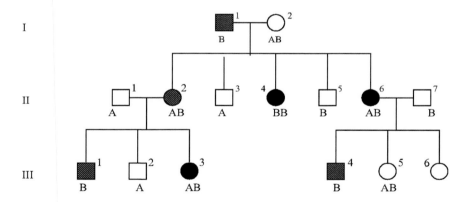

5 In this pedigree A and B represent alleles at a locus closely linked to the disease locus. Assuming no recombination, individual III6 can be informed that

(a) the condition in her family is showing autosomal dominant inheritance

(b) the disease allele is linked to allele B

(c) the disease is caused by allele B

(d) if she has inherited a BB genotype then she will be affected

(e) all sons born to her aunt II4 will be affected.

Answers

1 (a) false—if all heterozygotes show marked short stature then even with variable expression one of the parents would show a degree of short stature

(b) false—non-penetrance cannot apply if all heterozygotes show short stature

(c) true—this is the most likely explanation

(d) false—pseudodominant inheritance implies that the condition really shows recessive inheritance whereas the stem of the question states that inheritance is autosomal *dominant*

(e) false—pseudoautosomal implies that the locus is in the pseudoautosomal region on the X or Y chromosome and again the stem of the question clearly states that inheritance is *autosomal* dominant

2 (a) false—individuals with Klinefelter syndrome are male

(b) true—women with Turner syndrome show the same incidence of X-linked recessive disorders as men

(c) true—the woman could be homozygous having inherited a mutation from both parents

(d) true—she could have inherited two copies of her father's mutation

(e) false—the locus for colour blindness is on the X chromosome, not in the mitochondrial genome

3 (a) true—Prader-Willi syndrome results from loss of imprinted genes on the paternally derived number 15 chromosome. In maternal uniparental disomy both copies are inherited from the mother

(b) false—paternal uniparental disomy for chromosome 15 causes the Angelman syndrome

(c) false—this causes Angelman syndrome

(d) true—this is the most common cause of Prader-Willi syndrome

(e) false—only answers (a) and (d) are correct

4 (a) true—loci cannot be linked if they are on different chromosomes, as originally concluded by Mendel in his law of independent assortment

(b) false—linkage implies that the loci are closely adjacent but not that they interact at either the DNA or protein level

(c) false—as above

(d) true—the recombination fraction indicates the proportion of meioses in which two alleles in coupling will be separated by a crossover

(e) false—alleles at these loci *may* show linkage disequilibrium but this cannot be assumed without population studies

5 (a) false—the pattern of inheritance is X-linked dominant. There is no example of male to male transmission and males have only a single allele at the marker locus

(b) true—inspection of the pedigree indicates that the disease allele is segregating with allele B

(c) false—the disease is not *caused* by allele B, it is simply segregating with it

(d) true—she will inherit the mutant allele from her mother and will therefore be affected

(e) false—on average half of her sons will be affected depending on whether they inherit their grandfather's B allele or their grandmother's B allele

Identifying genes for Mendelian diseases

It can reasonably be argued that the most exciting development in medical research over the last 20 years has been the discovery of the genes that are responsible for serious Mendelian disorders such as Huntington disease, cystic fibrosis, and Duchenne muscular dystrophy. Previously genes could only be identified through knowledge of their protein product. This process, referred to as **functional cloning,** was applied very successfully to study the genes encoding proteins such as haemoglobin (the haemoglobinopathies—see Chapter 8), factor VIII (haemophilia A), and phenyl-alanine hydroxylase (phenylketonuria—p. 210).

Unfortunately, functional cloning strategies could not be used for conditions in which the molecular pathogenesis linking genotype and phenotype was unknown. In the mid 1980s this applied to most of the common Mendelian disorders for which patients were referred to genetics clinics. To address this issue a new approach, known as **positional cloning,** was devised. Instead of starting from knowledge of the disease process and the relevant protein as in functional cloning, positional cloning was based on molecular methods to locate (or *map*) and isolate the disease gene, followed by identification of its protein product (Fig. 5.1). As this represented a reversal of the traditional approach, positional cloning was sometimes referred to as the application of *reverse genetics*.

Over the last two decades, improvements in modern technology and knowledge gained from the Human Genome Project have combined to make positional cloning an extremely powerful method for identifying genes responsible for single-gene disorders. The basic

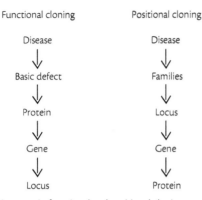

Fig. 5.1 Key steps in functional and positional cloning.

principles involved are outlined, followed by examples of how the processes of functional and positional cloning have been successfully employed to unravel the genetic basis of most of the more common single-gene disorders.

Functional cloning

Although different strategies have been applied, the general approach to functional cloning can be summarized in three stages:

- *Synthesis of a short oligonucleotide probe.* Based on a knowledge of the genetic code (p. 7) and the amino acid structure of the purified gene product, a short (hence 'oligo') probe is synthesized which can then be labelled and used for hybridization. A major problem is that because of the degeneracy of the genetic code (i.e. different nucleotide triplets code for the same amino acid), it is not possible to design an oligonucleotide probe which can be guaranteed to be a perfect match for the nucleotide sequence in the relevant gene. In early studies guesswork and fortune played an important role. Now a *degenerate oligonucleotide* is synthesized, consisting of a mixture of all of the possible nucleotide permutations. Ideally it is desirable that the probe should contain as many methionine and tryptophan residues as possible, as these are encoded by only a single codon.

- *Screening a DNA libarary.* The short oligonucleotide probe, suitably labelled, is now used to screen a DNA library constructed from a cell line in which the relevant gene is known to be expressed (e.g. immature red blood cells for the haemoglobin genes). Complementary DNA (cDNA) is DNA which has been synthesized from mRNA using reverse transcriptase. A cDNA library consists of a collection of cDNAs

generated from mRNA and then cloned or ligated into a suitable vector. Once established, the cDNA library is transferred to a nitrocellulose filter and screened by hybridization to the oligonucleotide synthesized from the protein product.

- *Screening a genomic library.* Once the cDNA for the relevant gene has been identified, it can be used as a probe to screen a genomic library constructed from total cellular DNA. In the original identification of the factor VIII gene, the genomic library was constructed using cells from an individual with four X chromosomes to enrich for the target sequences. Clones which contain the desired target sequences, i.e. the gene, are identified by hybridization to the cDNA and then analysed to obtain the full structure of the gene including exons, introns and promoter sequences.

An alternative and more contemporary approach for functional cloning involves the use of an antibody raised against the purified protein to screen a cDNA expression library. A cDNA expression library differs from a conventional DNA library in that the cDNAs from the relevant cell line are cloned into *expression vectors*, which contain promoters and regulatory sequences together with polymerase necessary for transcription and translation. Once identified on the basis of the protein product, the cDNA clone can be used to screen a genomic library to identify the relevant disease gene.

> **Key Point**
>
> In functional cloning a gene is identified through knowledge of the protein product.

Positional cloning

As the name implies, positional cloning involves the cloning or isolation of a gene through knowledge of its location in the human genome. As an aid to understanding, the procedure can be considered as consisting of two main stages: first, defining the approximate gene location, and second, identifying candidate genes within it (Fig. 5.2).

Mapping the candidate region

Ideally the candidate region should be as small as possible so that the number of genes to be analysed can be reduced to a minimum. The standard approach involves extensive linkage analysis utilizing as many

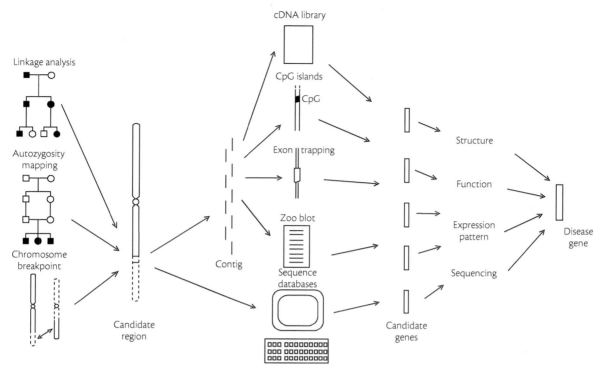

Fig. 5.2 Strategic approaches to positional cloning.

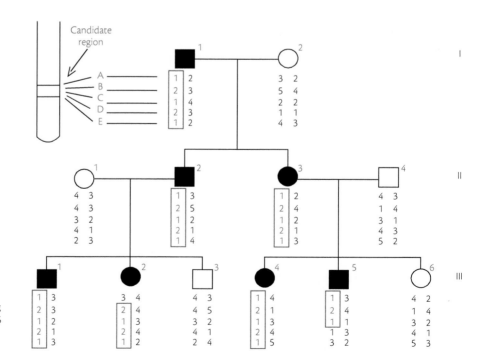

Fig. 5.3 Family tree showing how crossovers in III2 and III5 narrow the candidate region down.

affected families as possible. Alternative approaches include autozygosity mapping and the use of chromosome abnormalities.

Linkage analysis

Coarse linkage analysis is undertaken using a panel of polymorphic markers, traditionally microsatellites (p. 143), evenly distributed across the genome. As the genetic (linkage) distance of the human genome equals approximately 3000 cM, with 1 cM corresponding to approximately 1 Mb, a panel of 300 evenly distributed polymorphic markers should enable the candidate region to be narrowed down to around 10 cM, equivalent to 10 Mb. On average a region of this size would be expected to contain approximately 100 genes. (The human genome contains approximately 30 000 genes distributed over a physical distance of 3000 Mb with a genetic [linkage] distance of 3000 cM; see p. 90.)

Once an approximate location has been identified, further markers within this region are used in a search

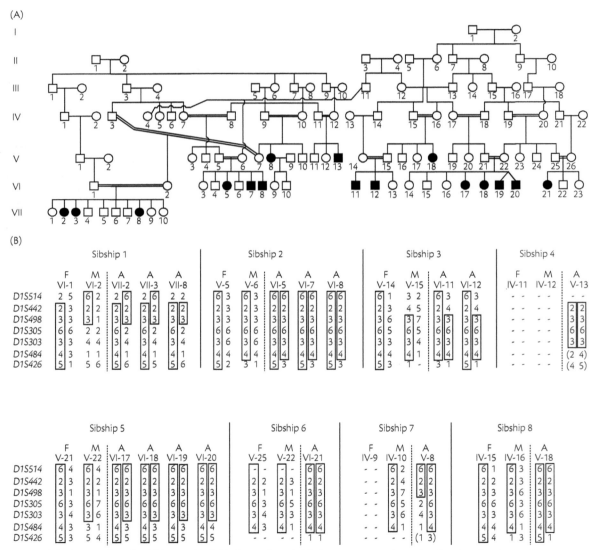

Fig. 5.4 (A) A large pedigree in which several individuals are affected with the rare autosomal recessive disorder pyknodysostosis. (B) Genotypes of affected individuals (A) and their parents (F = father, M = mother) for informative markers at the centromeric region of chromosome 1 where the disease locus was found to be located. Reproduced with permission from Gelb BD, Edelson JG, Desnick RJ (1995) Linkage of pycnodysostosis to chromosome 1q21 by homozygosity mapping. *Nature Genetics*, **10**, 235–237.

for recombinants to try to narrow the candidate region down to a physical size of 1 Mb or less. The underlying principle is illustrated in Fig. 5.3. Two recombinations have occurred within the candidate region in the meioses leading to the conceptions of III2 and III5. These indicate that the common region of overlap consists of only loci B and C, thereby implying that the disease gene is located in this narrow region. The ultimate goal is that by studying a large number of families it becomes possible to narrow the candidate region down to a sufficiently small size for more detailed physical analysis.

Autozygosity mapping

Markers or genes at a particular locus are described as **autozygous** if they are identical as a result of descent from a common ancestor. Autozygosity mapping is an extremely powerful method for mapping genes which cause autosomal recessive disorders by carrying out linkage analysis. The principle is illustrated in Fig. 5.4, which shows a large pedigree with multiple examples of consanguinity and several individuals affected by the rare autosomal recessive disorder, pyknodysostosis. If, as is likely, all of the affected individuals are homozygous by descent (i.e. autozygous) at the disease locus, then it is probable that they will also be homozygous by descent at closely adjacent loci. Thus if the family is analysed with multiple polymorphic markers distributed across the genome, and an autozygous region is

found which is shared by all affected individuals but by none who are unaffected, then it is extremely likely that the disease locus is present within the shared region. The mathematical power of this technique is such that a single inbred family with only three affected siblings or cousins can be sufficient to map a rare autosomal recessive disease locus.

Chromosome abnormalities

Just occasionally an individual is encountered who has both a chromosome abnormality and a single-gene disorder. Sometimes the association may be entirely fortuitous. Alternatively, the two may be causally related and in these rare situations this can provide a convenient short cut to home in on the candidate gene.

Chromosome abnormalities which have provided valuable clues about possible disease loci include balanced rearrangements, such as translocations and inversions, together with unbalanced abnormalities, such as deletions and ring chromosomes (see Chapter 2). A balanced rearrangement can cause a single-gene disorder if one of the breakpoints disrupts the disease locus or separates it from closely adjacent upstream control regions. This latter situation is referred to as a *position effect*. The resulting effect on gene function will be haploinsufficiency (p. 13). Haploinsufficiency can also result from a deletion which involves loss of the relevant disease locus. In theory it is also possible

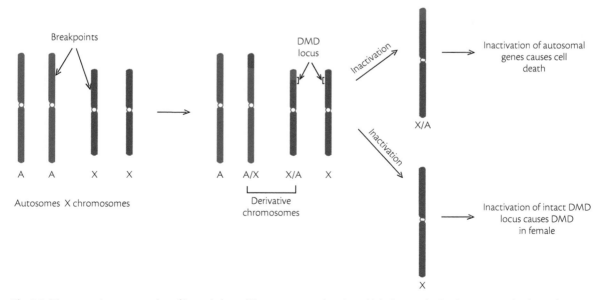

Fig. 5.5 Diagrammatic representation of how a balanced X-autosome translocation which disrupts the Duchenne muscular dystrophy (DMD) locus can result in a female being affected.

that a rearrangement could lead to a gain-of-function effect by bringing a gene into close proximity with a new promoter, but in practice this is uncommon.

Translocations between an autosome and an X chromosome have proved to be particularly useful for positional cloning. A reciprocal X-autosome translocation will result in the formation of two derivative chromosomes (Fig. 5.5), one of which consists mainly of an X chromosome and an attached segment of autosome. The other consists mainly of an autosome with a segment of the X chromosome. In a female, when the process of X chromosome inactivation occurs during early embryogenesis (see p. 78), all cells in which the derivative X-autosome is inactivated will lose activity from the attached autosomal segment as the inactivation signal extends across the chromosome. As a result, all of these cells will be at a major biological disadvantage and most will not survive. Thus the only cells that survive will be those in which the normal X chromosome is inactivated, i.e. the X-autosome derivative X is active. Thus, if a recessive gene on the X chromosome has been damaged at the translocation breakpoint, the woman will develop the disorder as she does not have any cells expressing the normal X chromosome to compensate. The study of an X-autosome translocation in a woman with Duchenne muscular dystrophy led not only to the mapping of the disease locus candidate region but also to the isolation of the actual Duchenne muscular dystrophy gene (p. 111).

Identifying candidate genes

Once the candidate region has been reduced to a size suitable for physical analysis, the search for the putative disease gene can begin. Several different methods have been employed with great success, as outlined in the discussion of various single-gene disorders that follows. A general approach can be summarized as consisting of three stages: construction of a contig, identification of expressed sequences, and finally identification of the disease gene.

Construction of a contig

A **contig** consists of a set of overlapping fragments of cloned DNA which span a small segment of the genome. In the context of physical mapping the contig should include the full segment of DNA which contains both the disease gene and the flanking linkage markers used to identify the candidate region. Originally contigs were constructed from libraries of cloned DNA in vectors such as **plasmids**, **phages**, and **cosmids**. A **vector** is a structure which can incorporate cloned DNA and which can replicate autonomously within a host cell thereby also replicating the incorporated DNA. Plasmids, phages, and cosmids all have the major disadvantage of being able to incorporate only small quantities of foreign DNA, so the construction of a contig using one of these vectors is extremely laborious. They have been superseded by **bacterial** and **yeast artificial chromosomes** (BACs and YACs) which can incorporate up to 300 kb and 1000 kb of DNA respectively. A YAC contig offers a suitable resource for isolating a gene of interest, although in practice ready-made contigs can now be downloaded from the human genome database.

Identifying expressed sequences

Several ingenious methods have been developed to identify genes in cloned DNA (Fig. 5.2), although some of these are becoming redundant as a result of the development of web-based databases from the Human Genome Project. Laboratory methods that have been successfully employed include the following:

◆ *Screening cDNA libraries.* Cloned genomic DNA from the candidate region, after insertion into a YAC, is used as a probe. The rationale is that any cDNA which matches genomic DNA from the YAC insert will bind to the YAC, enabling it to be isolated and analysed. This can be achieved by selectively amplifying complementary sequences within the genomic DNA and cDNA library and then cloning them.

◆ *Searching for CpG islands.* Stretches of CpG dinucleotides, known as CpG islands, are found close to the 5′ end of many housekeeping genes where they are thought to act as binding sites for transcription factors. CpG represents a cytosine-guanine dinucleotide, with the p representing the linkage phosphate. These clusters are unmethylated, which means that they can be cleaved by a methylation-sensitive restriction enzyme such as *Hpa*II. In the past these CpG islands were often referred to as HTF islands, with HTF standing for *Hpa*II tiny fragments. CpG islands can be selectively cloned from a DNA contig and used to screen cDNA libraries for expressed sequences.

◆ *Exon trapping.* This technique, also known as exon amplification, is aimed at isolating exons from genomic DNA by subcloning small portions of the DNA into an expression system which enables insert DNA to be transcribed into RNA followed by RNA splicing to yield an RNA transcript. Using reverse transcriptase this can then be converted to cDNA, which should in theory consist of exons from the original subcloned genomic fragment.

- *Zoo blotting.* This involves a search for DNA sequences that are conserved between species, the rationale being that such sequences are likely to represent part of an important gene. Genomic DNA from the candidate region is hybridized to DNA from a large number of species on a Southern blot. Any fragment which hybridizes to a large number of species is likely to contain coding sequences from one or more important genes.

Interrogating the human genome sequence databases

With the complete sequencing of the human genome, it is now usually no longer necessary to invest valuable laboratory resources in a search for expressed transcripts in the candidate region. Instead a direct search of the data emerging from the Human Genome Project can be made using a computer program known as a *genome browser*, of which Ensemble is one of the best-known examples (*www.ensemble.org*). This enables all of the known expressed sequences and genes in a specific region of the genome to be displayed with links to other genome sequences. Full information about each gene's sequence, structure, and homology is provided, in as much as this is known. The term *cloning in silico* has been applied to the application of this technology.

Identifying the disease gene

Having identified a number of genes within the candidate region, the next step is to try to decide which of these is most likely to be the true disease gene. This is usually achieved by assessing each gene on the basis of its structure, function, and expression pattern.

The structure of a gene may give a strong indication of its function. For example, if sequence analysis indicates that it contains several zinc finger motifs, this would suggest that the encoded protein is likely to be a transcription factor (p. 6). If the encoded protein has intracellular, extracellular, and transmembrane domains, this would point to it being a cell receptor involved in signal transduction (p. 168). Homology to other known genes, as established by reference to sequence databases, might indicate what type of protein the gene encodes or in which developmental pathway it is involved.

The expression pattern of a gene can be determined by a number of methods. These include tissue *in situ* hybridization and cDNA sampling. Tissue *in situ* hy-

TABLE 5.1 Examples of disease genes identified by positional cloning

Disease	MIM number	Inheritance	Locus	Gene	Gene product
Achondroplasia	100800	AD	4p16	FGFR3	Fibroblast growth factor receptor 3
Adult polycystic kidney disease	173900	AD	16p13	PKD1	Polycystin-1 (a cation channel)
Cystic fibrosis	219700	AR	7q31	CFTR	Cystic fibrosis transmembrane regulator (a chloride channel)
Duchenne muscular dystrophy	310200	XR	Xp21	DYS	Dystrophin (anchors cytoskeleton)
Familial adenomatous polyposis	175100	AD	5q21	APC	APC protein (a tumour suppressor)
Familial breast cancer	114480	AD	17q12	BRCA1	BRCA1/2 proteins—involved in DNA repair (tumour suppressors)
		AD	13q12	BRCA2	
Fragile X syndrome	309500	XR	Xq27.3	FMR1	FMRP (regulates mRNA translation)
Huntington disease	143100	AD	4p16.3	IT15	Huntingtin
Marfan syndrome	154700	AD	15q21	FBN	Fibrillin (in microfibrils)
Myotonic dystrophy	160900	AD	19q13	DMPK	Protein kinase
Neurofibromatosis					
Type 1	162200	AD	17q11.2	NF1	Neurofibromin (tumour suppressor)
Type 2	101000	AD	22q11-13	NF2	Merlin (tumour suppressor)
Retinoblastoma	180200	AD	13q14	RB1	Rb protein (tumour suppressor)
Tuberous sclerosis	191100	AD	9q34	TSC1	Hamartin (growth inhibitor)
			16p13	TSC2	Tuberin (growth inhibitor)
Von Hippel-Lindau	193300	AD	3p25	VHL	VHL protein (tumour suppressor)

bridization involves hybridization of a labelled probe derived from the potential candidate gene, against RNA in sections of tissue which have been mounted on a glass slide. The tissues in which the probe hybridizes give a clear indication of the spatial and temporal expression profile of the gene. cDNA sampling methods are based on computer-based comparison of the sequence of a segment of the candidate gene with the sequences of cDNAs from different tissue cDNA libraries. This approach does not give as clear an indication of the expression profile as tissue *in situ* hybridization, although new techniques have been developed based on analysis of very short sequence samples, which enable much more rapid and extensive identification of which tissue cDNA samples show expression of the candidate gene.

The final step in positional cloning is to undertake sequencing of the most plausible candidate gene looking for mutations that are present in patients but not in their unaffected relatives or in a large panel of control samples from the general population. Ideally it is desir-

able to find mutations such as frame-shift insertions and deletions, which are almost certainly going to be pathogenic, rather than missense substitutions of unpredictable effect.

Over the last 15–20 years the application of various positional cloning strategies has led to the identification of a very large number of genes responsible for a wide range of single-gene disorders (Table 5.1). Starting with the cloning of the gene for chronic granulomatous disease in 1986 (Box 5.1) scientists have now identified the genes responsible for almost all of the common disorders that were previously unexplained at both the DNA and protein levels. The different ways in which this was achieved are now illustrated using six important disorders as examples.

Key Point

In positional cloning a gene is identified by mapping the locus to a region of a chromosome and then analysing candidate genes within that region.

BOX 5.1 LANDMARK PUBLICATION: CLONING THE GENE FOR CHRONIC GRANULOMATOUS DISEASE

Chronic granulomatous disease (CGD, MIM 306400) is a rare X-linked recessive disorder in which white blood cells are unable to generate superoxide and other metabolites necessary for killing bacteria. Affected males present in early childhood with recurrent bacterial infection and abscesses. Regular treatment with prophylactic antibiotics such as trimethoprim has greatly improved the long-term prognosis.

Although rare, CGD holds a position of distinction in medical genetics because it was the first disease for which the gene and protein product were identified by positional cloning. Mapping of the disease locus to chromosome Xp21 was achieved by linkage analysis and through the discovery of a small number of boys with Duchenne muscular dystrophy (DMD—see p. 111) who had contiguous gene deletions at Xp21. Transcripts were obtained from leukaemic cells that had been treated to induce the NADPH-oxidase system which generates superoxide, so that the normal CGD gene was likely to be expressed. To identify the specific CGD transcripts, the cDNA from these cells was hybridized with RNA from a patient with a known CGD/DMD deletion, the rationale being that any cDNA which did not hybridize was likely to represent genes from the deletion. This was then hybridized to a Southern blot of bacteriophage clones derived from

Xp21. Two overlapping clones were identified and found to contain a 5-kb transcript which, on the basis of its tissue distribution and its abnormality in several CGD patients, was concluded to represent the CGD gene.

In subsequent studies it was established that the CGD gene, now known as *CYBB*, encodes the cytochrome-b β subunit, which is required for the activation of NADPH-oxidase. This was achieved by raising antibodies against a synthetic peptide produced from the CGD gene cDNA. These antibodies also reacted against a component of cytochrome-b.

This discovery of the CDG gene paved the way for reliable carrier detection and prenatal diagnosis, but of much greater importance it demonstrated the feasibility of positional cloning as a viable technique. CGD was simply the first of a large number of serious single-gene disorders in which the basic gene defect was identified through various ingenious positional cloning strategies.

Reference

Royer-Pokora B, Kunkel LM, Monaco AP *et al.* (1986) Cloning the gene for an inherited human disorder—chronic granulomatous disease—on the basis of its chromosomal location. *Nature*, **322**, 32–37.

Huntington disease (MIM 143100)

This condition takes its name from an American family doctor, George Huntington, who described several affected individuals in a large family in 1872. The prevalence in adults is around 1 in 10 000. The clinical features consist of slowly progressive intellectual decline and a movement disorder, characterized by involuntary jerky movements and twitching, known as chorea. The average age of onset is 40–45 years and the mean interval from onset to death is approximately 15 years. Unfortunately there is no effective treatment or cure.

Huntington disease shows autosomal dominant inheritance but with the rather unusual feature that when transmitted by males it shows anticipation, in that there is a tendency for onset to occur at a younger age. Another unusual feature is that the disease runs a similar course in homozygotes and heterozygoes.

Finding the gene for Huntington disease

Families with Huntington disease constitute one of the most common sources of referral to genetics clinics, so it is not surprising that this was one of the first disorders to be tackled by positional cloning strategies. What is surprising is that by remarkable good fortune the disease locus was mapped to chromosome 4p16 in an early linkage study. As indicated previously (p. 100), a modern linkage analysis involving a whole genome scan would use approximately 300 evenly spaced polymorphic markers. In the early 1980s, when this research began, only a small number of markers were available. By chance, the twelfth marker used, known as G8, was found to show close linkage to the disease locus.

This first report of linkage was published in 1983. The technical challenges posed by positional cloning are illustrated by the fact that the gene was not isolated for another 10 years. This was eventually painstakingly achieved by construction of a large genomic contig across the candidate region of over 2 Mb, followed by detailed haplotype analysis and a search for markers showing strong linkage disequilibrium (p. 90) with the disease locus. The disease gene was finally isolated by exon trapping (p. 102).

The gene which, when mutated, causes Huntington disease was given the rather unimaginative name of *IT15* (important transcript number 15) (Table 5.2). It is now known to contain 67 exons, which encode a 3144-amino-acid protein known as huntingtin. All individuals with Huntington disease have a CAG triplet repeat expansion in the first exon. Normal alleles contain up to 26 repeats. Alleles with the potential to expand to become disease alleles contain between 27 and 35 repeats. These are known as **mutable** alleles (Table 5.2). Alleles containing 36–39 repeats are associated with reduced penetrance as some heterozygotes never develop symptoms whereas others become affected in late life. Finally, alleles containing over 39 repeats always result in disease. Alleles of this size show marked meiotic instability.

Anticipation in Huntington disease is explained by an increase in the size of the triplet repeat expansion when transmitted from generation to generation. The fact that this is almost always associated with paternal

---

TABLE 5.2 Genetic characteristics of Huntington disease and fragile X syndrome

	Huntington disease	**Fragile X syndrome**
Disease locus	4p16	Xq27
Mode of transmission	Autosomal dominant	X-linked
Gene	*IT15*	*FMR1*
Gene product	Huntingtin	FMR protein
Trinucleotide repeat	CAG in first exon	CGG in 5′ untranslated region
Allele sizes	10–26 (normal)	6–54 (normal)
	27–35 (mutable)	55–200 (premutation)
	36–39 (reduced penetrance)	>200 (full mutation)
	> 40 (fully penetrant)	
Effect of full mutation	Toxic gain-of-function	Loss-of-function
Mitotic instability	No	Yes
Anticipation	Paternal transmission	Maternal transmission

transmission is probably simply a reflection of the vast number of sperm produced by a man in contrast to the small number of ova produced by a woman. The range of expansions seen in sperm is greater than that in ova, so statistically it is more likely that a large expansion will be transmitted by a sperm than an ovum.

Insight into molecular pathogenesis

The precise role of huntingtin remains unclear, although it is known to be widely expressed in the central nervous system. The expanded CAG repeat encodes a polyglutamine tract that is incorporated into the protein product. The abnormal protein tends to accumulate in the cell nucleus in the form of aggregates, which are neurotoxic and lead to neuronal death. Cytoplasmic accumulation leads to defects in axonal transport. This toxic gain-of-function mutational effect probably explains why homozygotes are usually no more severely affected than heterozygotes

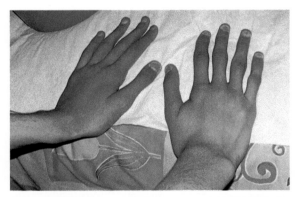

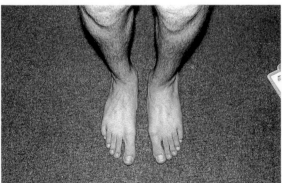

Fig. 5.6 Views of the hands and feet showing arachnodactyly in Marfan syndrome. Courtesy of Dr William Reardon, National Centre for Medical Genetics, Ireland.

> **Key Point**
>
> Huntington disease is caused by expansion of a CAG triplet repeat in the gene which encodes huntingtin. The mutant protein accumulates in the form of aggregates which are neurotoxic.

Marfan syndrome (MIM 154700)

Marfan syndrome, named after a French paediatrician who first reported the condition in 1896, affects approximately 1 in 10 000 individuals and is characterized by tall stature, arachnodactyly (long thin fingers and toes—Fig. 5.6), and weakness of connective tissue. This can manifest in many forms including increased joint laxity, a spinal curvature known as a scoliosis, dislocation of the lens, and aortic dilatation which can progress to form an aneurysm. If untreated, the cardiac complications can prove fatal. It has long been suspected that Abraham Lincoln was affected (see Box 14.3).

The pattern of inheritance is autosomal dominant, with close to complete penetrance and variable expression. Thus almost all individuals who are heterozygous for a pathogenic mutation are affected, but variation in severity within a family is well recognized. Homozygosity usually results in death in infancy or early childhood.

Finding the gene for Marfan syndrome

Several attempts were made to map the locus for Marfan syndrome, based on linkage analysis using markers for connective tissue candidate genes including members of the collagen gene family. All of these gave negative results. Although disappointing, this allowed an *exclusion map* to be drawn up which ruled out 75% of the autosomal genome. Analysis of the remaining 25% using polymorphic markers finally led to mapping of the disease locus to chromosome 15q21 in 1990.

At this point fortune intervened with the discovery that a component of extracellular microfibres, known as fibrillin, was deficient in many patients. A portion of the cDNA encoding fibrillin had been cloned and mapped to chromosome 15 by *in situ* hybridization. Thus the gene encoding fibrillin became an obvious candidate for the Marfan syndrome gene. Within 9 months of the initial mapping linkage report, confirmation that Marfan syndrome is caused by mutations in *FBN1* (the name given to the fibrillin gene) was forthcoming.

Subsequently it was shown that *FBN1* is a large gene with 65 exons. Over 200 different mutations have been reported. The large size of the gene and the marked mutational heterogeneity mean that mutation analysis in an individual family is a major undertaking.

Insights into molecular pathogenesis

Fibrillin is now known to be the major component of extracellular microfibrils, where it plays a vital role in maintaining the integrity of elastin and other connective tissues. Each protein molecule contains five distinct domains, two of which show homology to epidermal growth factor and are able to bind with calcium. Seventy percent of all identified mutations are missense and most of these involve one of the calcium-binding motifs. Most mutations are thought to exert a dominant negative effect by encoding abnormal fibrillin monomers, which interact with wild-type (normal) monomers to prevent the formation of normal microfibrillar aggregates. This in turn results in the formation of abnormal elastin fibres, leading to connective tissue weakness in crucial structures, such as the wall of the aorta and the suspensory ligament of the lens.

Key Point

Marfan syndrome is caused by mutations in the gene which encodes fibrillin. This is a major component of extracellular microfibrils.

Cystic fibrosis (MIM 219700)

Cystic fibrosis is the most common severe autosomal recessive disorder in children of western European origin, in whom it shows an incidence of approximately 1 in 2500. This implies a carrier frequency of 1 in 25 ($2pq = 2 \times 49/50 \times 1/50 = $ (approximately) $1/25$—see p. 136). The severe form of cystic fibrosis is characterized by recurrent pulmonary infection beginning in infancy or early childhood, leading eventually to irreversible lung damage and respiratory failure (Box 5.2). Most severely affected children also show pancreatic insufficiency with malabsorption. Other complications include insulin-dependent diabetes mellitus, hepatic cirrhosis, and nasal polyps. It is now recognized that cystic fibrosis can present in milder forms ranging from adult-onset chest infection to isolated male infertility caused by congenital bilateral absence of the vas deferens (CBAVD). Treatment with physiotherapy, antibiotics, and pancreatic enzyme supplements has dramatically improved life expectancy. However, despite intense research in gene therapy, no definitive cure has yet been developed.

The search for the cystic fibrosis gene

Complete lack of understanding of the molecular pathogenesis of the disease meant that positional cloning strategies held out the only short-term hope of identifying the cystic fibrosis gene. After 5 years of

BOX 5.2 CASE CÉLÈBRE: FRÉDÉRIC CHOPIN

Frédéric Chopin was born in Poland in 1810 but spent much of his adult life in Paris, where he achieved widespread acclaim for his skills as a piano recitalist and a composer. Sadly, much of Chopin's life was blighted by chronic illness, which at the time was attributed to a diagnosis of tuberculosis ('consumption'). However, following a detailed review of his medical history it has been suggested that he may instead have had cystic fibrosis. Although tall, he was frail and extremely thin with poor exercise tolerance and a tendency to develop heat prostration, possibly as a result of the severe salt loss that can occur in cystic fibrosis. Other features of his medical history consistent with this diagnosis were his aversion to fatty foods which resulted in abdominal pain and diarrhoea, his failure to father children, and the frequent chest infections from which he suffered in adult life. He eventually died at the age of 39 years from what appears to have been a combination of cardiac and respiratory failure. A point of note in his family history is that he had a sister who died at the age of 14 years as a result of recurrent chest infection.

With the benefit of hindsight a diagnosis of cystic fibrosis, which was unknown in the ninteenth century, is entirely plausible, although tuberculosis—perhaps superimposed on cystic fibrosis—certainly cannot be excluded. It is now well recognized that cystic fibrosis can present with milder adult onset symptoms and that siblings can be affected to different degrees. However, survival to the age of 39 years in the pre-antibiotic era would have been very unusual.

Reference

O'Shea JG (1987) Was Frédéric Chopin's illness actually cystic fibrosis? *Medical Journal of Australia*, **147**, 586–589.

effort, the disease locus was mapped to chromosome 7q31 by linkage analysis in 1985. Using other markers from this region it became clear that there was strong linkage disequilibrium between the cystic fibrosis mutation and adjacent loci, indicating that many of the cystic fibrosis genes in the population probably originated from a single founder mutation.

When the candidate region was reduced by linkage analysis to a size of around 500 kb, attempts were made to clone the cystic fibrosis gene by constructing phage and cosmid contigs and then searching for transcripts using the methods outlined earlier in this chapter (p. 102), together with other techniques including chromosome *walking* and *jumping*. These involve isolating sequences which are immediately or closely adjacent on the chromosome by screening genomic libraries for overlapping clones. Eventually a large contiguous segment of DNA was isolated which contained four genes. By using these as probes to detect transcripts from tissues that are affected in cystic fibrosis, one of these four was eventually identified as the cystic fibrosis gene in 1989.

Confirmation that the cystic fibrosis gene had been identified was provided by the discovery of a common mutation consisting of a deletion of 3 bp encoding phenylalanine at residue 508 in exon 10. This mutation, designated as ΔF508, was found to be present on 70% of all cystic fibrosis chromosomes and accounted for the strong linkage disequilibrium observed during the linkage phase of the gene mapping.

Insights into molecular pathogenesis

The cystic fibrosis gene is known as *CFTR* (cystic fibrosis transmembrane conductance regulator). It encodes a protein consisting of 1480 amino acids which functions as a chloride channel. The predicted protein structure includes two nucleotide binding fold (NBF) domains, which bind and cleave ATP to provide energy for transport, two transmembrane (TM) domains, which anchor the protein and constitute the chloride channel, and a single regulatory (R) domain, which undergoes phosphorylation leading to opening of the chloride channel.

Over 1000 different mutations have been identified in CFTR. Broadly these can be classified into five groups on the basis of their effects on protein synthesis or function:

- Reduced synthesis due to 'severe' mutations (e.g. nonsense, frame-shift, and splice-site), which produce a truncated or abnormal mRNA.

- Defective maturation due to mutations such as ΔF508, which prevent normal processing of the CFTR protein to the cell membrane.

- Abnormal activation caused by mutations involving an ATP-binding domain.

- Altered conductance caused by mutations involving the chloride channel.

- Defective regulation caused by mutations that impair the function of the regulatory domain.

A strong genotype–phenotype correlation has emerged. Levels of normal CFTR protein activity below 3%, as occurs in homozygotes for the common ΔF508

mutation, result in severe early-onset cystic fibrosis with pulmonary infection and pancreatic insufficiency. Levels greater than 3% result in milder phenotypes such as CBAVD. Some mutations, including ΔF508, show a clear association with the severe presentation but for others it is much more difficult to predict the disease outcome (Box 5.2).

> **Key Point**
>
> Cystic fibrosis is caused by mutations in the gene which encodes the cystic fibrosis transmembrane conductance regulator. This functions as a chloride channel.

Pyknodysostosis (MIM 265800)

Unlike cystic fibrosis, pyknodysostosis is extremely rare although it has achieved disproportionate notoriety because of the possibility that Toulouse Lautrec was affected (see Box 4.2). The reason that it is included here, along with much more common disorders, is because it provides an excellent example of the power of autozygosity mapping.

Pyknodysostosis is classified as a form of skeletal dysplasia. The skeleton shows generalized sclerosis with increased fragility. Clinical features include delayed closure of the anterior fontanelle, a small chin, and short stature. Inheritance is autosomal recessive.

Mapping the gene for pyknodysostosis

The gene for pyknodysostosis was mapped independently by two groups in 1995. In both instances this was achieved by linkage analysis based on single large inbred kindreds with multiple affected cases. In the first kindred, of Israeli Arab origin, autozygosity mapping was carried out using 285 microsatellite markers from across the genome (see Fig. 5.4). This suggested linkage to the pericentromeric region of chromosome 1. By using other markers for this region, mapping of the disease locus to chromosome 1q21 was confirmed with a very significant maximum LOD score of $10^{11.72}$. Subsequent haplotype analysis in the 16 affected individuals and their parents localized the gene to a region of 4 cM.

The second group used a slightly different approach in their analysis of a large Mexican family containing 10 affected individuals in four sibships. This involved looking for allelic variation in all affected homozygotes by using 363 polymorphic markers from across the genome, the rationale being that the disease locus will be positioned close to the marker loci at which there is the least allelic variation if all affected individuals are homozygous by descent. Once again this yielded an impressive maximum LOD score for the disease locus being at or close to chromosome 1q21.

The gene for pyknodysostosis

Further linkage analysis in the Israeli Arab kindred narrowed the candidate region down to a genetic distance of 2 cM, at which point a YAC contig was constructed spanning the relevant region. Coincidentally a gene known as *cathepsin S* had been mapped to this region, and as *cathepsin S* was known to encode a lysosomal enzyme expressed in osteoclasts during bone resorption, it was deemed to be an excellent candidate gene. By constructing PCR primers for part of the *cathepsin S* sequence and then searching for an identical site in the YAC contig, a technique known as *sequence tagged site* (STS) content mapping, it was possible to show that *cathepsin S* mapped within the candidate region. However, no pathogenic mutation was identified in this gene in affected members of the family.

Surmising that *cathepsin S* might be closely adjacent to the gene for another lysosomal enzyme *cathepsin K*, possibly as a result of an evolutionary tandem duplication event, the research group proceeded to confirm that *cathepsin K* mapped to the candidate region. They were then able to identify pathogenic mutations in a number of affected individuals from different families. Further evidence that *cathepsin K* is the pyknodysostosis gene was provided by transiently expressing normal and mutant *cathepsin K* alleles in a cell expression system and measuring transcript levels, by Northern blotting, and protein production, by immunoblotting.

Cathepsin K encodes a lysosomal enzyme, cysteine protease, which is involved in the degradation of bone matrix by osteoclasts. Histological examination of bone from patients with pyknodysostosis shows the osteoclasts to contain cytoplasmic vacuoles containing large collagen fibrils. Thus it is easy to see how loss-of-function mutations in *cathepsin K* can lead to an increase in bone density, i.e. osteosclerosis. Although pyknodysostosis is an extremely rare condition, the discovery of the disease-causing gene could have a much broader application if, for example, a method could be found for down-regulating the gene in conditions such as osteoporosis which are characterized by lack of bone density.

Key Point

Pyknodysostosis is caused by mutations in the gene that encodes the lysosomal enzyme cysteine protease. This is normally involved in the breakdown of bone matrix by osteoclasts.

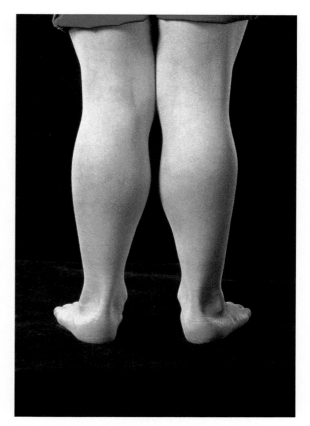

Fig. 5.7 The calves of a young boy with Duchenne muscular dystrophy showing pseudohypertrophy.

Duchenne muscular dystrophy (MIM 310200)

Duchenne muscular dystrophy affects approximately 1 in 3500 boys. It is characterized by the onset in early childhood of slowly progressive proximal muscle weakness with associated 'pseudohypertrophy' of the calves due to degenerative muscle changes (Fig. 5.7). Most boys are wheelchair bound by 10–12 years and death usually occurs by 20–25 years as a result of cardiorespiratory failure (Box 5.3). A small proportion of affected boys also have mild learning difficulties. There is no effective cure. The clinical features in Becker muscular dystrophy are very similar, but the rate of muscle degeneration is much slower with later age of onset and relatively normal life expectancy.

Inheritance is X-linked recessive. Most heterozygous females are asymptomatic although around 5–10% have mild muscle weakness. Until the mid 1980s there was no way of offering prenatal diagnosis, and carrier testing,

BOX 5.3 CASE HISTORY: DUCHENNE MUSCULAR DYSTROPHY

A family was first referred to the genetics clinic in 1977, shortly after a diagnosis of Duchenne muscular dystrophy had been made in their oldest son.

The parents had first became concerned when their son was aged 15 months because he had not yet started walking independently. They were reassured that this was not unusual and that the normal range for boys achieving independent locomotion is from 9 to 18 months. As predicted, he started walking a few months later and the parents' anxieties were eased until they noticed that he was struggling to keep up with his friends at playtime when he started school at the age of 4 years. The family doctor shared their concerns and referred their son to a paediatrician who carried out a thorough examination. This revealed that the little boy had a slightly unusual waddling gait with proximal limb muscle weakness and thick calves. He also had a positive Gower's sign, in that when trying to rise from the supine position on the floor he had to use his hands to climb up his legs, indicating that the quadriceps muscles were very weak.

The paediatrician arranged a number of investigations including assay of serum creatine kinase, a muscle enzyme released by damaged or dystrophic muscle. The little boy was found to have a grossly elevated level. The diagnosis of Duchenne muscular dystrophy was confirmed by the finding of increased variation in fibre diameter with internal nuclei and increased fat and connective tissue in a muscle biopsy sample.

At the genetics clinic a detailed family history was taken which revealed that the boy's mother had had an older brother who had died at the age of 15 years with a vague diagnosis of 'cerebral palsy'. No medical records could be obtained to confirm this. Creatine kinase analysis carried out on the mother indicated that she had a high level consistent with being a carrier. By this time the parents had had two more children, a boy and a girl. At their request creatine kinase assay was carried out on the younger brother. Fortunately this was normal.

The family were given full information about X-linked recessive inheritance and alerted to the fact that there was almost certainly a risk of 1 in 2 that any future son would be affected. They decided not to have further children, but the mother became pregnant unintentionally 2 years later and was seen again at the genetics clinic. She was keen to have a test in pregnancy to check that the baby was not affected, but at that time the only test available was fetal sexing by chromosome analysis following amniocentesis. After much discussion the couple chose to pursue this option. Unfortunately the fetus was shown to be male and the couple requested termination of pregnancy at 20 weeks gestation.

The genetics department maintained informal contact with the family over the following years and when the Duchenne muscular dystrophy gene was identified in 1988 a sample of blood was collected from the affected boy, who was now wheelchair bound, for mutation analysis. This showed that he had a large deletion involving seven exons and his mother was confirmed to be a carrier. The family requested that carrier testing be carried out on their daughter, but after discussion accepted that this would probably be better delayed until she herself could make her own informed decision as an adult. Over the years contact was maintained with the family through regular phone calls from one of the department's genetic counsellors. At the age of 19 years the daughter requested an appointment to arrange for carrier testing. She was now in a steady relationship and wanted to plan for the future. Both she and her partner had been very distressed by her affected brother's recent death and they had decided that if she was a carrier they would want to explore options for prenatal diagnosis. Unfortunately mutation testing showed that the daughter was a carrier. It was explained that reliable prenatal diagnosis could now be offered by testing DNA obtained from chorionic villi at around 11 weeks gestation (p. 274).

This case history amply illustrates how the development of reliable molecular analysis for the Duchenne muscular dystrophy gene has transformed the range of options available for families affected by the disease. In 1980 all that could be offered in a pregnancy was fetal sexing raising the spectre of termination of a male fetus at 20 weeks gestation on the basis of a possible risk of 1 in 2 for being affected. Now it is usually possible to offer precise prenatal diagnosis at 11–12 weeks gestation to women whose carrier status has been confirmed by molecular analysis. In many instances the development of mutation testing has enabled women, who would otherwise have opted not to have children, to have entirely healthy families.

Reference

Emery AEHE, Muntoni F (2003) *Duchenne muscular dystrophy*, 3rd edn. Oxford University Press, Oxford.

based on assay of the muscle enzyme creatine kinase, was unreliable. Its high incidence and pernicious nature meant that Duchenne muscular dystrophy was high on the list of disorders for which positional cloning offered the only hope for identifying the basic defect.

Finding the Duchenne muscular dystrophy gene

Three lines of evidence helped pinpoint the Duchenne muscular dystrophy locus to the middle of the short arm of the X chromosome at Xp21. These involved linkage analysis, the study of X-autosome translocations in affected females, and the identification of affected boys with contiguous gene deletions.

Linkage analysis using the very small number of markers available at that time initially mapped the locus to a large region of Xp. This approach also indicated that the loci for Duchenne muscular dystrophy and Becker muscular dystrophy were likely to be allelic. Coincidentally, reports of several affected females emerged in all of whom a balanced X-autosome translocation was identified with an X chromosome breakpoint at Xp21. Given that the pattern of cell survival in these women after X chromosome inactivation is skewed (p. 102), this pointed strongly to Xp21 as the position of the Duchenne muscular dystrophy locus. Confirmation was provided by the identification of tiny contiguous gene deletions at Xp21 in a small number of boys with several X-linked recessive disorders including Duchenne muscular dystrophy and chronic granulomatous disease, an immune disorder in which neutrophils cannot kill bacteria (Box 5.1).

The Duchenne muscular dystrophy gene was subsequently cloned through a combination of approaches. The first of these involved the competitive reassociation of DNA samples from a boy with a contiguous deletion and from a male with a 49, XXXXY karotype. The rationale was that any DNA which did not reassociate was probably derived from the deleted region. This led to the identification of a fragment of X chromosome DNA that was found to be deleted in other boys with Duchenne muscular dystrophy who did not have a cytogenetically visible deletion. Concurrently, a fragment from the Xp21 breakpoint in an X-autosome translocation from an affected female was cloned and found to contain chromosome 21 ribosomal RNA genes at one end and X chromosome material at the other end. It was correctly deduced that this X chromosome material represented part of the Duchenne muscular dystrophy gene.

Subclones from these putative gene fragments were examined for the presence of conserved sequences using a zoo blot (p. 103), and relevant subclones were then used as probes to isolate cDNA clones from muscle cDNA libraries. The entire transcribed sequence was eventually isolated in a series of overlapping cDNA clones. The Duchenne muscular dystrophy gene was found to be the longest known gene in humans, containing 79 exons with a 14-kb transcript and a genomic size of 2.5 Mb. Overall it occupies just over 1% of the entire X chromosome. This large size almost certainly accounts for the relatively high incidence of Duchenne muscular dystrophy.

It is now known that approximately 60% of patients with Duchenne muscular dystrophy and Becker muscular dystrophy have a deletion of one or more exons in the Duchenne muscular dystrophy gene. In Duchenne muscular dystrophy these are usually frame-shift, whereas in Becker muscular dystrophy they are in-frame. Around 5% of patients have a duplication. The remaining 30–35% of cases are caused by point mutations.

Insights into molecular pathogenesis

The Duchenne muscular dystrophy gene is now known to encode a large (247-kDa) protein, dystrophin, which is expressed in cardiac, skeletal, and smooth muscle. It is located at the sarcolemma of skeletal muscle where it interacts with a large number of other proteins in what is known as the dystrophin-associated protein complex. Mutations in the various genes which encode these cytoskeletal proteins cause several other forms of congenital and later onset muscular dystrophy. Dystrophin is thought to play an important role in anchoring the cytoskeleton to the cell membrane and in linking the sarcolemmal cytoskeleton to the extracellular matrix. Its large size may facilitate a role as a 'shock absorber' to protect the muscle membrane during muscle contraction.

Mutations which cause Duchenne muscular dystrophy lead to protein truncation or loss of the reading frame with little or no production of dystrophin. In contrast, mutations which maintain the translational reading frame, i.e. in-frame deletions, result in the production of a shortened but partially functional protein as seen in Becker muscular dystrophy. Quantitative defects in dystrophin protein in muscle biopsy specimens, as determined by immunofluorescence, can be used for diagnostic purposes and to distinguish between Duchenne and Becker muscular dystrophy.

Although vigorous efforts have been made to identify new therapeutic approaches based on gene therapy, none of these has yet proved successful. However, the discovery of the Duchenne muscular dystrophy gene has meant that genetic testing can now be offered to families in which one or more boys have been affected. As a direct result of the successful application of positional cloning, reliable DNA-based methods of carrier detection and prenatal diagnosis are now available enabling female relatives of affected boys to make informed reproductive choices (see Box 5.3).

Fragile X syndrome (MIM 309550)

This condition affects approximately 1 in 3000 males and is the most common inherited cause of mental retardation. (Down syndrome is more common, but is only rarely inherited—p. 55.) In addition to learning difficulties, which are usually severe, most affected males have a long face with a prominent chin and large ears. Other features include autistic behaviour, large testes after puberty, and increased joint laxity.

Fragile X syndrome came to attention in 1977 when it was realized that the condition was usually associated with the presence of a **fragile site** at the end of the long arm of the X chromosome at Xq27 (Fig. 5.8). This was consistent with the X-linked pattern of inheritance observed in many families. However, the inheritance pattern appeared somewhat idiosyncratic, in that some females were mildly affected and some males, who had clearly transmitted the condition through their daughters to affected grandsons, were unaffected. For many years ingenious theories were proposed to try to account for these observations, which essentially remained unexplained until the gene was identified in 1991.

Identifying the fragile X gene

The location of the fragile site obviously provided a strong pointer to the position of the disease locus and this was confirmed by linkage analysis. Using a YAC contig which spanned the fragile site, a CpG island was isolated which showed hypermethylation in affected but not in unaffected males. This was demonstrated by resistance to methylation sensitive restriction enzymes (p. 102). By subcloning the region adjacent to this CpG island, a 7-kb fragment was identified which contained not only the fragile X gene, designated *FMR1*, but also a small region which was increased in size due to an expanded CGG repeat in fragile X chromosomes. *FMR1* was subsequently found to contain 17 exons, which encode a protein of 632 amino acids, known as FMRP. This is expressed in brain, testes, lymphocytes, and placenta.

In more than 99% of patients with fragile X syndrome, the mutation involves expansion of the CGG repeat which is located in the 5′ untranslated region of *FMR1*. The normal gene contains 6–54 CGG triplets. Individuals with between 55 and 200 triplets have what is known as a **premutation** (see Table 5.2). These individuals are unaffected. Males with a premutation are known as normal transmitting males, and transmit the premutation in a stable form to all of their daughters, who, like their fathers, are not affected. When a premutation is transmitted by a female, there is a high risk

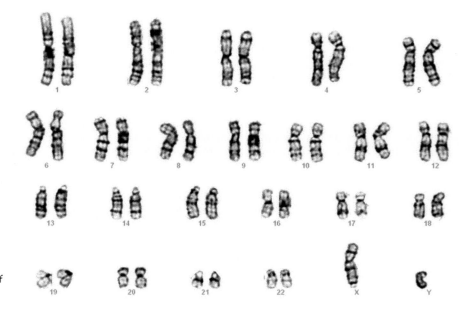

Fig. 5.8 Karyotype showing the presence of a fragile site at the end of the long arm of the X chromosome in a boy with fragile X syndrome. Courtesy of Applied Imaging Corp.

that it will be unstable during meiosis, resulting in an increase in size to between 200 and 2000 repeats. These represent full mutations. A male with a full mutation will be affected with the full fragile X syndrome. Approximately 50% of females with a full mutation show mild mental retardation.

This mutational mechanism underlying most cases of fragile X syndrome explains why some transmitting males are unaffected and why anticipation (p. 81) is observed when women with a premutation transmit *FMR1* in an unstable form to sons and daughters. It is not known why the triplet repeat expansion is unstable during meiosis in the female but not in the male, although it is well recognized that it shows post-zygotic instability in the lymphocytes of both affected males and females. Expansions more than 200 triplets in size result in hypermethylation of adjacent sequences including *FMR1*. This leads to transcriptional silencing. Normally FMRP acts as a regulator of mRNA translation in the brain, where it binds to 4% of all neuronal transcripts. Absence of FMRP probably alters the translational profile of many neuronally expressed mRNAs.

This type of mutational mechanism had never been observed previously in any species. As well as providing an explanation for the origin of fragile sites in chromosomes, the discovery of the CGG repeat site in *FMR1* meant that families with fragile X syndrome could be offered accurate carrier detection and prenatal diagnosis. Subsequently a test for the presence of FMRP in blood was developed using monoclonal antibodies. This meant that the <1% of cases due to point mutations in *FMR1* could be diagnosed reliably for the first time. Unfortunately no effective treatment has evolved from this discovery.

It is of interest that another gene located on the X chromosome, known as *MECP2* (methyl-CpG-binding protein 2), is normally involved in transcriptional suppression by binding to methylated CpG dinucleotides. Mutations in *MECP2* result in Rett syndrome (MIM—312750), in which affected females show psychomotor regression from late infancy onwards leading to mental retardation. This condition is thought to be lethal in males.

> **Key Point**
>
> Fragile X syndrome is caused by expansion of a CGG repeat in *FMR1* which encodes a regulator of mRNA translation in the brain.

Summary

The two main strategies used for identifying genes which cause human disease are functional and positional cloning. Functional cloning can be applied when the gene product is known. A short oligonucleotide probe is used to screen a cDNA library constructed from a tissue in which the gene is expressed. cDNA is then used to screen a genomic library for the target gene.

Positional cloning strategies have to be applied when the gene product is not known. A candidate region is localized by linkage analysis, autozygosity mapping, or information provided by a chromosome rearrangement in an affected individual. Candidate genes are then identified, usually by constructing a YAC contig which spans the candidate region, followed by cDNA library screening, searching for CpG islands, zoo blotting, and exon trapping. These methods are now being superseded by searching human genome databases. The disease gene is identified on the basis of its structure and expression pattern followed by sequencing and a search for mutations in affected individuals.

Positional cloning strategies have been extremely successful over the last 15 years, leading to improved understanding of most of the common single-gene disorders which were previously unexplained. This is discussed in the context of Huntington disease, Marfan syndrome, cystic fibrosis, Duchenne muscular dystrophy, and fragile X syndrome. The discovery of the genes which cause these disorders means that it is now possible to offer genetic testing for preclinical diagnosis, prenatal diagnosis, and carrier detection. There is cautious optimism that the discovery of these genes and their protein products will pave the way for successful gene therapy.

Further reading

Cox TM, Sinclair J (1997) *Molecular biology in medicine.* Blackwell Science, Oxford.

Shaw DJ (ed.) (1995) *Molecular genetics of human inherited disease.* Wiley, Chichester.

Strachan T, Read AP (2003) *Human molecular genetics 3.* Garland Science, New York.

Swallow DM, Edwards YH (ed.) (1997) *Protein dysfunction in human genetic disease.* Bios Scientific Publishers, Oxford.

Multiple choice questions

1 Strategies which can be used to map the candidate region in positional cloning include

   (a) screening a cDNA library

   (b) linkage analysis

   (c) autozygosity mapping

   (d) identifying chromosome breakpoints associated with the disease

   (e) exon trapping

2 Characteristics of Huntington disease include

   (a) autosomal dominant inheritance

   (b) homozygotes and heterozygotes are affected equally

   (c) the mutation involves expansion of a triplet repeat

   (d) anticipation when transmitted by a female

   (e) the mutation exerts a loss-of-function effect

3 Characteristics of cystic fibrosis include

   (a) autosomal recessive inheritance

   (b) a positive genotype–phenotype correlation

   (c) the most common mutation is an insertion of a codon for phenylalanine

   (d) the gene encodes a transcription factor

   (e) mutations exert a loss-of-function effect

4 Characteristics of Duchenne muscular dystrophy include

   (a) X-linked recessive inheritance

   (b) the basic defect lies in the gene which encodes creatine kinase

   (c) the most common mutations are in frame deletions

   (d) Becker muscular dystrophy is an alleleic condition

   (e) carrier testing is best carried out by mutation or linkage analysis

5 Characteristics of fragile X syndrome include

   (a) the gene is located at the end of the short arm of the X and Y chromosomes

   (b) males and females are affected equally

   (c) the mutation involves expansion of a triplet repeat

   (d) anticipation when transmitted by a male

   (e) the mutation exerts a gain-of-function effect

Answers

1 (a) false—screening a cDNA library can only be undertaken when genomic DNA has been obtained from the candidate region

 (b) true—this is still the method most commonly used

 (c) true—this method has been very successful for mapping rare autosomal recessive disorders

 (d) true—a chromosome breakpoint can provide a very useful clue to the position of the locus of a single-gene disorder

 (e) false—exon trapping is a technique for identifying exons from genomic DNA once the candidate region has been mapped

2 (a) true

 (b) true—Huntington disease is unusual in that homozygotes are usually no more severely affected than heterozygotes

 (c) true—expansion of a CAG repeat in the first exon

 (d) false—Huntington disease shows anticipation when transmitted by a male

 (e) false—the mutation exerts a toxic gain-of-function effect

3 (a) true—cystic fibrosis is one of the most common autosomal recessive disorders in western European populations

 (b) true—certain mutations, such as the common ΔF508 deletion, are associated with a severe presentation, whereas others result in milder presentations such as congenital absence of the vas deferens

 (c) false—the most common mutation is a deletion of a phenylalanine codon

 (d) false—the gene encodes a chloride channel

 (e) true—mutations all result in failure of chloride channel synthesis or function

4 (a) true

 (b) false—elevated levels of creatine kinase are a secondary effect of the gene defect

 (c) false—the most common mutations are frame-shift (out of frame) deletions

 (d) true—Becker muscular dystrophy is much milder and usually results from in-frame deletions

 (e) true—these are much more reliable than older methods based on creatine kinase assay

5 (a) false—the gene is located at the end of the long arm of the X chromosome

 (b) false—males are much more severely affected than females

 (c) true—expansion of a CGG repeat in the 5′ untranslated region

 (d) false—fragile X syndrome shows anticipation when transmitted by a female

 (e) false—the mutation exerts a loss-of-function effect

Polygenic inheritance and complex diseases

Up to this point most of the disorders discussed in this book can be explained on the basis of a straightforward genetic mechanism, in that they are caused either by an identified chromosome abnormality or a mutation in a single gene. One of the notable features of these conditions is that with a few exceptions they are rare. In contrast, there is a large group of disorders which are common, which show familial aggregation, and which cannot be explained by conventional Mendelian inheritance. These conditions are referred to as the *common* or *complex* diseases and their analysis, both at the molecular and at the clinical level, now constitutes the most active area of research in human and medical genetics.

In general the disorders that fall into this category can be divided into two groups: those that present at birth or early childhood, such as non-syndromal cardiac defects and Hirschsprung disease, and those that occur in later life, such as diabetes mellitus and schizophrenia.

Despite extensive research, the underlying molecular pathogenesis of most of these disorders remains unclear. Generally the prevailing evidence points to a complex and poorly understood interaction between genes at more than one locus (hence **oligogenic** or **polygenic** as opposed to **monogenic**) with environmental factors that may be involved before or after birth. The pattern of inheritance which applies to this group of disorders taking into account both environmental factors and multiple genes is known as **multifactorial**.

Identifying a multifactorial disorder

By definition, multifactorial disorders have a familial component that cannot be explained by simple Mendelian inheritance. These are the common diseases which 'run in families' but for which, with very few exceptions, no genetic tests are available. So what are the features of a disorder which enable scientists to deduce that there are likely to be underlying genetic and environmental components? Most such conclusions are reached on the basis of a combination of some or all of the following approaches.

Family studies

Essentially these consist of identifying families in which one or more individuals have a particular disorder and then studying the incidence of the disorder in other family members. Both of these steps require careful attention. In the first, the identification of affected individuals, it is important to try to achieve full ascertainment in a particular population. If only severely affected individuals are ascertained, perhaps through attendance at a hospital clinic, or if only families with more than two affected individuals are ascertained, perhaps through a genetics clinic, then this will introduce a degree of bias. Confirmation of the diagnosis in allegedly affected relatives is essential but can be difficult for reasons of confidentiality or because of problems with classification. This is particularly relevant in the study of psychiatric disorders, for which relatives may not have disclosed that they are undergoing treatment.

Having identified families, the next step is to determine the proportions of relatives affected. This is usually achieved by breaking down the results on the basis of the degree of relationship to the index case, i.e. the individual through whom the family was ascertained (Table 6.1). This requires precise information about the number of relatives and whether or not they are or

have been affected. Often an adjustment will have to be made to allow for variable or late age of onset, to cater for the fact that many relatives will not yet have lived through the period of risk.

Finally, the incidence, or more precisely the *lifetime expectation* or *morbid risk*, of the disorder is determined for various relatives. This can be quoted as a percentage or as a relative risk as compared to the general population risk. Relative risks are often denoted by λ_R, this being the risk to relative R as compared to the population risk. Thus if λ_S equals 10, then the risk that a sibling will develop the condition is 10 times greater than the incidence in the general population.

It is important to note that simply establishing that a condition shows familial clustering is not sufficient to confirm multifactorial inheritance. The disorder could be caused exclusively by shared environmental factors such as infection or nutritional deficiency. Thus confirmation of a genetic contribution requires further analysis in the form of either twin or adoption studies.

Twin studies

The analysis of concordance rates in twins can provide very valuable information about the relative contribution of genes and environment in causing a particular disorder. Twins are **concordant** if they are both affected or unaffected; they are **discordant** if only one is affected. Approximately one third of all twin pairs are **monozygotic**; the remaining two thirds are **dizygotic**. Monozygotic twins arise from a single fertilized zygote, which splits into two, and are therefore genetically identical, with the very rare exception of a post-zygotic (i.e. post-fertilization) mutation or non-disjunction resulting in only one of the twins being affected. Dizygotic twins arise when two ova are fertilized by different sperm; they are no more identical genetically than other siblings, and therefore share on average 50% of their genes.

If a condition is exclusively genetic, concordance rates are usually 100% in monozygotic twins but much lower in dizygotic twins (Table 6.2). If a condition is multifactorial, then the concordance rate is higher in monozygotic twins than in dizygotic twins but rarely as high as 100% because it is unlikely that monozygotic twins will share an identical environment, even *in utero*. Finally, if a disorder is exclusively environmental in origin, then monozygotic and dizygotic concordance rates will be approximately equal.

Twin studies can provide useful insight into the relative contributions of genetic and environmental factors. Higher concordance rates in monozygotic than

---

TABLE 6.1 Degrees of relationship

Degree	Proportion of genes shared (on average)	Examples of relationship
First	50% (1/2)	Parents, siblings (brothers and sisters), children
Second	25% (1/4)	Grandparents, grandchildren, uncles, aunts, half-siblings
Third	$12^{1}/_{2}$% (1/8)	Great-grandparents, greatgrandchildren, half-uncles, half-aunts, first cousins

TABLE 6.2 Expected concordance rates in monozygotic (MZ) and dizygotic(DZ) twins for different patterns of inheritance

Pattern of inheritance	MZ	DZ
Chromosome abnormality		
De novo, e.g. trisomy 21	100%	<1%
Inherited, e.g. unbalanced translocation	100%	10% (approximate)
Autosomal dominant		
New mutation	100%	<1%
Transmitted from a parent	100%	50%
X-linked recessive		25%
New mutation	100%	<1%
Transmitted from a carrier mother	100%	50% (if both are male)
Multifactorial		
High heritability	>50%	<50%
Low heritability	<50%	<<50%
Environmental—no genetic influence	Same as for DZ twins	Same as for MZ twins

dizygotic twins point strongly to multifactorial inheritance (Table 6.2). However, as with family studies there are some limitations, not least of which is the relative rarity of twins in the general population (approximately 1 in 80 of all pregnancies is a twin pregnancy). Other confounding factors are that concordant twin pairs are more likely to be ascertained than discordant twin pairs, and monozygotic twins probably often share more environmental factors than dizygotic twins. Studies of twins separated at birth can resolve some of these issues, but in practice the number of such twin pairs is extremely small.

Adoption studies

As indicated above the ideal scenario for distinguishing between the effects of genetics and environment ('nature and nurture') in causing a disease would involve the study of monozygotic twins separated at birth. For obvious reasons very few such studies have been undertaken. Instead, research has focused on adoption studies utilizing one or more of three different strategies.

◆ Study of the incidence of the disorder in children who were adopted away from an affected biological parent.

◆ Study of the incidence of the disorder in biological parents and other relatives of children who were adopted and subsequently developed the disease.

◆ Comparison of the incidence of the disorder in adopted children with affected biological parents and unaffected adoptive parents, with that in adopted children who have unaffected biological parents and affected adoptive parents.

In all of these approaches, a higher incidence in biological relatives as compared to the unrelated adoptive family implies that genetic factors contribute to the disease. Using the third strategy, if the incidence is greater in children adopted into an affected family then clearly this points strongly to environmental factors.

> **Key Point**
>
> Multifactorial inheritance implies that both genetic and environmental factors are implicated. This pattern of inheritance is suspected on the basis of family, twin, and adoption studies.

Oligogenic and polygenic inheritance

In genetics it has been recognized for many years that several human characteristics show a normal continuous distribution. Examples include height, weight, head circumference, skin colour, blood pressure, and possibly intelligence, for which a controversial nature vs nurture debate still rages. Thus if the heights of a large number of individuals are plotted their distribution takes the form of a normal (bell-shaped) curve. The shape of a normal curve is defined by its mean and standard deviation.

An observation which displays a normal distribution is said to have a **continuous** phenotype and the study of continuous traits is referred to a **quantitative** genetics. Correspondingly, loci which contribute to a continuous phenotype are known as **quantitative trait loci**.

Early geneticists were intrigued by how human characteristics such as height were inherited and noted how random segregation of different alleles at a relatively small number of loci could account for a continuous phenotype with a normal distribution (Fig. 6.1). These observations led to the concept of **polygenic inheritance** with the proposal that the relevant human characteristics are caused by the additive effects of a variable number of genes, known as **polygenes**, each of which makes a small contribution to the overall

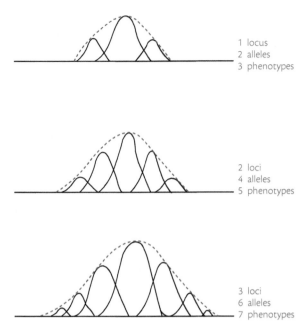

Fig. 6.1 How a trait determined by a small number of loci, each with two different alleles, can assume a continuous distribution. Variation at 1, 2 and 3 loci gives distributions of 1:2:1, 1:4:6:4:1 and 1:6:15:20:15:6:1 respectively. Adapted from McGuffin P, Owen MJ, Gottesman II (2002) *Psychiatric genetics and genomics*. Oxford University Press, Oxford.

phenotype. Support for the hypothesis of polygenic inheritance came from the study of conditions within families. For example, if a characteristic such as height is caused by interaction of say 20 genes at 10 loci (each locus has 2 alleles), then first-degree relatives would share on average 10 of these genes. They would be expected to show, and indeed were observed to show, a correlation of 0.5 or 50% for height. By extending this concept it became possible to draw up charts to predict the expected height of children based on the mid-parental height of their parents. Such charts are now used widely in clinical practice to monitor growth in childhood.

As an alternative to polygenic inheritance, and as a compromise between the two extremes of monogenic and polygenic inheritance, the concept of **oligogenic inheritance** has gained ground. This assumes that a trait or condition can be caused by genes at a relatively small number of loci and also allows for alleles at one locus to exert more influence than those at other loci. Thus this model caters for the influence of a major susceptibility locus, i.e. dominance, acting on an oligogenic background, with some genes influencing the

expression of others, a phenomenon referred to as **epistasis**. The concept of oligogenic inheritance has gained credibility as a result of research carried out on Hirschsprung disease, as described later in this chapter.

Before moving on to consider how the polygenic model can be adapted to account for multifactorial disease, mention should be made of the concept of **heritability**. This is defined as the proportion of total variance caused by additive genetic effects. Measurement of the heritability of a condition or trait gives an indication of the extent to which alleles at different loci contribute to the variability in phenotype. Thus if it is established that the heritability of a continuous phenotype, such as height or IQ, is high, this implies that genetic factors account for a large proportion of the observed differences in height or IQ seen in the population in which the heritability value was obtained. Heritability is sometimes denoted by H and is indicated as a fraction of 1, e.g. 0.6, or as a percentage, e.g. 60%.

Polygenes and environment—multifactorial inheritance

The development of the concept of polygenic inheritance served two purposes, in that as well as providing an explanation for how genes could generate a continuous phenotype, it also yielded insight into how an underlying disease susceptibility might be transmitted within families. The challenge then was to develop the concept to explain how an underlying normally distributed continuous susceptibility could account for a **discontinuous**, or dichotomous, disease phenotype, i.e. affected or not affected.

The most widely accepted explanation was proposed by Falconer in the 1960s and is known as the *liability-threshold model*. According to this model, the underlying disease susceptibility, which is known as **liability** and includes both genetic and environmental factors, is

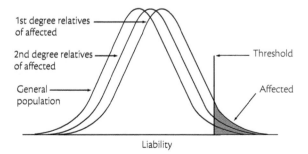

Fig. 6.2 The liability-threshold model. The liability curves for close relatives of affected persons are shifted to the right.

normally distributed in the population. Towards the right-hand side of this liability curve, a threshold value is reached beyond which all individuals are affected. In relatives of affected individuals the liability curves are shifted to the right, because of their shared genes and environment, so that risks to relatives are increased as compared to the general population (Fig. 6.2).

The model has a sound mathematical foundation based on the statistics of the normal distribution and is both attractive and plausible. For example, it is probably not too simplistic to hypothesize that palatal closure is under the control of perhaps 20 genes determining factors such as cell migration, cell adhesion, and tongue size. If too many of these genes are abnormal, or interact abnormally perhaps because of adverse environmental factors in the uterus, then the threshold value for defective palatal closure is reached and the baby is born with a cleft palate. First-degree relatives will on average share 10 of these genes and will therefore be at risk of inheriting a sufficient number of 'bad' genes to render them susceptible.

One of the particular attractions of the liability-threshold model is that it is consistent with some of the observations that arise from family studies. These can be summarized as follows:

- Risks are greatest amongst the closest relatives of the affected individual and decline sharply in more distant relatives. This is explained by different degrees of shift in the liability curve for different degrees of relationship (Fig. 6.2).

- Risks are increased if there is more than one affected individual in the family. The presence of multiple affected members implies that the family has an unusually large number of adverse factors contributing to its liability.

- For some conditions, notably cleft lip or palate, risks are greater for relatives of the most severely affected individuals. This implies that those with the most severe manifestations lie at the extreme end of the liability curve.

- If there is a marked difference in incidence between the sexes, then the risks will be greatest for individuals of the more susceptible sex who have affected relatives of the less susceptible sex. Thus, for example, if a disorder is more common in boys than in girls, then the risks for close relatives will be greatest for the male relatives of affected girls. This can be explained on the basis that girls must lie further along the liability curve than boys to be affected, so that their close relatives will also lie towards the

extreme of the curve with male relatives being more at risk than female relatives because of their sexually determined increased susceptibility.

A caveat—fact or fiction?

The liability-threshold model for multifactorial inheritance is plausible, attractive, and ingenious. It is widely quoted and reasonably easy to understand. It also makes an excellent subject for multiple choice questions. The problem is that it is totally unproven and, even with today's technology, unproveable. Until molecular research unravels the mysteries of multifactorial inheritance, the liability-threshold model provides an extremely useful framework for considering complex diseases, but is would be wrong to assume that its validity has been established. In the fullness of time it may well prove to be correct, but until then it should be remembered that it is just a model and not a proven scientific fact.

> **Key Point**
>
> The liability-threshold model for multifactorial inheritance has been developed to explain how an underlying continuous normally distributed susceptibility, known as liability, can generate a discontinuous phenotype (affected or not affected).

Approaches to finding susceptibility genes

The identification of genes which contribute to multifactorial disorders is an area of intense activity amongst the scientific community, where vast amounts are being invested both by the research funding bodies and by the pharmaceutical industry. The goal is to identify genes which play an important role in conveying susceptibility to the common disorders of adult life with a view to developing tests for their preclinical detection and new genetically based approaches for their prevention and treatment.

Unfortunately progress has been disappointing, to the extent that after more than a decade of major investment and research only a handful of susceptibility genes for multifactorial diseases have been identified. Three main strategies have been applied: linkage analysis, association studies, and linkage disequilibrium.

Linkage analysis

The use of linkage analysis in the context of single-gene disorders, both for risk assessment and to facilitate positional cloning, has been outlined in Chapters 4

(p. 91) and 5 (p. 100). This is referred to as **parametric** linkage analysis because the parameters of the analytical procedures are clear. These include factors such as mode of inheritance, penetrance, and gene frequency. For multifactorial disorders, none of these parameters is known; thus linkage analysis for multifactorial disorders is said to be **non-parametric** and predictably it is much more difficult and less specific.

The method most commonly employed involves analysis of affected sib pairs looking for areas of the genome that are shared by affected siblings more often than would be expected by chance. The underlying principle is illustrated in Fig. 6.3. On average sib pairs would be expected to share 2, 1 or 0 alleles at a particular locus in a ratio of 1 : 2 : 1. If a locus is identified at which large numbers of sib pairs with a particular disorder show statistically significant deviation from this ratio, then this indicates that alleles at this locus, or at a closely adjacent linked locus, are in some way implicated in causing the disease.

In a typical study, a sample of at least 200 affected sib pairs would be analysed along with their parents at around 300 evenly spaced loci using a panel of polymorphic microsatellites or single nucleotide polymorphisms (SNPs—p. 141). This yields a significant probability of detecting linkage. The mathematical analysis is undertaken using computer programs such as MAPMAKER/SIBS and GENEHUNTER. Areas of the genome identified in this way are referred to as *shared segments* and alleles at these loci are described as being *identical by descent* (IBS).

This type of linkage analysis is sometimes referred to as *whole genome scanning* and is relatively powerful in that it can detect large segments of the genome which contain one or more susceptibility loci. Unfortunately these segments are much too large to enable easy identification of the relevant gene(s). For this to be achieved it is necessary to move on to the next level of research which employs a combination of association and linkage disequilibrium analysis.

Association studies

The principle underlying this approach is relatively straightforward. If an allele at a particular locus occurs more often in individuals with a disorder than in an unaffected suitably matched control population, this indicates that the allele is associated with this disease. Before molecular analytic techniques became available, association studies focused on the use of biological markers such as the HLA system and the ABO blood groups. Occasionally these threw up an important association, such as that between the HLA B27 allele and ankylosing spondylitis, but more often weak associations emerged, the importance of which could not be interpreted. Now the standard approach involves the analysis of a large panel of markers, such as microsatellites or SNPs, often using microarray technology (p. 21), in as large a sample of individuals as possible.

A major advantage of association studies is that, unlike linkage analysis, families with single affected cases can be used. A major disadvantage is that an apparent association may not actually be causally relevant, as there are at least four possible explanations which have to be considered when a possible association emerges:

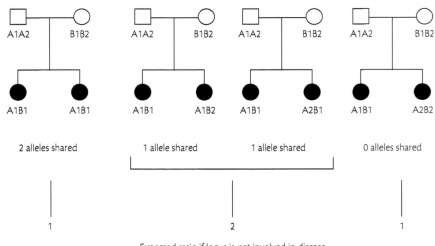

Fig. 6.3 The rationale behind the use of affected sib pairs in linkage analysis.

Expected ratio if locus is *not* involved in disease

- The association is a genuine causal association. This is the ideal situation, but in practice it is unlikely if the disease association is with an anonymous microsatellite or SNP. However, it is recognized that polymorphic variation in DNA such as SNPs can influence gene expression, so there is some justification for labelling the loci which harbour relevant polymorphic markers as quantitative trait loci. In practice it is desirable for an association to be plausible before it is accepted.

- The association is spurious because of population stratification. This means that both the disorder and the apparently associated allele are common in the population being studied. One way to test for this is to carry out a *transmission disequilibrium* test, which essentially involves showing whether the relevant allele has been transmitted more often by parents (who are unaffected) to their affected child than to their other unaffected children.

- The association is spurious because of a statistical error. If a sufficiently large number of markers are analysed it is quite possible that simply by chance, one or more will show an apparent statistical association with the disease. Careful allowance has to be made for this when undertaking the analysis.

- Linkage disequilibrum. As outlined in Chapter 4 (p. 90), this refers to the association of specific alleles at linked loci more or less often than would be expected by chance. Therefore if alleles at two linked loci are in disequilibrium, then it is possible that an apparent association between a marker and a multifactorial disease actually reflects an association between the allele with which it is in linkage disequilibrium and the disease. This possibility can be explored further and has proved to be an effective method for identifying susceptibility loci.

Linkage disequilibrum

Once an approximate localization has been made on the basis of linkage analysis or association studies, the next step usually involves an attempt to home in on the candidate region using linkage disequilibrium. The rationale for this approach is based on the hypothesis that most affected members of the population under study will have inherited their susceptibility allele from a common ancestor (i.e. identical by descent), so if the susceptibility allele arose by mutation in the not too distant past, then closely linked alleles adjacent to the susceptibility allele will also be identical.

Looked at another way, most if not all affected members of the population would be expected to have a short shared segment, equivalent to a common haplotype, surrounding their susceptibility allele. This is particularly likely if the population being studied is homogeneous in that it is small, stable, and isolated from immigration. Iceland represents an ideal location for this sort of study because of its relatively small population of approximately 270 000, long-term population stability, good medical facilities, and well-documented historical records. These factors prompted a genomics company, DeCode, to negotiate a contract with the Icelandic government to undertake precisely this type of research using DNA samples from the Icelandic population.

Thus in practice, once an approximate location has been identified, this region is analysed in depth using polymorphic DNA markers, usually SNPs, which map to the region to narrow down the area of interest. When this has been reduced to a size suitable for more detailed scrutiny, attempts are made to identify candidate genes and undertake mutation analysis. This approach has been greatly facilitated by the Human Genome Project and Internet-accessible genetic databases which provide details of known genes and expressed sequences which map to the relevant region.

Unfortunately, this is not as simple as it might seem. The technological requirements are enormous, and a huge amount of work is involved in carrying out assocation/linkage disequilibrium studies to cover the entire genome. If the population is heterogeneous, it is likely that there will have been several different founders of one or more susceptibility alleles so that linkage disequilibrum will be very difficult to detect. Different ethnic groups and populations may well develop the same disease for totally different reasons and therefore have very different genetic susceptibilities. Finally, the genetic changes which cause minor abnormalities in gene expression resulting in multifactorial disease may be very different from the more obvious mutations that cause single-gene disease. In an attempt to address at least some of these limitations, the UK government has recently launched a major Biobank project aimed at providing samples from half a million adults as a resource for genetic research into complex diseases. This initiative has not met with universal support (Box 6.1).

It is likely that all of these factors have contributed to the slow progress achieved to date. Nevertheless, considerable insight has been gained for some of the most common multifactorial disorders as outlined in the next section.

Key Point

Vigorous efforts are being made to identify genes which convey susceptibility for common multifactorial disorders. Methods used include linkage analysis, disease associations, and linkage disequilibrium. Progress to date has been very limited.

Examples of multifactorial disorders

Most conditions which are believed to show multifactorial inheritance can be classified either as congenital malformations or as acquired disorders of childhood or adult life (Table 6.3). All of these malformations can occur as part of multiple malformation syndromes, but when present in isolation in otherwise healthy children they usually show multifactorial inheritance. Similarly, many of the acquired diseases of adult life can be part of more complex medical diagnoses, but again in isolation they usually demonstrate multifactorial inheritance. Rare families in which there is

TABLE 6.3 Examples of disorders believed to show multifactorial inheritance

Present at birth or onset in infancy	Acquired with onset in childhood or adulthood
Cardiac defects, e.g. atrial septal defect, tetralogy of Fallot, ventricular septal defect	Alzheimer disease
Cleft lip/palate	Asthma
Congenital dislocation of the hip	Autism
Hirschsprung disease	Coronary artery disease
Neural tube defects	Depression—unipolar and bipolar
Pyloric stenosis	Diabetes mellitus—types 1 and 2
Talipes	Hypertension
	Inflammatory bowel disease, Crohn disease and ulcerative colitis
	Schizophrenia

BOX 6.1 CASE CÉLÈBRE: THE UK BIOBANK

Frustrated by the slow rate of progress in research into the common disorders of adult life, the UK Department of Health together with the Medical Research Council and the Wellcome Trust have awarded £45 million for the establishment of a national resource for biomedical research. This is to be known as the UK Biobank. It will be based on samples of blood collected from 500 000 adults aged 45–69 years who will also provide information about their medical history, lifestyle, and diet. The health of those who participate will be monitored for a period of at least 10 years, enabling information to be gathered about the development and progress of chronic disorders such as cerebrovascular accidents (strokes), heart disease, and diabetes.

The underlying rationale is that this initiative will enable scientists to carry out association/linkage disequilibrium studies and other research on what will be the largest population resource in the world. The long-term goal is to improve understanding of the biology of these disorders and develop ' improved diagnostic tools, prevention strategies and tailor made treatments'.

Given the difficulties encountered in this type of research, it is perhaps not too surprising that this initiative has not been greeted with universal enthusi-

asm. A report from the House of Commons Science and Technology Committee in 2003 was particularly critical, arguing that the project is speculative and politically driven and that the money could be better spent elsewhere. The MRC countered that the goal of improving human health requires planning and funding on a long-term basis, and that the Biobank is in the national interest.

Ultimately only time will reveal whether the Biobank is indeed a worthwhile long-term investment. If problems such as population heterogeneity and complex statistical analysis can be resolved, leading to a better understanding of the common disorders, then there can be little doubt that the establishment of the Biobank will have been an extremely wise decision. Given a choice between funding research into chronic ill-health or other initiatives, such as nuclear weapons and space exploration, most individuals would probably opt for the former even if in the short-term the rewards are minimal.

Reference

Details of the UK Biobank can be found on its website, *www.ukbiobank.ac.uk*

a particularly high frequency of a multifactorial disorder suggestive of single-gene inheritance have proved especially valuable for identifying susceptibility loci.

Alzheimer disease

Alzheimer disease is the commonest cause of both senile and pre-senile dementia, with a prevalence of around 20% in individuals over 80 years of age. Clinically it manifests as progressive loss of memory, emotional disturbance, and loss of intellectual skills

---

BOX 6.2 CASE CÉLÈBRE: IRIS MURDOCH

Iris Murdoch was one of the great literary figures of the twentieth century. She was born in 1919 in Dublin of Anglo-Irish parents but lived most of her life in England where she studied classics, ancient history, and philosophy at Oxford University, before going on to lecture in philosophy at Oxford for 25 years. Her first published work was on Jean-Paul Sartre, the French novelist and philosopher, and she subsequently wrote several other works on philosophy including *Metaphysics as a Guide to Morals* and *The Sovereignty of Good and Other Concepts*. However, she is best remembered for her 26 novels, one of which, *The Sea! The Sea!*, won the Booker prize in 1978.

Although her mother died as a result of Alzheimer disease, neither Iris Murdoch nor her husband, who was a professor of English at Oxford, had any inkling that the condition could be genetic until she herself developed early signs at the age of 75 years. Her husband's account of the illness makes sad reading. He describes Alzheimer disease as being 'like an insidious fog' and it is depressing to learn of his wife's intellectual decline from literary genius to a childlike state in which her favourite television programme was the *Teletubbies*. The illness culminated in her death a few months short of her 80th birthday in 1999.

Iris Murdoch has been the subject of several biographies, one of which, by her husband, John Bayley, forms the basis of the film *Iris* starring Judi Dench, Jim Broadbent, and Kate Winslett, all of whom received Oscar nominations. The natural history of her intellectual decline as depicted in the film is typical of that seen in many older patients.

Reference

Bayley J (1998) *Iris. A memory of Iris Murdoch*. George Duckworth and Co, London.

(Box 6.2). The neuropathological findings consist of neurofibrillary tangles made up of tau protein, which are neurotoxic, and senile plaques consisting of amyloid fibres known as AB which are derived from amyloid precursor protein (APP).

Support for a genetic contribution to Alzheimer disease originally came from both family and twin studies. The predicted lifetime risk of developing Alzheimer disease for first-degree relatives of an affected individual is increased 3–4-fold as compared to controls. Twin concordance rates in different studies vary from 30 to 80% for monozygotic twins and from 10 to 40% for dizygotic twins. Heritability estimates vary from 0.44 to 0.8.

Autosomal dominant Alzheimer disease (MIM 104300)

Review of family histories revealed that in a small subset of families showing pre-senile onset (before 65 years), inheritance was suggestive of autosomal

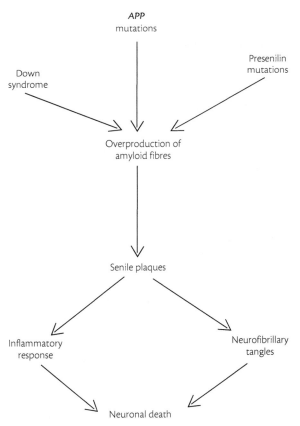

Fig. 6.4 Simplified diagram to explain the pathogenesis of Alzheimer disease. Adapted from McGuffin P, Owen MJ, Gottesman II (2002) *Psychiatric genetics and genomics*. Oxford University Press, Oxford.

dominant transmission. The first gene to be implicated was *APP* on chromosome 21. This encodes the amyloid precursor protein. The small number of mutations identified in *APP* lead to an increase in production of APP and hence AB, this being the chief constituent of the characteristic senile plaques. This observation is thought to explain the high incidence of Alzheimer disease seen in older adults with Down syndrome (p. 55).

It soon became apparent that mutations in *APP* accounted for only 5% of the rare autosomal dominant families. Subsequently, linkage analysis identified susceptibility loci on chromosomes 14 and 1 where genes now known as Presenilin-1 (*PS1*) and Presenilin-2 (*PS2*) were isolated. Mutations in these genes have been shown to account for approximately 70% and <5% respectively of all presenile-onset autosomal dominant families. *PS1* and *PS2* are thought to act by enhancing APP processing, leading to overproduction of AB which in turn leads to the formation of senile plaques and neurofibrillary tangles (Fig. 6.4).

Polygenic Alzheimer disease

The pattern of inheritance in most families with Alzheimer disease is much more consistent with polygenic than with autosomal dominant inheritance, in that most affected individuals develop symptoms after the age of 65 years and offspring risks are much lower than 50%. Linkage analysis in these families focused attention on chromosome 19 where a polymorphism at the APOE locus was found to show a strong association with late-onset Alzheimer disease. Apolipoprotein E (ApoE) is synthesized in the liver and brain and is involved in lipid metabolism and tissue repair. There are three common alleles at the APOE locus: E2, E3, and E4. The frequency of the E4 allele is significantly increased in patients with Alzheimer disease, to the extent that the odds ratio for developing Alzheimer dis-

ease is increased 12-fold for E4E4 homozygotes as compared with E3E3 homozygotes. This has proved to be one of the strongest known disease associations, with estimates of the lifetime risk for developing Alzheimer disease being around 35% for E4E4 men and 50% for E4E4 women (Table 6.4). The discovery of this association has spawned a thriving but controversial genetic susceptibility testing industry (p. 130). Linkage analysis suggests that there are at least four other genes which convey susceptibility to late-onset Alzheimer disease, but as yet these have not been identified.

> ### Key Point
>
> In most families Alzheimer disease shows multifactorial inheritance. Autosomal dominant early-onset Alzheimer disease can be caused by mutations in at least three different genes.

Coronary artery disease

Coronary artery disease in the form of atherosclerosis kills almost 1 million people a year in the USA, where it is now the principal cause of death. It is more common in men than in women, and six times more common in Americans than in Japanese. However, when Japanese emigrate to the USA their rate of coronary artery disease triples, thereby illustrating the importance of environmental factors. These include smoking, diet, and lack of exercise.

Based on studies such as those of the rates in Japanese immigrants, it has been estimated that genetic factors account for approximately 50% of the risk of atherosclerosis with a heritability of 0.5–0.6 for premature coronary artery disease, defined as onset before age 55 years. Risks to brothers of male index cases with premature coronary artery disease are increased by a factor of 3–5, and to both brothers and sisters of female index cases by a factor of 7. These observations suggest that women require a greater genetic susceptibility than men to be affected. Concordance rates vary from 40 to 65% for monozygotic twins and from 15 to 30% for dizygotic twins.

Coronary artery disease is a well-recognized complication of several single-gene disorders involving lipid metabolism. These include familial hypercholesterolaemia (p. 213), and various other rare forms of familial hyperlipidaemia. Over 20 genes have been proposed as candidates for polygenic coronary artery disease. These include genes which control lipid metabolism, blood pressure, clotting, and fibrinolysis. One particular polymorphism which has been studied at length is that

TABLE 6.4 Lifetime risk for developing Alzheimer disease depending upon ApoE4 genotype

Genotype	Risk Males (%)	Females (%)
Unknown	6.3	12
No E4 allele	4.6	9.3
One E4 allele	12	23
Two E4 alleles	35	53

Adapted from McGuffin P, Owen MJ, Gottesman II (2002) *Psychiatric genetics and genomics.* Oxford University Press, Oxford.

of an insertion (I)/deletion (D) in intron 16 of *ACE*, the gene which encodes the enzyme which converts angiotensin I into angiotensin II. Homozygotes for the DD genotype have higher plasma and tissue levels of the enzyme than those with the ID or II genotype. They also show a higher incidence of coronary artery disease, with one study suggesting that in the absence of other risk factors such as smoking, hypertension, diabetes, and obesity, the ACE DD genotype accounts for 35% of cases of myocardial infarction.

Several companies are now marketing kits aimed at determining susceptibility to coronary artery disease based on assay of cholesterol and tests for predisposing genetic changes such as the ACE deletion/insertion polymorphism. Reduction of high levels of cholesterol by diet and with the use of drugs which inhibit cholesterol synthesis, such as 3-hydroxy-3 methylglutaryl coenzyme A inhibitors (statins), is widely recommended in older men and in those with a strong family history. The role of genetic susceptibility testing is much more controversial (p. 130).

Diabetes mellitus

Two main types of diabetes mellitus are recognized. Both show multifactorial inheritance.

Type 1 diabetes—insulin-dependent diabetes mellitus (IDDM)

This relatively rare form of diabetes, with a prevalence of around 1 in 200, usually presents in childhood or early adult life. It is caused by autoimmune destruction of the pancreatic β-cells which produce insulin. Both family and twin studies are consistent with multifactorial inheritance. The recurrence risk for siblings and offspring of an affected individual is 5–6%. Monozygotic and dizygotic twin concordance rates are approximately 30–40% and 5–10% respectively.

The search for susceptibility genes has identified two definite loci. The first of these, the HLA system, accounts for 30–40% of total genetic susceptibility. Around 95% of all affected individuals have the HLA-DR3 and/or HLA-DR4 antigens, whereas these are found in only 50% of the general population. This locus has been designated as *IDDM1*. The second locus, *IDDM2*, includes the insulin gene on chromosome 11 together with a closely adjacent upstream region which contains multiple copies of a 14-bp microsatellite repeat sequence. A short number of repeats conveys susceptibility to type 1 diabetes, probably by reducing the expression of the insulin gene in the fetal thymus gland, thereby reducing subsequent immunological tolerance to insulin and insulin-producing cells in the pancreas.

Together *IDDM1* and *IDDM2* account for around 50% of the total genetic predisposition to type I diabetes. Many attempts have been made to identify other susceptibility loci based on whole genome linkage analysis using very large numbers of affected sib pairs. Although many possible susceptibility regions have been identified, these are very large and no other specific susceptibility genes have been isolated.

Type 2 diabetes—maturity onset diabetes

Type 2 diabetes affects approximately 5% of adults after the age of 45 years and can often be treated by diet or oral hypoglycaemic agents, as most patients retain a degree of endogenous insulin production. Risks to first-degree relatives are 10–15%, i.e. two to three times the general population risk, and the concordance rate for monozygotic twins is 90%, so the evidence in favour of a genetic aetiology is strong.

Progress in finding genes which contribute to the common late-onset form of maturity onset diabetes has been slow. Linkage and association studies have identified over 20 different loci, but often the original findings have not been replicated with subsequent studies giving conflicting results. At present the genes that convey susceptibility to type 2 diabetes remain elusive and there appears to be little likelihood of successfully identifying these genes in the near future.

In some families with early-onset non-insulin-dependent diabetes, inheritance is clearly autosomal dominant. These families have a condition which is distinct from type 2 diabetes, known as maturity-onset diabetes of the young (MODY—MIM 125850). Mutations have been identified in several genes including the hepatocyte nuclear factor-4-*a* gene (*MODY1*), the glucokinase gene (*MODY2*), the hepatic transcription factor-1 gene (*MODY3*), and the islet duodenum homeobox-1 (*IDX1*) gene (*MODY4*—p. 170). Most of the genes that cause MODY are transcription factors that regulate expression of the insulin gene or pancreatic development.

Hirschsprung disease

Hirschsprung disease is caused by absence of ganglion cells in the distal part of the colon and rectum. It is divided into long and short segment types (L-HSCR and S-HSCR), depending on the presence or absence of disease proximal to the sigmoid colon. Presentation is usually at birth or in early infancy with acute intestinal obstruction and abdominal distension.

Approximately 30% of all cases occur in association with other abnormalities or as part of a condition such as Down syndrome. Multifactorial inheritance has long been suspected for the remaining 70% of isolated non-

TABLE 6.5 Risks to siblings in non-syndromal Hirschsprung disease

Sex of affected sibling	Recurrence risk for	
	Brothers (%)	Sisters (%)
Short segment (S-HSCR)		
Male	4.7	0.6
Female	8.1	2.9
Long segment (L-HSCR)		
Male	16.1	11.1
Female	18.2	9.1

Data adapted from Harper PS (2001) *Practical genetic counselling*, 5th edn. Arnold, London.

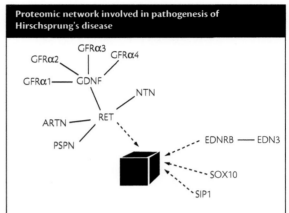

Proteomic network involved in pathogenesis of Hirschsprung's disease

Solid lines connecting proteins=networked interactions. Dashed arrows=putative pathogenetic relations of individual proteins and subnetworks to Hirschsprung's disease phenotype ("black box").

RET=receptor tyrosine kinase, NTN-neurturin, GDNF=neurotropic factor derived from glial-cell lines, GFRα=GDNF receptor, ARTN-artemin, PSPN=persephin, EDNRB=G-protein-coupled endothelin B receptor, EDN3-endothelin 3, SOX10=sex-determining-factor-related box, SIP1=survival-of-motor-neurons interacting protein-1.

Fig. 6.5 Simplified diagram to show the proposed pathogenesis of Hirschsprung disease. The diagram shows various receptors and ligands believed to be involved. Reproduced with permission from McCabe ERB (2002) Hirschsprung's disease: dissecting complexity in a pathogenetic network. *Lancet*, **359**, 1169–1170.

syndromal cases, based mainly on the results of family studies. These have yielded recurrence risks for siblings varying from around 2% for sisters of a boy with short-segment disease to as high as 17% for brothers of a girl with long-segment disease (Table 6.5). Note that these observations are consistent with the liability-threshold model given that the male to female ratio for affected individuals is 4 : 1 and that L-HSCR represents a much more severe form of the condition than S-HSCR.

In contrast to most multifactorial conditions, recent research has shed considerable light on the underlying molecular pathogenesis. It has emerged that the proto-oncogene *RET* (p. 198) is a major susceptibility locus, with heterozygous mutations identified in around 50% of all familial cases and up to 75% of all children with L-HSCR. These mutations exert a loss-of-function effect. *RET* encodes a transmembrane tyrosine kinase receptor that moderates cell signalling in the embryonic enteric nervous system. Not all family members with a *RET* mutation are affected, indicating that penetrance of *RET* mutations is incomplete.

Mutations have also been identified in several other genes in much smaller numbers of affected children. These include the RET ligands *GDNF* and *NTN*, together with the endothelin gene, *EDNRB*, and its ligand *EDN3*. *EDNRB* is also known to be involved in the normal development and migration of the neural crest derived intestinal nervous system. To date mutations have been identified in a total of nine genes involved in neuronal development and unknown genes at two other loci are believed to act as modifiers of *RET* expression.

It is not known how all these genes and their encoded products interact to cause Hirschsprung disease, a fact succinctly acknowledged in a recent review which

depicts the various genes with arrows pointing to a large black box (Fig. 6.5). However, Hirschsprung disease is the only multifactorial condition in which the major underlying genetic pathways have been elucidated. The relatively small number of genes involved has resulted in Hirschsprung disease being categorized as an oligogenic disorder. The term **synergistic heterozygosity** has been coined to refer to the interaction of a small number of mutant alleles at different loci.

> **Key Point**
>
> At least nine genes have been implicated in conveying susceptibility for Hirschsprung disease. *RET* is the most important of these, with mutations in 50% of familial cases.

Neural tube defects

This term embraces a group of serious disorders in which part of the neural tube fails to close completely during early embryogenesis. Recognized clinical presentations include anencephaly, occipital encephalocoele, and lumbosacral spina bifida. Some neural tube defects occur as part of a multiple malformation syndrome or chromosomal abnormality, but most occur in isolation and are classified as non-syndromal. Evidence that genetic factors are involved is strong. Family studies show that in high-incidence areas the risks for first- and second-degree relatives are approximately 4% and 1% respectively. The heritability has been estimated to be approximately 0.6.

Research in mice has identified over 60 genes in which mutations can cause neural tube defects, but with very few exceptions none of these studies have been replicated in humans. The only consistent finding to emerge is that a common polymorphism in *MTHFR* involving a C → T mutation at nucleotide 677 conveys a 2–4-fold increase in risk for a neural tube defect if the mother is homozygous, rising to a 6–7-fold increase in risk if both the mother and baby are homozygous. *MTHFR* encodes the enzyme methylenetetrahydrofolate reductase, which catalyses a step in the conversion of tetrahydrofolate to 5-methyltetrahydrofolate, which in turn acts as a co-substrate for the methylation of homocysteine to form methionine. Homozygosity for the 677C → T polymorphism is associated with reduced red cell folate levels and

it has been estimated that this polymorphism accounts for up to 50% of the protective effect of periconceptional folic acid supplementation (Box 6.3).

Although progress in identifying loci that convey susceptibility for nueral tube defects has been slow, the demonstration that periconceptional folic acid supplementation can prevent up to 80% of all cases has focused attention on the importance of trying to identify environmental factors which interact with genetic susceptibility and can be modified accordingly. Unfortunately very few such factors have been identified for any of the other multifactorial disorders. This applies particularly to psychiatric disorders such as depression and schizophrenia.

> ### Key Point
>
> Periconceptional folic acid supplementation can prevent approximately 80% of all non-syndromal neural tube defects.

BOX 6.3 LANDMARK PUBLICATION: FOLIC ACID AND NEURAL TUBE DEFECTS

It had long been recognized that nutritional factors might play a role in causing neural tube defects, but it was not until 1980 that serious attention was paid to the possibility of preventing this group of disorders using dietary modification. This was the year in which Smithells *et al.* published the preliminary results of their research into vitamin supplementation.

Women who had delivered a baby with a neural tube defect were invited to participate in a study that involved taking a multivitamin preparation before and during the early stages of their subsequent pregnancy. Amongst the 178 pregnancies conceived by fully supplemented women, one infant was affected with a neural tube defect. In contrast, the 260 pregnancies conceived by mothers who were not supplemented included a total of 13 affected with a neural tube defect, an incidence of 5%.

At the time of this study, neural tube defects had an incidence approaching 1 in 200 in some parts of the UK, so it is not surprising that these results generated great interest and publicity. Predictably, they were also greeted with considerable scepticism as there was much concern that the unsupplemented mothers did not constitute a valid control group. Prompted by the pressing need to resolve the situation and clarify

which, if any, of the vitamins in the multivitamin cocktail was effective, the Medical Research Council organized a major multicentre trial recruiting women from throughout the UK. The results pointed strongly to folic acid being the effective preventive agent.

On the basis of these and other studies, low-dose folic acid supplementation is now recommended for all pregnant women in the UK, although experience to date indicates that compliance is poor. In North America this problem has been addressed by introducing mandatory fortification of flour. It is still not absolutely clear how the beneficial effects of folic acid supplementation are mediated, but experience with the *MTHFR* polymorphism suggests that dietary supplementation can compensate for subtle genetically determined variations in folate metabolism which can be of crucial importance during embryogenesis.

References

Smithells RW, Sheppard S, Schorah CJ *et al.* (1980) Possible prevention of neural-tube defects by periconceptional vitamin supplementation. *Lancet*, **i**, 339–340.

MRC Vitamin Study Research Group (1991) Prevention of neural tube defects: results of the Medical Research Council vitamin study. *Lancet*, **338**, 131–137.

Schizophrenia

There is strong evidence that genetic factors play a major role in causing schizophrenia. Family studies yield risks of 10–12% for first-degree relatives, 2–4% for second-degree relatives, and around 2% for third-degree relatives (Table 6.6). Concordance rates vary from 50 to 70% for monozygotic twins and from 10 to 15% for dizygotic twins. Adoption studies of all types (p. 119) have consistently shown an increased incidence of schizophrenia in biological relatives as compared with adoptive relatives. Thus, despite problems with classification and diagnosis, the evidence for a major genetic aetiological contribution is overwhelming. Estimates of heritability vary from 0.8 to 0.85.

Unfortunately, progress in identifying susceptibility alleles has been slow. Several chromosomal regions have been implicated, based either on the results of whole genome linkage analysis or because individuals with chromosome abnormalities (such as the deletion of chromosome 22 which causes the DiGeorge syndrome—p. 65) have developed schizophrenia-like symptoms (Fig. 6.6). However, no specific susceptibility genes have been identified, although several studies indicate that *DTNBP1* and *NRG1* are strong candidates.

DTNBP1 encodes a protein known as dysbindin, which is thought to play a role in synaptic function in the brain although its precise function is unknown. *NRG1* encodes neuroreglin 1, which induces myelination. Both of these genes have been identified through several independent linkage and association/linkage disequilibrium studies, but no specific pathogenic mutations have been identified in either gene. Thus, while

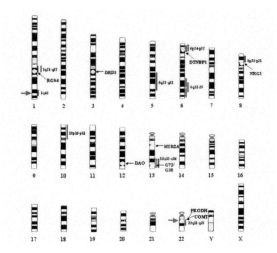

Fig. 6.6 Ideogram showing major chromosomal regions implicated by linkage studies in schizophrenia. Blue lines indicate areas for which suggestive evidence of linkage has been found in more than one data set. Red lines indicate regions where evidence of linkage has achieved genome-wide significance. Red arrows indicate sites of chromosome abnormalities associated with schizophrenia. Yellow circles indicate location of possible schizophrenia susceptibility loci. Reproduced with permission from O'DonovanMC, Williams NM, Owen MJ (2003) Recent advances in the genetics of schizophrenia. *Human Molecular Genetics*, **12**, R125–134.

the evidence for a causal role is strong, absolute proof that one or both of these genes conveys susceptibility to schizophrenia is lacking.

Susceptibility testing for multifactorial disorders

The identification of susceptibility loci using the methods outlined earlier in this chapter has raised the possibility that susceptibility testing for common multifactorial disorders should be made available to the general public. At present, this is offered on a small scale by commercial companies based mainly in the UK and the USA. These companies offer 'predictive genomic profiling' for conditions such as osteoporosis, asthma, hypertension, and coronary artery disease. Some also claim to be able to predict the body's ability to process toxic agents such as nicotine and alcohol. Following assessment by one of these companies, a journalist wrote in the *Daily Telegraph* (25 October 2003) that

> I won't be getting a blood clot in the near future, and I don't have the polymorphism associated both with Alzheimer's and high cholesterol. I'm blessed with a

TABLE 6.6	Family risks in schizophrenia

Relationship to index case	Lifetime risk for development of schizophrenia (%)
Identical twin	50
Non-identical twin	15
Sibling	9
Half sibling	6
Child	13
Grandchild	5
Nephew/niece	4
First cousin	2

Data adapted from McGuffin P, Owen MJ, Gottesman II (2002) *Psychiatric genetics and genomics*. Oxford University Press, Oxford.

genetic constitution that laughs loud and long in the face of drugs such as nicotine and—oh, happy, happy, words—alcohol.

Clearly this makes for good journalism, and presumably also good commercial profit, but an objective assessment of the state of the art suggests that the interpretation of these results might well be over-optimistic. This can be illustrated by considering the association of ApoE4 with Alzheimer disease, which is one of the strongest known associations in multifactorial disease. The lifetime risk of developing Alzheimer disease for an E4/E4 homozygote is 35% for men and 50% for women, in contrast to 4.5% for a man and 9.3% for a woman with no E4 allele. However, much of this risk is for onset in advanced old age and at present there is no reliable way of predicting when the disease is likely to start. In addition 50% of E4/E4 homozygotes will never develop Alzheimer disease and 5–10% of the population will develop the disease even though they do not possess a single copy of the E4 allele. The association of ankylosing spondylitis with the HLA B27 allele represents another example of a disease association which is not thought to be sufficiently strong to justify its use for susceptibility testing in the general population (p. 145).

These observations raise very real issues about the value of these tests and whether they should be used in clinical practice or be made available commercially for genetic profiling. A positive result might have adverse implications for employment, insurance and an anxious individual's mental state. In the light of these concerns, the UK Alzheimer's Disease Genetics Consortium, the American College of Medical Genetics, and the American Society of Human Genetics have all recommended that ApoE testing should not be used for routine clinical diagnosis or predictive testing in Alzheimer disease. The only scenario in which it is envisaged that ApoE testing might be acceptable would be if a therapy should be developed based on knowledge of an individual's ApoE4 status.

The use of empiric recurrence risks

The lack of understanding of the basic mechanisms involved in multifactorial inheritance means that it is not usually possible to derive recurrence risks on a theoretical basis. If it is known with certainty that inheritance is determined by an underlying normally distributed liability, then in theory recurrence risks can be based on statistics of the normal distribution and knowledge of the general population incidence (I). In this situation it has been shown that theoretical recur-

TABLE 6.7 Approximate empiric recurrence risks used for counselling in multifactorial disorders

Disorder	Recurrence risk for Siblings (%)	Offspring (%)
Autism	3–5	Not known
Cardiac defects	2–4	2–6
Cleft lip/palate	2–6	4
Depression		
Unipolar	10–15	10–15
Bipolar	7–10	7–10
Diabetes		
Type 1	5–6	5–6
Type 2	10–15	10–15
Neural tube defects	4	4
Schizophrenia	9	13

rence risks for first-, second-, and third-degree relatives are $I^{1/2}$, $I^{1/4}$ and $I^{7/8}$ respectively. For a disorder with an incidence of 1 in 1000, the risks for first-, second-, and third-degree relatives will be approximately 1 in 30, 1 in 200, and 1 in 500 respectively.

In reality it is not possible to be confident about the genetic mechanisms underlying most multifactorial disorders, so instead of theoretical risks, use is made of the observed incidence figures in various relatives as obtained in family studies. These are referred to as **empiric risks** because they are based on observation rather than theory. Examples of empiric risks used when counselling for some of the more common multifactorial disorders are given in Table 6.7.

Summary

Multifactorial is the term given to the mode of transmission shown by a large number of disorders which show familial clustering but which is not in accord with any recognized pattern of single gene inheritance. These disorders include several common congenital malformations and acquired disorders of childhood and adult life. The underlying genetic mechanism is thought to involve interaction of relatively large numbers of genes—hence oligogenic or polygenic—with environmental factors. A liability-threshold model has been proposed to explain how a discontinuous phenotype can be generated from an underlying continuous genetic/environmental liability.

Disorders which show multifactorial inheritance can be identified by a combination of family, twin and adoption studies. Family studies involve estimation of the incidence of the condition in close relatives. These also provide empiric recurrence risks which can be used for genetic counselling. Twin studies rely on comparison of the concordance rates in monozygotic and dizygotic twin pairs. Adoption studies involve comparison of the incidence of the disorder in biological and adoptive relatives.

Major research is being undertaken to identify susceptibility genes for the common multifactorial disorders. The approaches used include linkage analysis, association studies and the study of linkage disequilibrium. Linkage based on whole genome scanning can identify genes of major effect but these are usually in such a large segment of the genome that isolation is very difficult. Linkage disequilibrium and association studies can identify genes of minor effect over small distances, but these studies are much more technically demanding.

Using these approaches small numbers of susceptibility loci/genes have been identified for some of the common multifactorial disorders including Alzheimer disease, type I diabetes mellitus and neural tube defects. However, with the exception of Hirschsprung disease in which considerable progress has been made, the nature of the genetic contribution to most multifactorial disorders remains unknown.

Further reading

Bishop T, Sham P (eds) (2000) *Analysis of multifactorial disease.* Bios Scientific Publishers, Oxford.

Falconer DS, Mackay TFC (1996) *Introduction to quantitative genetics.* Longman, Harlow.

King RA, Rotter JI, Motulsky AG (ed.) (2002) *The genetic basis of common diseases,* 2nd edn. Oxford University Press, Oxford.

McGuffin P, Owen MJ, Gottesman II (ed.) (2002) *Psychiatric genetics and genomics.* Oxford University Press, Oxford.

Post SG, Whitehouse PJ (ed.) (1998) *Genetic testing for Alzheimer disease: ethical and clinical issues.* Johns Hopkins University Press, Baltimore.

Multiple choice questions

1 The following are correct examples of degrees of relationship:

(a) monozygotic twins—first degree

(b) dizygotic twins—second degree

(c) parent and child—first degree

(d) grandparent and grandchild—second degree

(e) first cousins—third degree

2 According to the liability/threshold model of multifactorial inheritance

(a) the recurrence risk is greater for relatives of the most severely affected cases

(b) the recurrence risk is not influenced by the number of affected relatives

(c) if the disorder is more common in girls than in boys, then the recurrence risk is greater for the relatives of affected girls than for the relatives of affected boys

(d) the liability curve for second degree relatives lies to the right of the curve for first degree relatives

(e) the liability curve for the general population lies to the left of the curve for affected individuals

3 Monozygotic twins

(a) result when two sperm fertilize an ovum which then splits in two

(b) would be expected to show 100% concordance for trisomy 21

(c) would be expected to show a higher concordance for a multifactorial disorder than dizygotic twins

(d) are more likely to be affected with a multifactorial disorder than dizygotic twins

(e) are a possible cause of chimaerism

4 A 45-year-old man is concerned that he may be at increased risk of developing Alzheimer disease because his mother developed the condition at the age of 82 years. It would be correct to tell him that

(a) his family history suggests autosomal dominant inheritance

(b) DNA mutation analysis is not available for either him or his children

(c) his risk is increased because his older brother with Down syndrome has developed the condition

(d) If ApoE testing shows that he does not have an E4 allele then he will definitely not develop the condition

(e) There is an increased risk that he will develop the condition but this risk is not high

5 Scientists studying a disorder which is known to show multifactorial inheritance find that it appears to be more common in individuals with blood group A. This could be because

(a) blood group A is very common in the population being studied

(b) the allele which encodes blood group A is in linkage disequilibrium with an allele which causes the disease

(c) the blood group A antigen actually causes the disease

(d) blood group A is more common in monozygotic than in dizygotic twins

(e) blood group B protects against developing the disease

Answers

1 (a) false—monozygotic twins share 100% of their genes, not 50% as applies for first-degree relatives

   (b) false—dizygotic twins are first-degree relatives

   (c) true—a parent and child share 50% of their genes

   (d) true—a grandchild has four grandparents and therefore shares on average 25% of each grandparent's genes

   (e) true—first cousins share on average 1/8 of their genes

2 (a) true—according to the theory the most severely affected individuals will lie to the extreme right of the liability curve

   (b) false—the more affected relatives, then the greater the risk

   (c) false—boys must have more 'bad genes' to be affected than girls. Therefore their relatives are at greater risk

   (d) false—the curve for second degree relatives lies to the left of the curve for first-degree relatives (Fig. 6.2)

   (e) true—see Fig. 6.2

3 (a) false—monozygotic twins result from the fertilization of a single ovum by a single sperm forming a zygote when then splits into two

   (b) true—with the rare exception of mosaic Down syndrome caused by a post-fertilization mitotic non-disjunction

   (c) true—this is one of the ways to demonstrate multifactorial inheritance

   (d) false—simply being a twin of whatever zygosity does not convey an increased risk for developing a multifactorial disorder

   (e) false—chimaeras have two cell lines derived from *different* zygotes

4 (a) false—autosomal dominant forms tend to show 'early' pre-senile onset

   (b) true—specific mutation analysis is not available for the polygenic/multifactorial form of Alzheimer disease

   (c) false—most people with Down syndrome develop Alzheimer disease after the age of 50 years because of a gene dosage effect. This does not mean that other relatives are at increased risk

   (d) false—there is still a small risk that he will develop the condition (see Table 6.4)

   (e) true—his risk is increased by a factor of 3–4 but is still relatively low compared to risks for autosomal dominant families

5 (a) true—this is referred to as population stratification

   (b) true—linkage disequilibrium is thought to be one of the commonest explanations for a disease association

   (c) true—unlikely but possible

   (d) false—blood groups have no association with twinning

   (e) true—it could be the lack of blood group B which conveys susceptibility rather than the presence of blood group A

Genes and populations

Certain genetic conditions are more common in some populations than in others. In the case of multifactorial acquired disorders of adult life, as described in the previous chapter, this can be explained by exposure to different environmental factors such as diet and lifestyle. But it is difficult to envisage how dietary factors could account for the striking differences seen in the frequencies of many single-gene disorders. Notable examples include cystic fibrosis (p. 107), sickle-cell disease (p. 156), and α/β-thalassaemia (p. 159), which all show particularly high incidences in specific ethnic groups (cystic fibrosis in the white population, sickle-cell disease in the black population, and thalassaemia in Mediterranean and Asian populations).

Thus the study of the distribution of genes in populations, which is referred to as **population genetics**, is not simply an interesting exercise for human geneticists. For biologists, an understanding of why a condition is common can help shed light on its underlying pathogenesis. For medical geneticists, knowledge of the carrier frequency in a population is essential for calculating risks in genetic counselling. Finally, for public health physicians, information about the prevalence of particular conditions and their carrier frequencies is vital for planning carrier screening programmes and allocating health care resources.

In this chapter we begin by reviewing the Hardy–Weinberg principle, a concept that is fundamental to almost all aspects of population genetics. This is followed by consideration of factors which can influence gene frequency leading to a condition being more common in one population than in another. This in turn leads on to an account of some of the more common polymorphisms observed in human populations. Finally, the clinical relevance of population genetics is emphasized by outlining criteria for carrier screening programmes and how these have met with varying success.

The Hardy–Weinberg principle

This was proposed independently in 1909, by an English mathematician, Hardy, and a German physician, Weinberg, to explain why dominant conditions do not become steadily more common in populations. According to the Hardy-Weinberg principle, in an ideal or model population the *relative proportions* of different genotypes remain constant from one generation to another. Thus, as a population increases in size, the numbers of individuals with a dominant or a recessive trait will increase, but the relative proportions of the different genotypes will remain constant. In this situation the population is said to be in **Hardy-Weinberg equilibrium**. Note that this only applies to a model population and that there are several factors which can disturb Hardy-Weinberg equilibrium. These are considered later.

According to the Hardy-Weinberg principle, if there are two alleles, A1 and A2, at a locus with frequencies of p for A1 and q for A2, then if the population is in equilibrium the frequencies of the different genotypes

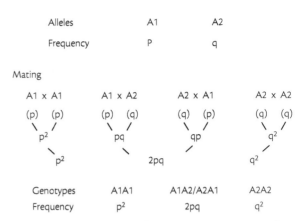

Fig. 7.1 Genotype frequencies following random segregation of alleles A1 and A2.

will be p^2 for A1A1, $2pq$ for A1A2 and A2A1, and q^2 for A2A2 (Fig. 7.1). The frequency of A1A1 offspring is the probability that an A1 allele will meet up with another A1 allele, i.e. p^2. Similarly, the frequency of A2A2 offspring is q^2. The frequency of heterozygous offspring is the sum of the probability that a sperm with A1 will fertilize an egg with A2 (i.e. pq), plus the probability that a sperm with A2 will fertilize an egg with A1 (i.e. pq), giving a total of 2pq. For the mathematically inclined this represents a simple expansion of the binomial, i.e. $(p+q)^2 = p^2 + 2pq + q^2$.

TABLE 7.1   Maintenance of Hardy–Weinberg equilibrium in a population

Genotype	AA	Aa	aa
Frequency	p^2	$2pq$	q^2

At equilibrium		Genotypic probability in offspring			Genotypic frequency in next generation		
Mating type	Frequency	AA	Aa	aa	AA	Aa	aa
AA × AA	p^4	1	0	0	p^4	0	0
AA × Aa	$2p^3q$	$1/2$	$1/2$	0	p^3q	p^3q	0
AA × aa	p^2q^2	0	1	0	0	p^2q^2	0
Aa × AA	$2p^3q$	$1/2$	$1/2$	0	p^3q	p^3q	0
Aa × Aa	$4p^2q^2$	$1/4$	$1/2$	$1/4$	p^2q^2	$2p^2p^2$	p^2q^2
Aa × aa	$2pq^3$	0	$1/2$	$1/2$	0	pq^3	pq^3
aa × AA	p^2q^2	0	1	0	0	p^2q^2	0
aa × Aa	$2pq^3$	0	$1/2$	$1/2$	0	pq^3	pq^3
aa × aa	q^4	0	0	1	0	0	q^4

Total frequency of
AA = $p^4 + 2p^3q + p^2q^2 = p^2(p^2 + 2pq + q^2)$ = p^2
Aa = $2p^3q + 4p^2q^2 + 2pq^3 = 2pq(p^2 + 2pq + q^2)$ = $2pq$
aa = $q^4 + 2pq^3 + p^2q^2 = q^2(p^2 + 2pq + q^2)$ = q^2

Adapted from Khoury MJ, Beaty TH, Cohen BH (1993) *Fundamentals of genetic epidemiology*. Oxford University Press, New York.

TABLE 7.2 Summary of genotype frequences assuming Hardy–Weinberg equiplibrium

Autosomal disorders		
Gene frequency	A1	p
	A2	q
Genotype frequencies	A1A1	p^2
	A1A2	2pq
	A2A2	q^2
X-linked disorders		
Gene frequency	X1	p
	X2	q
Genotype frequencies Males	X1	p
	X2	q
Females	X1X1	p^2
	X1X2	2pq
	X2X2	q^2

Intuitively it is apparent that as a population reproduces, the alleles will be reproduced proportionately as long as there are no external factors such as selection for or against a particular genotype. Thus the genotype proportions will remain as p^2 for A1A1, 2pq for A1A2/A2A1, and q^2 for A2A2. This can be demonstrated mathematically by drawing up a table to show the frequencies of different genotypes in successive generations (Table 7.1).

For an X-linked trait with alleles X1 and X2, with frequencies of p and q respectively, the frequencies of X1 and X2 in males will simply be p and q. The frequencies of the X1X1, X1X2/X2X1, and X2X2 genotypes in females will be p^2, 2pq, and q^2 respectively (Table 7.2).

Using the Hardy–Weinberg principle to determine carrier frequencies

If it can be assumed that a population is in Hardy–Weinberg equilibrium for two alleles at a particular locus, then knowledge of the genotype distributions (p^2, 2pq, q^2) can be used to determine carrier frequencies. This is mainly relevant for autosomal recessive and X-linked recessive disorders in which, by definition, carriers are phenotypically normal, so that there is no way of determining carrier frequencies simply by counting known carriers in a defined population.

Autosomal recessive inheritance

If the incidence of an autosomal recessive disorder is known in a population, then it is a relatively simple task to determine the carrier frequency. The incidence of the disorder equals the incidence of affected homoygotes, i.e. q^2. The gene frequency, q, equals the square root of the disease incidence, and the carrier frequency equals 2pq, where p + q = 1. Therefore if we are considering a condition such as cystic fibrosis, which affects approximately 1 in 2500 children of western European origin, $q^2 = 1/2500$ and q = 1/50. The carrier frequency equals 2pq = 2 × 49/50 × 1/50, which is approximately 1/25 (assuming 49/50 = 1). Thus a quick estimate of the carrier frequency can be made by doubling the square root of the disease incidence (I), i.e. carrier frequency = $2\sqrt{I}$.

To check that a calculation is correct, try reversing the situation and imagine that you are calculating the disease incidence based on knowledge of the carrier frequency. The chance that both members of a partnership will be carriers is 1/25 × 1/25, which equals 1/625. On average a quarter of the children born to such partnerships will be affected. Thus the expected incidence of the disorder is 1/625 × 1/4, which brings us back to the disease incidence of 1 in 2500 that we started with. If you get a different answer then you have made a mistake somewhere!

Approximate values for gene frequency, disease incidence, and carrier frequency for various autosomal recessive disorders are given in Table 7.3. In theory these values are only valid if the disorder is in Hardy–Weinberg equilibrium. For practical purposes in genetic counselling this is usually assumed to be the case. An example of how knowledge of the carrier frequency can be used in calculating genetic risks is given in Chapter 13 (p. 249).

TABLE 7.3 Approximate incidence and carrier frequency for autosomal recessive disorders

Incidence (q^2)	Carrier frequency (2pq)	Disorder
1/400	1/10	Haemochromatosis
		Sickle cell disease
		α- and β-thalassaemia
1/2000	1/22	Cystic fibrosis
1/3600	1/30	Tay-Sachs disease
1/5000	1/35	Congenital adrenal hyperplasia
1/10 000	1/50	Oculocutaneous albinism
		Phenylketonuria
		Spinal muscular atrophy
1/40 000	1/100	Friedreich ataxia
1/160 000	1/200	Hurler syndrome
1/250 000	1/250	Ataxia telangiectasia

X-linked recessive inheritance

As with autosomal recessive disorders, a quick estimate of the carrier frequency in females (2pq) can be made based on knowledge of the disease incidence in males (q). Thus for a condition such as haemophilia A with an incidence in males of 1 in 5000, the carrier frequency equals $2 \times p \times 1/5000$, which is approximately 1 in 2500. The incidence of affected females due to homozygosity equals q^2, i.e. $(1/5000)^2 = 1$ in 25 million. A quick check that this is correct is provided by considering the probability that an affected man will partner a carrier woman $(1/5000 \times 1/2500)$. On average, half of their daughters will be homozygous affected, giving an overall incidence of affected homozygotes of $1/5000 \times 1/2500 \times 1/2$, which equals $(1/5000)^2$ or 1 in 25 million.

> #### Key Point
>
> If a population is in Hardy–Weinberg equilibrium, the carrier frequency approximates to twice the square root of the disease incidence for an autosomal recessive disorder and to twice the incidence in males for an X-linked recessive disorder.

Why eugenics doesn't work

Before considering some of the factors that can disturb Hardy–Weinberg equilibrium, it is worth pausing for a moment to reflect on how we can apply this principle to demonstrate the fallacy behind the application of eugenics principles to 'genetically cleanse' a population. **Eugenics**, which can be defined as the science and philosophy of trying to improve the hereditary characteristics of a population, deservedly gained a horrendous reputation during the first half of the twentieth century, as discussed in Chapter 14.

Imagine that, in a moment of madness, a government decides that oculocutaneous albinism (MIM 203100) is an undesirable trait and that albinos should not be allowed to reproduce. Oculocutaneous albinism is an autosomal recessive condition with an incidence of 1 in 10 000, which numbers amongst its list of distinguished sufferers both Noah of biblical fame and the Reverend Dr Spooner, an Oxford don, who had the endearing habit of reciprocally transposing the first letters of words—hence 'Spoonerism'. Famous examples include complaining about students who had 'hissed my mystery lecture' and who had not worked sufficiently diligently to the extent that they had 'tasted two worms'.

If we assume that the albinos in society would be foolish enough to comply with this ridiculous government edict, then consider how long it would take for the gene frequency to show a significant decline.

If the population is in Hardy–Weinberg equilibrium, the ratio of carriers to homozygous affected will be $2pq : q^2$, which reduces to $2p : q$. If the incidence of affected homozygotes is 1 in 10 000, then $q^2 = 1/10\,000$, $q = 1/100$, and $p = 99/100$. Therefore the ratio of carriers to homozygotes equals 198 : 1. Another way of looking at this is that amongst 10 000 people, there will be 1 affected homozygote and 198 carriers. If the affected individual fails to reproduce, then two mutant alleles will be lost, but the 198 carriers will continue to reproduce normally. By developing this theme it can be shown that it will actually take 100 generations, i.e. approximately 2000 years, to reduce the gene frequency by 50%.

Factors that can influence the Hardy–Weinberg distribution

The Hardy–Weinberg principle applies to an ideal or model population in which no internal factors are operating that could disturb the distribution of alleles within that population. This means therefore that there should be no new mutational events, no selection for or against a particular genotype, no chance fluctuation because of small population size, no selection of partners on the basis of genotype, and no migration of new genes into the gene pool. Thus the ideal population should be large and stable with random mating and free of new mutations. In reality no such population exists, although Hardy–Weinberg equilibrium is usually assumed when calculating carrier frequencies for genetic counselling. The following factors which can disturb the Hardy–Weinberg distribution are now considered:

- unequal mutation–selection equilibrium
- selection for heterozygotes (heterozygote advantage)
- founder effect and genetic drift
- non-random mating
- migration and gene flow.

Mutation–selection equilibrium

In large, randomly mating populations the two factors which are most likely to influence gene frequencies are new mutations and selection against affected homozygotes. In practice these are often assumed to balance out, particularly if a carrier test becomes available which shows that the observed carrier frequency

TABLE 7.4 Formulae for calculating mutation rates assuming mutation-selection equilibrium

Inheritance	Formula
Autosomal dominant	$\mu = I(1-f)/2$
Autosomal recessive	$\mu = I(1-f)$
X-linked recessive	$\mu = I^M(1-f)/3$

μ, mutation rate; I, incidence; I^M, incidence in males; f, reproductive fitness.

equals the carrier frequency that would be expected if the population is in equilibrium.

In these situations, by applying a little algebra it is possible to derive simple formulae which can be used to determine the rates at which mutations occur (Table 7.4). In the following discussion μ denotes the mutation rate, I denotes the disease incidence, and f denotes the reproductive **fitness** of affected individuals, as defined by the ability to have children as compared with the general population

Autosomal dominant inheritance

If the disease is in equilibrium then new mutations (2μ per generation) are being balanced by the loss of mutant alleles through selection against affected individuals. Therefore

$2\mu = I(1-f)$ or $\mu = I(1-f)/2$

For example, for a condition with an incidence of 1 in 10 000 and fitness 0.5 (indicating that affected individuals have on average half as many children as unaffected) the mutation rate equals $I(1-f)/2 = 1$ in 40 000. This means that 1 in 40 000 gametes carries a new mutation.

Autosomal recessive inheritance

For an autosomal recessive disorder in equilibrium,

$2\mu = I(1-f) \times 2$ or $\mu = I(1-f)$

In this situation two mutant alleles are lost with each affected homozygote who fails to reproduce. Thus a factor of 2 is introduced on the right-hand side of the equation. For a condition with incidence 1 in 10 000 and fitness 0.5, the mutation rate = $I(1-f) = 1$ in 20 000.

X-linked recessive inheritance

For an X-linked recessive disorder in equilibrium

$3\mu = I^M(1-f)$ or $\mu = I^M(1-f)/3$

In this situation three X chromosomes are transmitted per couple per generation, so there are three opportunities for a mutation to occur. I^M represents the incidence of the condition in males.

Duchenne muscular dystrophy (p. 109) is an example of an X-linked recessive condition in which affected boys rarely reproduce. Thus reproductive fitness equals

zero. The disease affects approximately 1 in 3000 boys, so the mutation rate is $I^M(1-f)/3 = 1$ in 9000. This implies that 1 in 9000 gametes carries a new Duchenne muscular dystrophy mutation. Therefore in an average male ejaculate containing 100 million sperm, approximately 11 000 will carry a new Duchenne muscular dystrophy mutation. This high mutation rate is explained by the large size of the Duchenne muscular dystrophy gene (p. 111).

Selection for heterozygotes (heterozygote advantage)

When a condition shows a high incidence in a large population there are two possible explanations. Either the mutation rate is very high, as is likely to apply for Duchenne muscular dystrophy, or carriers have a biological advantage over non-carriers resulting in increased reproductive fitness. This is referred to as **heterozygote advantage**. Heterozygote advantage is thought to explain, at least in part, the relatively high gene frequencies observed for the haemoglobinopathies, cystic fibrosis, and Tay–Sachs disease.

In the case of the sickle-cell trait, heterozygote advantage is explained by resistance to infection with falciparum malaria (p. 157). This was first suspected on the basis of the almost identical geographical distribution shown by sickle-cell disease and malaria. Confirmation came with the demonstration that the incidence of malaria parasitaemia was lower in children with sickle-cell trait than in those without the trait. It was also shown that following inoculation of malaria parasites into adult volunteers with and without the sickle-cell trait, parasitaemia only became established in the latter group. More recently, molecular analysis of haplotypes adjacent to the β-globin locus has indicated that there were probably three or four original sickle-cell mutations that became established because of the selective pressure conveyed by resistance to malaria.

The mechanism underlying resistance to malaria in the sickle-cell trait is thought to involve the preferential removal of red cells which undergo sickling on invasion by the malaria parasite. The explanation for heterozygote advantage in the thalassaemias is not so well understood, but the closely overlapping distributions of malaria with both α- and β-thalassaemia are consistent with resistance to malaria being the probable explanation for their high gene frequencies in specific populations (p. 159).

The explanation for heterozygote advantage in cystic fibrosis and Tay–Sachs disease probably also lies in resistance to infection. Typhoid fever was endemic in Europe in the middle ages. *Salmonella typhi*, the causative organism, utilizes the cystic fibrosis chloride channel (p. 108) to enter gastrointestinal epithelial cells. Mice which are heterozygous for the common ΔF508 cystic fibrosis mutation have been shown to have much lower levels of *Salmonella typhi* in their gastrointestinal mucosa than homozygous normal mice. Thus the prevailing evidence points to relative resistance to typhoid fever as the probable explanation for the high incidence of cystic fibrosis in western European populations.

There is some evidence that carriers of Tay–Sachs disease (p. 215) have increased immunity to tuberculosis, although how this is mediated is not known. Tuberculosis was extremely common in those parts of eastern Europe from which much of the world's Ashkenazi Jewish population originates. Three mutations account for 99% of all Tay–Sachs disease mutations in this population, which suggests that these 'founder' mutations conveyed a biological advantage in carriers, which has led to the high carrier frequency of 1 in 30.

The founder effect and genetic drift

The **founder effect** refers to the mechanism whereby a population shows a relatively high frequency for a particular mutation because one member of the original small ancestral population was a carrier. A high gene frequency can become established either because of heterozygote advantage, as already discussed, or because of chance fluctuations in the number of offspring inheriting the mutation when the population was small. The term **genetic drift** refers to changes in gene frequency due to random or chance fluctuations occurring from generation to generation. The impact in large populations is minimal, but in a small population large changes in gene frequency can occur within a few generations simply as a result of the lottery of Mendelian segregation.

Many examples of founder effect and drift have been observed. A specific mutation which causes variegate porphyria in many white South Africans can be traced back to a single Dutch couple who emigrated from the Netherlands and married in Cape Town in 1688 (p. 218). Approximately 2% of the Ashkenazi Jewish population in Australia and the USA carry a specific muation in either *BRCA1* (185delAG) or *BRCA2* (6174delT), thereby accounting for the relatively high incidence of familial carcinoma of the breast and ovary (p. 200) identified in this group. Autosomal recessive disorders resulting from a founder effect are particularly common in genetic isolates, i.e. populations with a restricted choice of partners for social, geographical, or religious reasons. Examples include the Ellis van Creveld syndrome (MIM 225500) in the Old-Order Amish in Pennsylvania and cartilage–hair hypoplasia (MIM 250250) in Finland.

Non-random mating

The Hardy–Weinberg principle assumes that there will be random segregation of alleles with random mating in the population. Two patterns of non-random mating can disturb Hardy–Weinberg equilibrium. These are **assortative** mating and **consanguinity**. In assortative mating individuals are attracted to each other on the basis of shared characteristics. Assortative mating sometimes occurs between people who share a disability such as poor vision or hearing loss. This can result in the clustering of mutant alleles within small nuclear families, leading to an increase in the overall proportion of homozygotes. Marriage between close relatives (i.e. consanguinity—p. 74) can have a similar outcome. This is why estimation of gene frequencies based on the incidence of affected individuals in small inbred communities can give falsely high results.

Migration and gene flow

The introduction of new alleles into a population as a result of migration and intermarrige, a phenomenon known as **gene flow**, leads to a new gene frequency in the hybrid population. This is the explanation proposed for the subtle gradation observed for the distribution of the B blood group which shows its highest frequency in eastern Europe and western Asia with much lower frequencies in western Europe, Africa, the USA, and Australasia.

A more dramatic example of gene migration is illustrated by the distribution of the common C282Y (Cys282Tyr) mutation, which accounts for most cases of haemochromatosis (MIM 235200). In this autosomal recessive disorder there is increased absorption of iron from the intestine, which eventually leads to damage to the liver, pancreas, and heart where the excess iron is stored. The disorder occurs almost exclusively in the white population. The common mutation occurs on a single background haplotype, consistent with a single original mutation which became established, probably because of the benefits of moderately increased iron absorption in carriers in times of famine and bloodletting.

Analysis of the distribution of the common C282Y allele in Europe (Fig. 7.2) indicates that it probably arose in Scandinavia and spread throughout Europe as

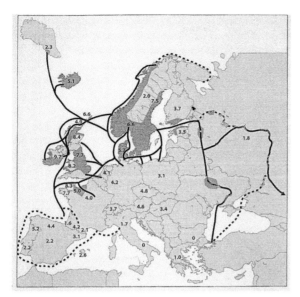

Fig. 7.2 Shows the pattern of gene flow for the common C282Y mutation which causes haemochromatosis. Solid and dashed lines denote major and minor travel routes used by the Vikings. Figures indicate the local frequency of the C282Y mutation. Reproduced with permission from Milman N, Pedersen P (2003) Evidence that the Cys282Tyr mutation of the HFE gene originated from a population in Southern Scandinavia and spread with the Vikings. *Clinical Genetics*, 64, 36–47. Published by Blackwell Publishing Ltd.

a result of Viking conquests. The high gene frequency of haemochromatosis in Europe can probably be attributed to an original founder effect with drift in a small ancestral Scandinavian population followed by heterozygote advantage on a large scale coupled with migration and intermarriage.

> **Key Point**
>
> Hardy–Weinberg equilibrium can be disturbed by several factors including mutation, selection for or against a particular genotype, small population size, non-random mating and gene flow. Estimates of gene frequencies based on the incidence of affected individuals in small inbred communities will give falsely high values.

Polymorphism

In its widest sense the term polymorphism refers to the existence of several different forms. In genetics it has a rather more specific meaning. A **genetic polymorphism** refers to variation in a gene, chromosome, or protein resulting in the existence of two or more forms, each of which has a frequency greater than that

which could be maintained by recurrent mutation alone. By convention, alleles which show a frequency of 1% or greater are described as polymorphisms, whereas those with a frequency of less than 1% are referred to as *rare variants*.

Genetic polymorphisms are extremely common. Amongst non-coding nucleotides, which constitute around 98% of the total genome, polymorphism occurs at 1 in 500 to 1 in 1000 bp. At a protein level, 20–30% of all loci are polymorphic. This high degree of variation at the molecular level is thought to account for the extreme diversity seen in the human species, and probably also accounts for a number of factors very relevant to medicine, including variable drug sensitivity (p. 225) and susceptibility to the common diseases of adult life (p. 121).

There is a tendency to use the term 'polymorphic' exclusively to refer to a locus which shows allelic variation that is of no medical significance. It is certainly true that most human polymorphisms are clinically silent. However, some clearly represent pathogenic mutations, such as ΔF508 in cystic fibrosis (p. 108) and βGlu6Val in sickle-cell disease (p. 156). Others become apparent only in certain circumstances, such as following blood transfusion (e.g. the ABO blood groups) or exposure to a toxic agent (e.g. G6PD deficiency— p. 227). It is also possible that polymorphisms in non-coding regions which appear to be of no importance may exert subtle effects on gene expression and interaction which are relevant to the pathogenesis of multifactorial disorders. This is one of the reasons why attention is being focused on the study of single nucleotide polymorphisms (SNPs) in the study of polygenic disorders (p. 122).

Genetic polymorphism can be identified at three levels: DNA, chromosomal, and protein. Each of these is now considered.

DNA sequence polymorphisms

The high degree of polymorphic variation seen in non-coding regions of the genome probably reflects the fact that no evolutionary disadvantage has been conferred. It is certainly possible that some non-coding polymorphisms have become established because of an evolutionary or biological advantage but at present this is difficult to confirm at a scientific level.

Single nucleotide polymorphisms (SNPs)

As the name implies, a SNP (usually pronounced 'snip') involves a change in a single nucleotide. Usually this is a substitution but it can also take the form of an insertion or a deletion. Over 4 million SNPs have been identified and as these are distributed throughout the genome they provide extremely useful markers for association studies in the search for loci conveying susceptibility to the common diseases (p. 122). Automated methods have been developed for identifying large numbers of SNPs in a single analysis using microarray technology. This procedure is referred to as *high-throughput genotyping*.

Restriction fragment length polymorphisms (RFLPs)

RFLPs are generated by SNPs that occur at restriction sites identifiable by Southern blotting (p. 15). A restriction site represents the short sequence of DNA which is recognised by a restriction enzyme resulting in the cleavage of the DNA strand. The presence or absence of a cleavage site can be detected using the appropriate restriction enzyme and a probe for a closely adjacent DNA sequence (Fig. 7.3). The different band patterns seen on a Southern blot represent the RFLPs. RFLPs were amongst the first genetic markers used for linkage analysis and for gene-tracking to facilitate genetic counselling (p. 91).

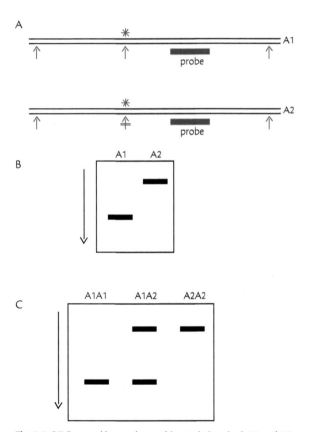

Fig. 7.3 RFLP caused by a polymorphic restriction site *. A1 and A2 are homologous chromosomes. The restriction enzyme cleaves chromosome A1 but not A2. B shows the appearance on Southern blot if each chromosome could be analysed separately. C shows the appearance on Southern blot for the different possible genotypes.

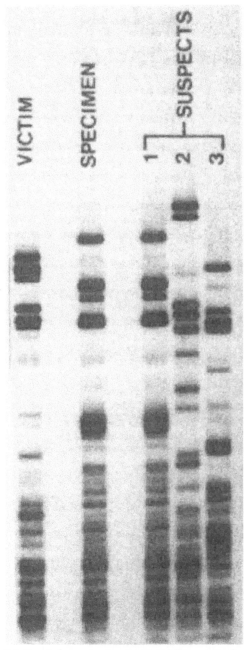

Fig. 7.4 Genetic fingerprint showing minisatellite analysis of samples from three suspects, the victim and the perpetrator of a crime. Suspect number 1 is clearly the perpetrator. Photograph kindly provided by Cellmark Diagnostics, Abingdon, UK.

Minisatellites

Minisatellites represent a form of length polymorphism caused by insertion in tandem (i.e. one after another) of a 10–100-bp core sequence of DNA. They are highly polymorphic and are sometimes referred to as **variable number of tandem repeat** (VNTR) polymorphisms. Minisatellites are widely scattered across the genome and can be identified at a large number of loci in a single Southern blot using a core sequence probe. This generates a unique pattern for each individual which is referred to as a *genetic fingerprint*. Alternatively, a single VNTR polymorphism can be studied using a probe which hybridizes to a unique locus specific DNA sequence immediately adjacent to the tandem repeat.

Genetic, or DNA, fingerprinting has proved to be an extremely powerful test for identifying family relationships as in paternity testing and immigration appeals,

when alleged close relatives have been denied access to the country in which their parents or children have settled. However, its widest and most valuable application has been in the field of forensic medicine, as a unique DNA fingerprint can be obtained from small amounts of body tissue such as semen and blood (Fig. 7.4) (Box 7.1).

Microsatellites

Microsatellites, which are also known as *simple sequence repeats* (SSRs), are similar to minisatellites in that they consist of multiple copies of tandemly repeated sequences. However, they differ in that the repeated core sequence contains only two, three, or four base pairs, known as di-, tri-, or tetranucleotides respectively. Over 10 000 polymorphic microsatellites have been identified. These have been extremely useful for mapping studies in positional cloning (p. 100) and as

---

BOX 7.1 CASE CÉLÈBRE: THE NARBOROUGH MURDERS

In 1987 the small town of Narborough in Leicestershire, in the Midlands of England, found itself the focus of unwelcome world attention, for it was here that mass population genetic fingerprinting was undertaken for the first time. This sad tale began in 1983 when a 15-year-old Narborough girl, Lynda Mann, was brutally raped and murdered close to her home. Three years later another 15-year-old Narborough girl, Dawn Ashworth, suffered the same fate.

After the second murder, a 17-year-old kitchen porter confessed to the crime during police investigation. One year previously a research scientist at the nearby University of Leicester, Alec Jeffreys (now Sir Alec Jeffreys), had discovered the technique known as genetic fingerprinting, which is based on the simultaneous analysis of polymorphic minisatellites at multiple loci. Before the young man's trial it was decided to use genetic fingerprinting to try to confirm or refute his guilt. Samples of semen from vaginal swabs from both victims were analysed and compared with DNA from the alleged perpetrator. Somewhat surprisingly, it emerged that his DNA did not match that of the semen samples obtained from either of the murdered girls. However, the DNA fingerprints did show conclusively that both girls had been murdered by the same man.

In an attempt to flush out the true murderer, who understandably was suspected to live in the vicinity, the police requested that all local men aged between 16 and 34 years volunteer to provide blood samples

for DNA analysis. Several months later over 98% of the relevant male population had given blood but the culprit remained unidentified. As the police doggedly tried to track down the remaining 2%, a bakery worker revealed in conversation that he had given a blood sample on behalf of a colleague at work who had persuaded him to trade identity for the purpose. When this news reached the police, the work colleague, Colin Pitchfork, was arrested and duly confessed to both crimes. His DNA profile was found to match that of the semen stains and at the subsequent trial he received a double life sentence.

This first dramatic example of the application of mass genetic fingerprinting to identify a murderer illustrates the power of a technique that has revolutionized the practice of forensic medicine throughout the world. In practice genetic profiling using minisatellites is no longer the method of choice, as the analysis by Southern blotting requires a large quantity of DNA. Instead, around 10 highly polymorphic single-locus microsatellites are used as these can be typed by PCR and allow a more precise mathematical calculation of probability.

Further details of the Narborough murders can be obtained in the book by Joseph Wambaugh referenced below.

Reference

Wambaugh J (1984) *The blooding*. Bantam Press, London.

markers linked to disease loci in genetic counselling (p. 91). Microsatellites have the major advantage that because of their small size they can be analysed by PCR rather than the more demanding technique of Southern blotting required for the analysis of minisatellites.

Chromosome polymorphisms

Although normal human chromosomes from different individuals generally have an identical appearance, subtle abnormalities can sometimes be observed using particular staining procedures (p. 29). Specifically, there is a band of heterochromatin adjacent to the centromere of chromosomes 1, 9, 16, and Y, which shows polymorphic variation in size and position. Heterochromatin is composed mainly of repetitive non-coding DNA and stains darkly using a special technique known as C (centromeric heterochromatin) banding.

This polymorphic (also known as *heteromorphic*) band of heterochromatin is of no medical significance. Historically its main claim to fame is that linkage of a red blood cell antigen, known as the Duffy blood group, to a large heterochromatic region on chromosome 1 was demonstrated in 1968. This was the first locus to be mapped to an autosome in humans.

Protein polymorphisms

Structural variation in polypeptides accounts for much of the physiological and biochemical diversity manifest in the human species. Some of the more commonly encountered examples are now considered.

The major histocompatability complex (MHC)

The MHC consists of three groups, or *classes*, of genes located within a 4-Mb cluster on the short arm of chromosome 6 (Fig. 7.5). The class I and II loci constitute the human leukocyte antigen (HLA) genes. These encode cell surface proteins that bind to an antigenic peptide to generate an immune response. The HLA class I and

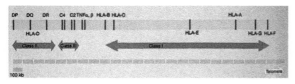

Fig. 7.5 Diagrammatic representation of the major histocompatibility locus on chromosome 6p. TNF = tumour necrosis factor. The locus for 21-hydroxylase deficiency is positioned close to the C2 and C4 complement components. Reproduced with permission from Lewin B (2000) *Genes VII*. Oxford University Press, New York.

II genes are the most polymorphic known coding sequences in the human genome, with over 200 alleles at the A locus and over 400 at the B locus. It is thought that this allelic heterogeneity has been generated by point mutations, gene conversions, and reciprocal recombination events. The HLA region shows strong linkage disequilibrium, indicating that there may have been strong evolutionary selection for certain combinations of alleles, possibly because of protection conveyed to common infectious agents. The HLA alleles on a particular chromosome are referred to as a **haplotype**. Usually a haplotype is inherited as an intact unit, although rarely a crossover can occur between the class I and class II clusters, which are at opposite ends of the major histocompatibility complex. Haplotype also refers in a more general sense to the pattern of alleles found at linked loci on a chromosome.

HLA and transplantation genetics

Compatibility for HLA alleles (*histocompatibility*) is a key factor in determining the successful outcome of bone marrow and renal transplantation. The ideal donor should be HLA identical. There is one chance in four that siblings will share identical HLA haplotypes, whereas it is very unlikely that this will apply to a parent and child, with the extremely rare exception of uniparental disomy (p. 83). The importance attached to having an HLA-identical sibling for bone marrow or stem cell transplantation has prompted some parents of children with conditions such as β-thalassaemia (p. 161) to request pre-implantation diagnosis for a future child whose tissue could be used to treat an affected older sibling. This has led to considerable controversy over the ethics of planning a 'designer baby'.

HLA and disease associations

Several autoimmune diseases have been found to show strong associations with specific HLA alleles. Examples include ankylosing spondylitis with B27, coeliac disease with DQ2, and insulin-dependent diabetes with DR3 and DR4 (p. 127). The underlying explanation for these disease associations is not known. One possibility is that the immune response generated by a foreign antigen–HLA protein complex cross-reacts with the host tissue.

None of the HLA disease associations has been deemed to be of sufficient strength to be useful for predictive diagnostic purposes. This can be illustrated by considering the relatively strong association that exists between ankylosing spondylitis and the B27 allele.

Approximately 95% of all white individuals with ankylosing spondylitis are B27 positive, in contrast to 10% of the general population. The incidence of ankylosing spondylitis is approximately 1 in 1000, so in an average population sample of 1000 people, 1 will have ankylosing spondylitis and 100 will be B27 positive. Thus only 1% of individuals who test positive for B27 would be expected to have or develop ankylosing spondylitis. For the other 99% a positive B27 result would simply generate unjustified concern.

Finally, an important distinction should be made between an HLA or disease association and true linkage between a disease locus and a single-gene disorder. The loci for two autosomal recessive conditions, 21-hydroxylase deficiency (p. 179) and haemochromatosis (p. 140), are tightly linked to the MHC on chromosome 6, with the 21-hydroxylase locus actually positioned within the class III cluster (Fig. 7.5). As well as being linked to the HLA loci, both of these conditions also show linkage disequilibrium with specific HLA alleles (21-hydroxylase deficiency with A3 and haemochromatosis with A8). This suggests that these conditions are caused by a small number of founder mutations that arose on specific HLA chromosomes with which they are still in disequilibrium. Linkage disequilibrium therefore provides another explanation for how a disease association can arise. This is the rationale behind the use of association and linkage disequilibrium studies in the search for susceptibility genes for multifactorial diseases (p. 122).

The ABO blood groups

The common human ABO blood groups are determined by the activity of a glycosyltransferase encoded by a polymorphic gene on chromosome 9q34. The A allele encodes transferase A which adds N-acetylgalactosamine to a red blood cell surface protein known as H or O antigen. The B allele encodes transferase B, which adds D-galactose to the H antigen. The O allele contains a single frame-shift deletion in the coding region leading to absence of transferase activity (Fig. 7.6).

The A and B alleles are **codominant** in that they are both expressed, whereas the O allele is recessive. Thus there are six potential genotypes and four possible phenotypes (i.e. blood groups): A, B, AB, and O (Table 7.5). Curiously, absence of an antigen is associated with the formation of antibodies against that antigen, possibly because of cross-reaction with antigens in environmental pathogens. This means that individuals who lack blood group A cannot be transfused with group A blood. Similarly, individuals who lack blood group B cannot be transfused with group B blood. People with blood group

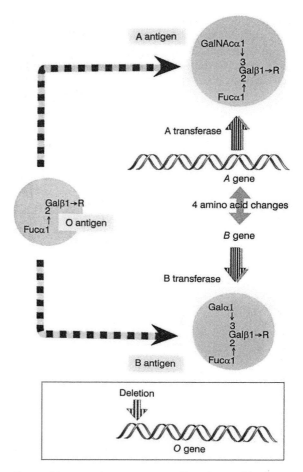

Fig. 7.6 Diagrammatic representation of how the ABO blood groups are determined. Reproduced with permission from Lewin B, (2000) *Genes VII*. Oxford University Press, New York.

TABLE 7.5 ABO blood group antigens and antibodies

Genotype	Phenotype	Antigen	Antibodies
AA	A	A	anti-B
AO	A	A	anti-B
BB	B	B	anti-A
BO	B	B	anti-A
AB	AB	A + B	none
OO	O	–	anti-A and anti-B

AB are sometimes referred to as universal recipients, as they can safely receive blood from a donor of any ABO group. Conversely, blood from persons of group O can be given to recipients of any ABO group. They are therefore referred to as universal donors.

The Rhesus blood group system

The Rhesus blood group antigens are encoded by two genes, *RHD* and *RHCE*, located in tandem on chromosome 1. An individual is Rhesus (Rh) D positive if there is at least one functional copy of *RHD* (i.e. RhDD or Dd). Deletion of both copies of *RHD* results in an individual being Rhesus D negative (ie Rhdd). The Rhesus C and E blood groups are determined by *RHCE*. Translation of a full-length mRNA transcript produces either the E or the e antigen, depending upon the presence or absence of a single point mutation in amino acid 226. RhC or c antigens are produced from shorter transcripts due to splicing out of various exons, with the difference between C and c being due to four point mutations.

Clinically the importance of the Rhesus blood groups is based on compatibility for blood transfusion and a condition known as haemolytic disease of the newborn. For both purposes the D antigen is approximately 20 times more potent that the C or E antigens. Individuals who are Rh D positive, i.e. RhDD or RhDd, can receive blood from Rh D positive or Rh D negative donors. Individuals who are Rh D negative (i.e. Rhdd) can only receive blood from Rh D negative donors. The incidence of the Rh D negative phenotype (Rhdd) varies from 20–30% in Europeans to 7% in African-Americans and less than 1% in Chinese and Japanese people.

> ### Key Point
>
> Individuals with an AB Rhesus positive blood type can receive blood from any donor. Those with an O Rhesus negative blood group can donate blood to any recipient.

Haemolytic disease of the newborn

An individual who is Rh D negative will form anti-D antibodies if exposed to Rh D positive blood cells. During pregnancy, small quantities of fetal red blood cells cross the placenta into the maternal circulation. If an Rh D negative mother is carrying an Rh D positive fetus, then at some point in the pregnancy she will become sensitized to the Rh D antigen. This does not usually impact upon the first pregnancy, but in a subsequent pregnancy with an Rh D positive fetus maternal anti-D antibodies will cross the placenta and destroy

BOX 7.2 LANDMARK PUBLICATION: PREVENTION OF RHESUS HAEMOLYTIC DISEASE

It had been recognized for many years that Rh-negative women were at risk of producing Rh antibodies following the delivery of an Rh-positive infant. It had also been noted that if there was also incompatibility at the ABO locus, this afforded a degree of protection. This led to the hypothesis that the protective effect of the ABO incompatibility was mediated by rapid destruction of the fetal red cells by maternal anti-A or anti-B antibodies, so that the antigenic red cells were removed from the maternal circulation before an anti-D immune response could be mounted.

This hypothesis was explored by giving injections of Rh-positive blood to Rh-negative male volunteers, consisting of blood donors in Liverpool (UK) and inmates at the Maryland State Penitentiary in Baltimore (USA). The volunteers were divided into an untreated control group and a study group treated with injections of Rh antibody. The results clearly showed that Rh antibody would prevent an immune response to the Rh antigen. Clinical trials were then carried out in several groups of pregnant women who were given injections of Rh immunoglobulin followiL' delivery of their infants. Once again it was apparent that this was an extremely effective means of preventing Rh iso-immunization, in that the women failed to develop Rh antibodies in subsequent Rh-positive pregnancies.

It s now standard procedure in every pregnancy that the mother's and baby's blood groups are checked and that all Rh-negative mothers giving birth to Rh-positive infants are given an injection of anti-D immunoglobulin. As a result, the incidence of haemolytic disease of the newborn has declined dramatically. This was the first example of a 'genetic' disorder that could be treated and prevented. At the time it was hoped that this would act as a model for the prevention of other genetic and immunological conditions. Unfortunately this hope has not as yet been fulfilled.

Reference

Clarke CA, Donohue WTA, McConnell RB *et al.* (1963) Further experimental studies on the prevention of Rh haemolytic disease. *British Medical Journal*, **1**, 979–984.

the fetal red blood cells. This rapidly leads to severe fetal anaemia with jaundice due to haemolysis.

The anaemia causes widespread oedema resulting in severe hydrops fetalis, similar to that seen in homozygous severe α-thalassaemia (p. 160) and Turner syndrome (p. 59). This often proves fatal. If the baby survives to birth, there is a risk of severe brain damage due to the high level of circulating unconjugated bilirubin. The pattern of encephalopathy caused by high levels of bilirubin is referred to as *kernicterus*.

Until the 1960s haemolytic disease of the newborn was a common cause of pregnancy loss and neonatal mortality. Research at that time showed that the maternal antibody response could be prevented by intramuscular injection of a small quantity of anti-D gamma-globulin at the time of delivery of a Rhesus D positive infant (Box 7.2). This combines with the fetal red blood cells and blocks the mother's immune response. This is now a standard treatment throughout the world. Haemolytic disease of the newborn can still occur as a result of other blood group incompatibilities, including ABO and Rhesus C or E, but these are rare and the outcome is usually relatively mild. The prevention of haemolytic disease of the newborn ranks as one of the major medical achievements of the twentieth century.

> ### Key Point
>
> All Rhesus-negative women who deliver or miscarry a Rhesus-positive infant should be given anti-D gamma-globulin to prevent haemolytic disease in a future pregnancy.

Colour blindness

Amongst western Europeans, approximately 8% of men and 0.7% of women are 'colour blind' in that they cannot accurately distinguish between red and green. Males with a defect in the red (protan) photoreceptors have normal blue and green cones but abnormal red cones with a spectral sensitivity that is shifted towards that of green cones. They are said to have protanopia if the condition is severe or protanomaly if it is mild. Men with a defect in the green (deutan) photoreceptors have normal blue and red cones but abnormal green cones, with a spectral sensitivity shifted towards that of the red cones. They are said to have deuteranopia if the condition is severe or deuteranomaly if it is mild. Deutan defects are approximately three times more common than protan defects.

These common and relatively harmless forms of colour blindness are caused by defects in the contiguous red and green pigment genes located at chromosome Xq28. These genes are highly homologous, a factor which predisposes to misalignment in meiosis, with a high frequency of unequal recombination (Fig. 7.7). The commonest molecular defect seen in deuteranomaly is a 5′ green–red 3′ hybrid gene. Complete deletion of a green pigment gene causes deuteranopia. Conversely, a 5′ red–green 3′ hybrid gene causes protanomaly and deletion of a red pigment gene causes protanopia. It has been shown that when there are more than two genes in the array, only the first two genes are expressed. This explains why the presence of a normal green pigment gene in the third position in the array does not compensate for defects in the upstream green–red hybrid gene (Fig. 7.7).

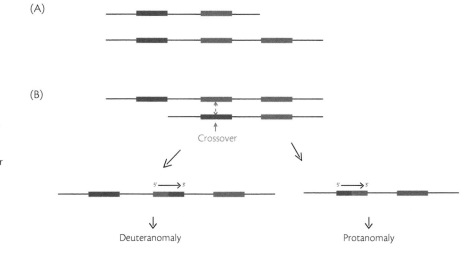

Fig. 7.7 Diagrammatic representation of the red and green pigment genes on chromosome Xq28: (A) the normal alignment with one or two green pigment genes. (B) the outcome of unequal crossing over generating a green vision defect (deuteranomaly) and a red vision defect (protanomaly).

TABLE 7.6 Criteria for population carrier screening

Participation	Disease	Test
Voluntary	Common	Simple
Informed consent	Severe	Inexpensive
Counselling and support	Intervention available	Reliable with high sensitivity and specificity

Population screening for carriers

As discussed elsewhere (p. 74), most members of the human race carry at least one autosomal recessive disorder. It is clearly impossible to design a programme for the detection of all carriers of every condition, but a strong case can be made for offering carrier detection for those conditions which are common in specific populations. Obvious examples are the haemoglobinopathies, cystic fibrosis, and Tay–Sachs disease.

There is general agreement that any population screening programme for carriers of an autosomal recessive disorder should fulfil several important criteria (Table 7.6). The relevant disorder should be sufficiently common to justify the expense of setting up the programme and the potential anxiety generated within the population. It should also be sufficiently serious to warrant a subsequent offer of intervention in the form of prenatal diagnosis for couples who are both found to be carriers. The programme should be well planned, with equity of access. Full information and support should be readily available for the target population. Participation should be entirely voluntary and available only to adults. Finally, the test itself should be simple, inexpensive and reliable as determined by sensitivity and specificity (Table 7.7).

Tay–Sachs disease

Tay–Sachs disease (p. 215) is a serious neurodegenerative disorder of early childhood caused by deficiency of hexosaminidase A, an enzyme encoded by the *HEXA* gene on chromosome 15. Affected children rarely survive beyond the age of four years and there is no effective treatment. The incidence in the Ashkenazi Jewish population used to be approximately 1 in 3600 with a carrier frequency of 1 in 30 (i.e. $q^2 = 1/3600$, $q = 1/60$, $2pq = 1/30$). The incidence in non-Jewish populations is much lower (<1 in 100 000).

Carriers can be detected by enzyme assay, as Tay–Sachs disease is unusual amongst enzyme deficiency disorders in that there is a clear distinction between enzyme levels in affected children, normal homozygotes, and heterozygotes. Carrier states can be confirmed by DNA analysis for one of the three common mutations that account for 99% of all carriers in the Ashkenazi Jewish population.

Since carrier screening for Tay–Sachs disease was first offered to Ashkenazi Jewish populations, over 50 000 carriers and 1400 affected pregnancies have been identified. The programme fulfils all of the criteria listed in Table 7.6 and has been well received by the target population. This is illustrated by a reduction of around 95% in the incidence of Tay–Sachs disease in the American Ashkenazi Jewish population over the last 30 years.

Cystic fibrosis

In contrast to Tay–Sachs disease, cystic fibrosis (p. 107) offers a good example of a condition which does not fulfil all of the criteria required to justify a carrier detection screening programme. Although it is common in specific populations, notably white people of western European origin, the disease course is variable and unpredictable. There is marked mutational heterogeneity, with over 1000 different pathogenic mutations reported, so carrier detection can be difficult and expensive. Finally, there is understandable concern about the ethics of offering termination of pregnancy for a condition in which average life expectancy has increased from 5 years in 1955 to a projected 50+ years

TABLE 7.7 Sensitivity and specificity

	Carrier	Not a carrier
Positive test result	a	b
Negative test result	c	d
Sensitivity = proportion of carriers detected		= a/(a + c)
Specificity = proportion of non-carriers detected		= d/(b + d)
False positives = proportion of incorrect positive results		= b/(a + b)
False negative = proportion of incorrect negative results		= c/(c + d)

for children diagnosed in infancy and treated vigorously thereafter.

These concerns are illustrated by the results of several pilot carrier screening studies carried out in Europe and the USA. These have shown that there is little enthusiasm for cystic fibrosis carrier testing amongst non-pregnant adults, with take-up rates of 10–20%. However, these increase to 70–80% in pregnancy or if an invitation to undergo testing is made by a respected family doctor or other health care professional.

In 2001 the American College of Obstetrics and Gynaecology recommended that cystic fibrosis carrier screening should be offered to:

individuals with a family history of cystic fibrosis, reproductive partners of persons who have cystic fibrosis, and couples in whom one or both partners

are Caucasian (carrier frequency 1/20 to 1/30) and are planning a pregnancy or seeking prenatal care.

The laboratory tests for carrier screening are based on detection of the most common 25–30 mutations that account for approximately 90% of all carriers. This means that a negative test result reduces an individual's risk for being a carrier from 1 in 25 to 1 in 241 (Table 7.8). Thus if one member of a partnership is shown to be a carrier with his or her partner testing negative, then the probability that they will have an affected child is $1 \times 1/241 \times 1/4 = 1$ in 964, or roughly 1 in 1000.

Several explanations have been offered to explain the public's apparent lack of interest in cystic fibrosis carrier testing. Knowledge of the condition is limited, the long-term prognosis is unpredictable, there is concern about implications for insurance, and the test has a sensitivity of only 90%. Follow-up studies have revealed that of those who undergo testing, up to 50% cannot recall the full details of their results when questioned a few months later.

The haemoglobinopathies

Sickle-cell disease and α/β-thalassaemia represent the most common autosomal recessive disorders in the world and are a major cause of morbidity and mortality in many of the world's poorer nations (p. 156). The responses to early screening programmes illustrate the spectrum of outcomes which can be observed in different populations under different circumstances.

TABLE 7.8 Bayesian calculation to determine carrier risk when mutation screening is negative

Probability	Carrier	Not a carrier
Prior	1/25	24/25
Conditional (negative mutation screen)	1/10	1
Joint	1/250	240/250
Odds	1 to 240	

Posterior probability for carrier status = 1/241.
For an explanation of Bayesian calculations, see Box 13.2.

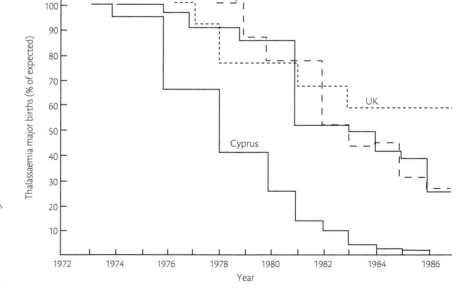

Fig. 7.8 Decline in the birth rates for β-thalassaemia major following the introduction of national carrier screening programmes. Reproduced with permission from from Modell B, Kuliev AM, Wagner M (1991) *Community genetics services in Europe. Report on a survey.* WHO Regional Publication Series No. 38, WHO Regional Office for Europe, Copenhagen.

With the support of the World Health Organization, screening for carriers of β-thalassaemia was introduced in several Mediterranean countries in the mid-1970s. The results over a 10-year period are indicated in Fig. 7.8. In Greece and Italy there was a 70% decrease in the birth of homozygous affected infants. The remarkable decline of almost 99% in Cyprus was almost certainly attributable to the active support of the Orthodox Church, which strongly encouraged carrier testing before marriage. In general it seems that these programmes were well organized with motivated staff and an informed target population to whom screening was acceptable. However, there is concern that the 99% decline observed in Cyprus could reflect a degree of coercion, which, if correct, would clearly conflict with the basic principle of voluntary participation.

The potential counter-productive effects of un-welcome authoritarian pressure are amply illustrated by the attempts to introduce compulsory sickle-cell carrier screening in parts of the USA in the 1970s. Confusion between the carrier state, also known as the trait, and the homozygous affected disease state, coupled with illogical discrimination against carriers by employers and insurance companies, rapidly led to hostility and the abandonment of the screening programmes. Subsequent offers of carrier testing with informed counselling and voluntary participation have been well received.

Population screening—good or bad?

If the outcomes of the screening programmes for Tay–Sachs disease, cystic fibrosis, and the haemoglobinopathies are representative of the general public's attitude to carrier screening, then it is clear that there are a number of important lessons to be learned. Specifically the target population has to be motivated and well informed, the disease has to be serious, and the test has to be accurate. Some populations may be more compliant than others. Attempts to impose screening on an unwilling population will almost inevitably be unsuccessful.

In defence of population screening it can be argued that, when well conducted, it enhances personal choice and is likely to lead to a decrease in human suffering through a reduction in the number of severely affected children. This in turn can help release valuable and scarce resources for other health care needs. On the other hand, carrier detection conveys a risk of stigmatization, with implications for personal anxiety, employment and insurance. In addition there is a danger that unreasonable pressure to participate will be applied for misguided eugenic purposes or simply to try to save money. The importance of informed consent and voluntary participation cannot be over-emphasized.

> **Key Point**
>
> Population screening to identify carriers of common autosomal recessive disorders should fulfil certain important criteria. Specifically the programme must be acceptable to the target population.

Summary

According to the Hardy–Weinberg principle, in a model population with no disturbing factors, the relative proportions of different genotypes remain constant from generation to generation. For a locus with two alleles, A1 and A2 with gene frequencies of p and q, the frequencies of the genotypes A1A1, A1A2 and A2A2 will be p^2, $2pq$, and q^2 respectively. This can be used to determine the carrier frequency of an autosomal recessive disorder if the incidence (I) of affected persons is known. The carrier frequency ($2pq$) is approximately to $2\sqrt{I}$.

Factors which can influence the Hardy–Weinberg distribution include disturbance of the mutation-selection equilibrium, heterozygote advantage, founder effect with drift, assortative mating, consanguinity, and gene flow. If a population is in equilibrium then mutation rates can be calculated using the formula $\mu = I\,(1-f)/2$ for autosomal dominant conditions, $\mu = I\,(1-f)$ for autosomal recessive conditions, and $\mu = I^M(1-f)/3$ for X-linked recessive conditions.

Genetic polymorphism refers to the existence of different alleles each with a frequency of 1% or greater. This is extremely common and accounts for the striking diversity seen in the human population. Polymorphism can be demonstrated at the DNA, chromosomal, and protein level. DNA polymorphisms include SNPs, RFLPs, minisatellites, and microsatellites. These have proved to be extremely useful for linkage analysis and in forensic medicine. Examples of protein polymorphisms include the HLA system, the ABO and Rh blood groups, and red–green colour blindness.

Carrier screening is now offered for conditions such as Tay–Sachs disease, cystic fibrosis, sickle-cell anaemia, and β-thalassaemia, each of which is common in specific ethnic groups. These programmes have met with mixed success. Important factors include the acceptability of the test to the target population, the severity of the relevant disease and the accuracy of the test. Informed consent and voluntary participation are essential components of any population screening programme.

Further reading

Hartl DL, Clark AG (1997) *Principles of population genetics*, 3rd edn. Sinauer Associates, Sunderland, MA.

Khoury MJ, Beaty TH, Cohen BH (1993) *Fundamentals of genetic epidemiology*. Oxford University Press, New York.

Vogel F, Motulsky, AG (1997) *Human genetics, problems and approaches*, 3rd edn. Springer-Verlag, Berlin.

Multiple choice questions

1 An autosomal recessive disorder in Hardy–Weinberg equilibrium affects 1 in 6400 babies. Therefore

  (a) the gene frequency of the mutant allele is 1 in 80

  (b) the gene frequency of the normal (wild-type) allele is 39 in 40

  (c) the carrier frequency is approximately 1 in 40

  (d) the probability that the mother of an affected child is a carrier is 1 in 40

  (e) the probability that the unaffected sibling of an affected person is a carrier is 1 in 2

2 Factors that can interfere with Hardy–Weinberg equilibrium include

  (a) small population size

  (b) large population size

  (c) non-assortative mating

  (d) consanguinity

  (e) heterozygote advantage

3 If a man with blood group AB Rh positive has children with a woman with blood group O Rh negative, then

  (a) on average half of their children will have blood group A

  (b) on average half of their children will have blood group AB

  (c) all of their children will have an O allele

  (d) all of their children will definitely be Rhesus positive

  (e) there is a risk of Rhesus isoimmunization in the mother following delivery

4 In an emergency, with no facilities to carry out tests to confirm compatability, it would be safe to give blood

  (a) from an O Rh-positive donor to an A Rh-positive recipient

  (b) from an A Rh-positive donor to an A Rh-negative recipient

  (c) from an AB Rh-negative donor to an O Rh-negative recipient

  (d) from a B Rh-negative donor to an AB Rh-positive recipient

  (e) from an O Rh-negative donor to an AB-positive recipient

5 Population carrier screening

  (a) should be compulsory to ensure that it is effective

  (b) should be carried out in early childhood

  (c) can identify all carriers of cystic fibrosis

  (d) is based on biochemical assay in cystic fibrosis

  (e) is offered for Tay–Sachs disease in selected populations

Answers

1 (a) true—$q^2 = 1/6400$. Therefore $q = 1/80$

 (b) false—if $q = 1/80$, then $p = 79/80$

 (c) true—the carrier frequency is $2pq$, which equals approximately $2q$, which equals $1/40$

 (c) false—the mother is an obligatory carrier as she has had an affected child

 (d) false—the probability that an *unaffected* sibling is a carrier is 2/3 (p. 73).

2 (a) true—a small population can lead to a founder effect with drift

 (b) false—large size is one of the characteristics of a model population

 (c) false—non-assortative mating equals random mating. It is *assortative* mating that can disturb Hardy–Weinberg equilibrium

 (d) true—consanguinity can lead to an increase in the numbers of affected homozygotes

 (e) true—in a model population there should be no selection for or against any particular genotype

3 (a) true—the mating is AB × OO, so on average half of the children will have an AO genotype

 (b) false—none of the children will have an AB genotype

 (c) true—each child must inherit an O allele from its mother

 (d) false—the full Rhesus status of the father is not known. He could be +/+ or +/–

 (e) true—each baby's blood group should be checked after delivery and the mother given anti-D if the baby is Rh positive

4 (a) true—blood group O can be given to anyone as long as there is no Rhesus incompatibility

 (b) false—Rh positive blood cannot be given to an Rh negative recipient

 (c) false—only group O blood can be given to a group O recipient

 (d) true—a person with blood group AB Rh positive is a universal recipient

 (e) true—this is an example of a universal donor giving to a universal recipient

5 (a) false—participation must be voluntary

 (b) false—only adults should participate

 (c) false—mutation testing/screening only detects around 85–90% of all carriers

 (d) false—it is based on DNA mutation analysis

 (e) true—this has been well received by the Ashkenazi Jewish population

Genes and haemoglobin

Inherited disorders of the structure and function of haemoglobin, known collectively as the *haemoglobinopathies*, hold a distinguished and prominent role in the annals of human and medical genetics. Historically, haemoglobin was one of the first human proteins for which the underlying amino acid structure and molecular basis were elucidated. Sickle-cell disease, one of the commonest forms of haemoglobinopathy, is the prime example of an inherited disorder for which heterozygote advantage and its underlying mechanism have been demonstrated. Knowledge of the sickle-cell mutation was exploited by the research group which developed the polymerase chain reaction (p. 16), and sickle-cell disease was the first inherited condition for which prenatal diagnosis was undertaken using linkage analysis (p. 91).

However, the importance of the haemoglobinopathies in medical genetics lies not in their historical significance but in their enormous impact on human morbidity and mortality. In 1994 it was estimated that approximately 5% of the word's total population, now 6 billion, carries a haemoglobin mutation and that over 350 000 children are born each year with a serious disorder of haemoglobin structure (e.g. sickle-cell disease) or synthesis (e.g. thalassaemia). Such is the impact of these disorders, both at an individual level and on a global scale, that the World Health Organization has actively promoted several national population screening programmes, with the dual goals of informing reproductive choice and, thereby, reducing the number of severely affected children. These programmes have met with variable success, as discussed in Chapter 7 (p. 149).

Haemoglobin structure and function

The haemoglobin molecule is made up of four poly-peptide chains, normally two α and two non-α, each of which has a single haem moiety consisting of an iron atom located at the centre of a porphyrin ring. This tetramer is spherical in structure, with the globin chains folded so that the four haem groups lie in surface clefts equidistant from each other (Fig. 8.1). The tetramer is held together by bonds between the α and non-α chains, and the quarternary structure changes as oxygen is taken up by oxygenation of each haem group.

Normally the oxyhaemoglobin dissociation curve has a sigmoid shape, which means that large amounts of oxygen are taken up or released by a small increase or decrease in oxygen tension. This is facilitated by inter-action of the haem groups, so that when one haem group in a tetramer becomes oxygenated, conforma-tional changes result in increased oxygen affinity of the closely adjacent non-oxygenated haem groups. Abnormal forms of haemoglobin which lack α chains show abnormal subunit interaction, with a different pattern of oxyhaemoglobin dissociation curve that pre-vents oxygen release at normal physiological oxygen tensions. Fetal haemoglobin, as discussed below, has greater oxygen affinity than that of adults because of reduced binding to the tetramer of 2,3-diphospho-glycerate, which normally stabilizes the deoxygen-ated form of haemoglobin, thus lowering its oxygen affinity.

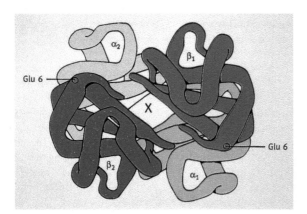

Fig. 8.1. Three-dimensional structure of the haemoglobin molecule. Glu 6 marks the site of the amino acid substitution which causes sickle-cell disease. X marks the site where 2,3-diphosphoglycerate binds in the deoxygenated state. Reproduced with permission from Elliott WH, Elliott DC (2001) *Biochemistry and molecular biology*, 2nd edn. Oxford University Press, Oxford.

Haemoglobin genes and chains

The human haemoglobin genes show considerable homology and almost certainly arose from a single common ancestral gene. They are located in two clus-ters, an α or α-like complex on chromosome 16 and a β or β-like complex on chromosome 11 (Fig. 8.2). Each of these clusters contains at least one pseudogene. **Pseudogenes** closely resemble functional adjacent genes but contain mutations which have rendered them inactive. Both they and the almost identical func-tional genes are thought to have originated from a common ancestral gene.

The α-like cluster is located close to the end of chro-mosome 16 at 16p13.3 and consists of three pseudo-genes (ψζ, ψα2 and ψα1), three functional genes (α1, α2, and ζ) and one gene (θ) of unknown function. The ζ gene encodes an α-like ζ (zeta) chain in early embry-onic life; α1 and α2, which arose from a duplication event approximately 60 million years ago, encode α-globin chains from late embryonic life onwards. Thus each normal individual has four active α-globin genes encoding α-globin chains.

The β-like gene cluster is located on the short arm of chromosome 11 at 11p15.5 and is spread over a region of approximately 60 kb. It consists of a single pseudo-gene (ψβ) and five functional genes (ε, Gγ, Aγ, δ, and β), which encode the ε, γ, δ, and β chains respectively. The γ chains are encoded mainly in fetal life, by Gγ and Aγ, when they combine with α chains to form fetal haemo-globin (Hb F). β-globin chain synthesis begins soon after birth. Normal adult haemoglobin (Hb A) consists of two α-globin and two β-globin chains. The α-globin and β-globin chains consist of 141 and 146 amino acids respectively. Adults also have a small quantity (2–3%) of Hb A₂, which consists of two α-globin chains and two δ-globin chains (Table 8.1).

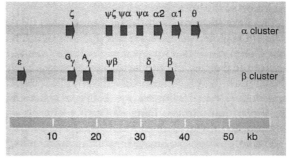

Fig. 8.2. Structure of the α and β globin gene clusters on chromosomes 16 and 11. Reproduced with permission from Lewin B (2000) *Genes VII*. Oxford University Press, New York.

TABLE 8.1	Human haemoglobin		
Haemoglobin	Chain structure	Expression pattern	% in normal adults
Gower I	$\zeta_2\epsilon_2$	Embryonic	0
Gower II	$\alpha_2\epsilon_2$	Embryonic	0
Portland	$\zeta_2\gamma_2$	Embryonic	0
Fetal (Hb F)	$\alpha_2\gamma_2$	Fetal	1
Adult (Hb A)[a]	$\alpha_2\beta_2$	Infancy →	97
Adult (Hb A2)	$\alpha_2\delta_2$	Infancy →	2–3

[a] A small amount of Hb A is glycosylated with a glucose moiety at the N-terminus. This is known as Hb A_{1c}. Hb A_{1c} is raised in individuals with diabetes mellitus, in whom it gives an indication of the quality of diabetic control over preceding months (p. 183).

Each globin gene has a similar structure consisting of a 5′ promoter region, a cap site, a 5′ untranslated region, an initiation codon, three exons, two introns, a termination codon, a 3′ untranslated region, and a poly(A) signal tail. The α- and β-like gene clusters share several features in common. They both have upstream locus control regions (LCRα and LCRβ), which regulate transcription of the actively expressed genes within the relevant cluster. The genes in each cluster share temporal and spatial expression patterns, which are regulated by the relevant locus control regions and more closely adjacent enhancer sequences and promoter regions. In both clusters the genes are arranged sequentially in order of their temporal developmental expression, a characteristic which they share with the *HOX* gene family clusters (p. 169). Spatially, the globin genes are expressed initially in the yolk sac and then by the liver in embryonic life, and the spleen and bone marrow in fetal life.

The different temporal expression patterns of the α- and β-globin chains explain the different ages of onset of the various forms of haemoglobinopathy. Severe α-thalassaemia presents *in utero* or at birth, whereas β-thalassaemia and sickle-cell disease do not manifest until the synthesis of β-globin chains begins in early infancy.

Key Point

Fetal haemoglobin (Hb F) consists of two α- and two γ-chains. Adult haemoglobin (Hb A) contains two α- and two β-globin chains. Adults also have a small amount of Hb A2, which is made up of two α- and two δ-globin chains.

The spectrum of globin gene mutations

Globin biosynthesis is a complex process involving transcription to produce a messenger RNA precursor (pre-mRNA); post-transcriptional processing with 5′-end capping and methylation; translation, which proceeds through three separate phases of initiation, elongation, and termination; and finally interaction of the haemoglobin chains to form mature haemoglobin. Given this degree of complexity, it is not surprising that the process is error-prone. Generally disorders of haemoglobin are divided into two main groups consisting of those due to structural abnormalities and those that arise because of reduced globin synthesis.

Abnormalities of haemoglobin structure

Over 700 abnormal haemoglobins have been described, although not all of these are pathogenic. They can be classified on the basis of the type of underlying mutation or according to their clinical consequences (Table 8.2). Originally haemoglobin variants were identified by protein electrophoresis, but this has largely been superseded by direct DNA sequencing. Most of the recognized variants are the result of point mutations leading to single amino acid substitutions. Some of these are clinically silent, but others result in clinical problems through a variety of effects on the haemoglobin molecule, which include:

◆ instability of the tetramer

◆ formation of an abnormal three-dimensional structure

◆ prevention of ferric iron reduction

◆ prevention of normal haem binding.

Several other mutational mechanisms have also been shown to result in structural haemoglobin variants (Table 8.2). These include frame-shift and in-frame deletions/mutations, chain termination mutations, and mispairing of homologous sequences leading to unequal crossing-over and the formation of fusion δβ genes. This latter mechanism accounts for a rare haemoglobin variant, Hb Lepore. Several different forms of Hb Lepore have been identified, together with complementary forms of Hb anti-Lepore. They have all arisen through mispairing of the closely homologous δ- and β-chains with subsequent unequal crossing-over. The resulting δβ fusion chain contains 5′ δ polypeptides and 3′ β polypeptides. The complementary anti-Lepore fusion chain consists of 5′ β polypeptides and 3′ δ

TABLE 8.2 Examples of variant structural haemoglobins and their clinical effects

Abnormal haemoglobin	Mutation	Outcome and clinical effects
Hb S	Single nucleotide substitution (βGlu6Val)	Hb polymerizes causing sickling and haemolytic anaemia
Hb C	Single nucleotide substitution (βGlu6Lys)	Hb cystallizes causing mild haemolytic anaemia
Hb M (Boston)	Single nucleotide substitution (αHis58Tyr)	Hb has low O_2 affinity causing asymptomatic cyanosis associated with methaemoglobinaemia (hence Hb M)
Hb E	Single nucleotide substitution (βGlu26Lys) → activation of cryptic donor splice site	Reduced synthesis of β chains causing very mild anaemia
Hb Tak	Frameshift insertion → elongated β chain	Hb has high O_2 affinity causing polycythaemia
Hb Gun Hill	Deletion (β - codons 91–95)	Unstable Hb tetramer leading to haemolytic anaemia
Hb Constant Spring	Substitution in stop codon → elongated α chain	Unstable α chain causing mild thalassaemia
Hb Lepore	Unequal crossing over between δ and β chains	Reduced synthesis of the δβ fusion chain

polypeptides, and the corresponding haemoglobin is known as Hb anti-Lepore or Hb Miyada.

Note that when letters of the alphabet had been exhausted variant haemoglobins were generally named after the hometown of the first reported patient—e.g. Hb Leiden, Hb Louisville, Hb Madrid, etc.). Sometimes the name of the family in which the variant was first identified was used, and occasionally the hospital at which the variant was first recognized (e.g. Hb Barts).

Abnormalities of haemoglobin synthesis

Inherited disorders associated with reduced or absent synthesis of one or more of the normal globin chains are known collectively as the thalassaemias. (Thalassa is the Greek for sea, and the disorders were so named because of their concentration around the Mediterranean.) The thalassaemias are classified on the basis of the globin genes involved (e.g. α-, β-, and δβ-thalassaemia) and whether gene expression is completely (e.g. β°) or only partially (e.g. β⁺) suppressed. As a group, the thalassaemias constitute the most common autosomal recessive disorders in the world.

The mutational basis of the common α- and β- forms of thalassaemia is complex and heterogeneous. α-thalassaemia is usually caused by large deletions, whereas β-thalassaemia shows marked mutational heterogeneity with point mutations and small deletions or insertions. These can influence the synthesis of β-globin chains at various stages from transcription through RNA processing, cleavage, and polyadenylation to translation. The most common β-thalassaemia mutation found in individuals originating from the Mediterranean region results in the creation of a new acceptor AG splice site in the first intron of the β-globin gene. In contrast, the most common β-thalassaemia mutation found in the Indian subcontinent is a small 619-bp deletion involving the 3′ end of the β-globin gene.

> **Key Point**
>
> Mutations in the haemoglobin genes can be subdivided on the basis of whether they affect haemoglobin structure or haemoglobin synthesis.

Sickle-cell disease

Sickle-cell disease (MIM 603903) is one of the most common inherited disorders in the world, with an estimated incidence in African-Americans of approximately 1 in 625. The disorder was first described in 1910 and was shown to be due to an abnormality in haemoglobin in 1949 (Box 8.1). The molecular basis and underlying pathophysiology are well understood but, despite extensive research, treatment remains largely supportive rather than curative. Sickle-cell disease is still a major cause of ill-health in the black population throughout the world.

Genetics and epidemiology

Sickle-cell disease shows autosomal recessive inheritance. Carriers, who are described as having the *sickle-cell trait*, are generally entirely healthy, although they can develop clinical problems, such as vaso-occlusive episodes (see below), in conditions of very low oxygen saturation such as may be encountered in deep-sea diving, flying in unpressurized aircraft, or during general anaesthesia.

The frequency of the sickle-cell trait is high in all populations originating from equatorial Africa, where

BOX 8.1 LANDMARK PUBLICATION: THE BASIC DEFECT IN SICKLE-CELL DISEASE

The technological advances of the last 50 years have been such that it is now difficult to contemplate how little used to be understood about the basic cause of most inherited disorders. In many ways sickle-cell disease, because of the ready accessibility of blood from affected individuals, served as the prototype for how inherited disorders might be investigated. This can be illustrated by a review of two major landmark publications.

The first, by Linus Pauling and colleagues at the California Institute of Technology in Pasadena, was published in 1949. Having observed that red blood cells underwent sickling if the haemoglobin was deoxygenated, they compared haemoglobin from affected and unaffected individuals by electrophoresis. The results clearly indicated that these haemoglobins behaved differently. They also showed that haemoglobin from people with sickle-cell trait was made up of two types, one identical to that found in normal individuals and the other identical to that found in patients with sickle-cell disease. Finally, Pauling and his colleagues compared the haem moieties from normal and sickle-cell haemoglobin and found them to be identical. Taken together, these observations strongly suggested that the basic defects in sickle-cell disease resided in the globin component of sickle-cell haemoglobin.

The next major step in the understanding of sickle-cell disease was reported by Ingram, from Cambridge, in 1957. He separated the polypeptides in normal and

sickle-cell haemoglobins by digestion with trypsin and then subjected them to a combination of electrophoresis and partition chromatography. One peptide showed a consistently different pattern and on subsequent analysis was found to differ in just one amino acid, i.e. glutamic acid in normal Hb A and valine in Hb S. Based on these results Ingram proposed that replacement of a single base pair in the DNA of the globin gene could account for the observed amino acid change and the ensuing clinical problems seen in sickle-cell disease. This was the first time that a single-gene mutation had been shown to cause an amino acid change in an inherited Mendelian disorder.

Against the backdrop of modern science these achievements seem tame by comparison. However, in their day they were groundbreaking, paving the way for the extraordinary revolution in molecular biology that followed. Linus Pauling in particular is widely regarded as one of the founding fathers of molecular biology. Amongst his many distinctions he is the only individual ever to receive two unshared Nobel prizes, for Chemistry in 1954 and for Peace in 1962.

References

Ingram VM (1957) Gene differences in human haemoglobin: the chemical difference between normal and sickle-cell haemoglobin. *Nature*, **180**, 326–328.

Pauling L, Itano HA, Singer SJ, Wells IC (1949) Sickle-cell anemia, a molecular disease. *Science*, **110**, 543–548.

carriers have a relative resistance to falciparum malaria. Carrier frequencies as high as 20% have been found in countries such as Kenya and Uganda. Lower carrier frequencies of 5–10% are seen in the Middle East and in countries around the Mediterranean where malaria is endemic. The carrier frequency in African-Americans is approximately 8%, consistent with the incidence of homozygotes of 1 in 625 (i.e. $1/12.5 \times 1/12.5 \times 1/4 = 1/625$—see p. 136)

Molecular pathogenesis

The sickle-cell mutation involves a base change of A to T in the second nucleotide of the sixth codon in the β-globin gene, resulting in a substitution of valine for glutamic acid (i.e. GAG to GTG) (Fig. 8.3). This leads to the formation of an $\alpha_2\beta_2$ tetramer which is unstable in the deoxygenated form. When the oxygen saturation falls below 85%, the $\alpha_2\beta_2$ tetramers polymerize to form par-

	Codon		
	5	6	7
Hb A	C C T	G A G	G A G
	(Pro)	(Glu)	(Glu)
Hb S	C C T	G T̲ G	G A G
	(Pro)	(Val)	(Glu)
Hb C	C C T	A̲ A G	G A G
	(Pro)	(Lys)	(Glu)

Fig. 8.3. The A to T mutation in the β-globin gene which causes sickle-cell disease. The G to A mutation which causes HbC is also shown.

allel rod-like structures, causing the red blood cells to become sickle-shaped.

Sickling of the red blood cells has two main consequences:

◆ the sickled red cells are fragile, with a shortened survival time, resulting in a chronic haemolytic anaemia

◆ they clump together to form aggregates, which can lead to occlusion of the peripheral circulation.

This tendency to increased viscosity and peripheral vaso-occlusion accounts for most of the more serious complications of sickle-cell disease.

Laboratory diagnosis

Sickled cells may be present in a peripheral blood smear (Fig. 8.4). A definitive diagnosis is made by haemoglobin electrophoresis, which shows a high level of Hb S, (>80%), or by direct mutation analysis for the Hb S mutation using a PCR-based technique. Carriers can be detected by a simple 'sickling' test which

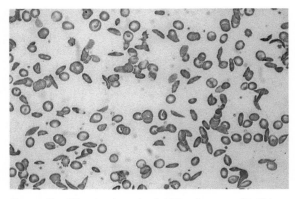

Fig. 8.4 Blood film showing red-cell sickling. Courtesy of Dr Claire Chapman, Leicester Royal Infirmary, Leicester.

BOX 8.2 CASE HISTORY: SICKLE-CELL DISEASE

The following case history, taken from the first documented report of a patient with sickle-cell disease, was published in 1910.

A 20-year-old student from the island of Grenada in the Caribbean sought medical advice because of a 5-week history of cough and fever. During childhood he had suffered from yaws, an infectious illness caused by a spirochaete. This is common in the tropics and causes pustular granulomatous lesions which heal slowly, leaving ulcers and scars. On leaving school at the age of 17 years he had gone to study in Chicago, where he began to experience palpitations and shortness of breath with a general lack of energy. He also noted that on occasions his sclerae had a yellowish tinge.

On examination it was confirmed that he did indeed have yellow sclerae, an indication of jaundice, which in his case was almost certainly caused by haemolysis. His mucous membranes were described as pale, this being a typical feature of anaemia. He also showed clear signs of chest infection, together with evidence of cardiac enlargement.

Investigations included a full blood count, which revealed a level of haemoglobin of 40% (i.e. <6 g/100 ml) with microcytes and nucleated red cells. The blood film showed a large number of 'thick elongated, sickle-shaped and crescent-shaped' forms. A trace of bile was observed in the patient's urine. No infectious agent that would account for the clinical findings could be identified.

The patient was treated with rest and 'nourishing food' for a period of 4 weeks, following which he was discharged with a haemoglobin level of 58% (8.4 g/100 ml). Over the next year or two he was seen on several occasions with recurrent bronchitis, persistent anaemia, a painful swollen knee, and finally with what was almost certainly a haemolytic crisis when he experienced pain in his back and limbs in association with pallor, jaundice, shortness of breath, and pyrexia. Following this the patient was lost to follow-up. Subsequent research has revealed that he returned to Grenada, where he practised as a dentist until his death at the age of 32 years. He is buried in the Catholic cemetery in Sauteurs, close to the cliff edge where the indigenous native Carib Indians committed mass suicide rather than submit to subjugation by invading Europeans.

With the benefit of hindsight, we can see that this man had relatively typical features of sickle-cell disease with chronic anaemia, haemolysis, and recurrent infection. It says much for his stoical courage that he managed to qualify as a dentist and survive to the age of 32 years without the benefits of either a diagnosis or any effective treatment.

Reference

Herrick JB (1910) Peculiar elongated and sickle-shaped red blood cells in a case of severe anaemia. *Archives of Internal Medicine*, **6**, 517–521.

involves inspection for sickle-cells in a peripheral blood film exposed to low oxygen saturation, or more reliably by either haemoglobin electrophoresis or direct mutation analysis.

Clinical features

Onset is usually in infancy as β-globin gene expression overtakes γ-globin gene expression. Affected children present with anaemia and jaundice, due to haemolysis, which persists throughout life. The disease course is marked by recurrent painful *sickle-cell crises* caused by anaemia, haemolysis, and vaso-occlusive ischaemia, particularly involving the abdominal viscera and the long bones (Box 8.2). Other organs including the brain and lungs can also be affected. Splenic infarction results in increased susceptibility to infection with bacteria such as *Streptococcus pneumonia* and *Haemophilus influenza*. The long-term outlook is unpredictable. Some affected individuals succumb to infection in early life, particularly in areas of socio-economic deprivation, whereas others survive well into middle age. Concomitant high levels of Hb F tend to be associated with a milder disease course.

Treatment and prevention

Several approaches have proved effective in reducing morbidity and mortality. These include immunization against pneumococcal infection, regular prophylactic treatment with penicillin to prevent infection, and folic acid supplementation to prevent folate deficiency which can exacerbate the anaemia. Crises are treated with fluid replacement, oxygen, and analgesics. Drugs such as hydroxyurea and butyrate lead to an increase in fetal haemoglobin production, with a reduction in the incidence of painful crises. Hydroxyurea has been approved specifically for the treatment of sickle-cell disease. Individuals with the sickle-cell trait, i.e. carriers, do not require any treatment other than advice about avoiding activities, such as deep sea diving and flying in unpressurized aircraft, which could expose them to low levels of oxygen saturation.

Bone marrow transplantation has been carried out in a relatively small number of patients with sickle-cell disease, with success rates of around 90%. Pre-existing organ damage due to vaso-occlusion is associated with a poorer outcome. One of the major problems in sickle-cell disease is that it can be very difficult to predict the outcome in a young child, making it equally difficult to determine whether bone marrow transplantation is indicated. Once it has become clear that the disease is following a severe course, the success rate for bone marrow transplantation becomes suboptimal. Thus ideally a method is needed for predicting future disease severity so that criteria for treatment can be established.

In theory, the births of many affected children could be prevented through population carrier screening programmes. However, as discussed in Chapter 7 (p. 150), these have often been received with hostility and indifference because of poor planning and suggestions of racist overtones. Prenatal diagnosis, based on direct mutation detection, can be offered when both parents are known to be carriers.

> **Key Point**
>
> The sickle-cell mutation results in a substitution of valine for glutamic acid in the sixth codon of the β-globin gene. Homozygotes have sickle-cell disease in which red blood cells sickle when exposed to low levels of oxygen saturation. This causes haemolytic anaemia and painful vaso-occlusive sickle-cell crises.

α-Thalassaemia

α-Thalassaemia (MIM 141800) is a major cause of ill health in South-east Asia, in parts of which the carrier frequency is as high as 1 in 5. The condition is also common in those parts of Africa and in Mediterranean regions where malaria is endemic. The high carrier frequencies are thought to reflect heterozygote advantage mediated by relative resistance to severe malaria, particularly the cerebral form, but the underlying mechanism for this remains unclear.

Genetics and molecular pathogenesis

α-Thalassaemia is caused by a deficiency of α-globin chain synthesis, with the most common underlying mutational mechanism being deletion of one or both of the contiguous α-globin genes. These deletions are caused by unequal crossing-over between homologous sequences in the α-globin gene cluster (Fig. 8.5).

Normally each individual should have four functional α-globin genes. Loss of one gene (αα/α–) constitutes a silent carrier state of no clinical importance (Table 8.3). Loss of two genes (α–/α– or αα/– –) results in mild anaemia and is referred to as α-thalassaemia trait. Loss of three genes (α–/– –) results in the formation of β_4 tetramers, known as Hb H, and accordingly is referred to as Hb H disease. Finally, loss of all four α-globin genes (– –/– –) causes a lethal condition known as hydrops fetalis in which the fetus cannot make any

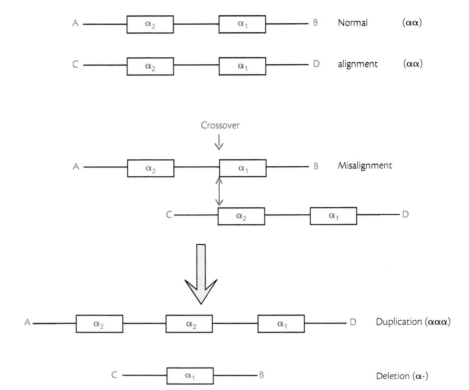

Fig. 8.5 How misalignment with unequal crossing over in meiosis can generate an α-globin gene deletion.

fetal haemoglobin, but instead makes haemoglobin consisting of γ_4 tetramers known as Hb Barts. The $\alpha\alpha/--$ genotype is found mainly amongst South-east Asians, so that homozygosity for the $--$ form of thalassaemia trait is seen almost exclusively in this population.

Clinical features

Individuals with α-thalassaemia trait ($\alpha\alpha/--$ or $\alpha-/\alpha-$) are usually asymptomatic, with only a mild anaemia and haemoglobin levels of 10–12 g/100 ml (normal range 12–14 g/100 ml). In contrast, individuals with Hb H disease ($\alpha-/--$) have a more severe anaemia with haemoglobin levels of 7–10 g/100 ml, associated with infection, haemolysis, and splenomegaly. Loss of all four α-globin genes presents in mid-pregnancy with severe anaemia and fluid overload, giving the clinical picture of hydrops fetalis. This presentation almost always results in death *in utero*.

The diagnosis of the various forms of α-thalassaemia is suspected on the basis of abnormal haematological

TABLE 8.3 Classification of α-thalassaemia			
Condition	**Number of functional α-globin genes**	**Main type of haemoglobin**	**Clinical outcome**
Normal	4 ($\alpha\alpha/\alpha\alpha$)	HbA ($\alpha_2\beta_2$)	Normal
Silent carrier or α^+-thalassaemia trait	3 ($\alpha\alpha/\alpha-$)	HbA ($\alpha_2\beta_2$)	Normal
α^0-thalassaemia trait	2 ($\alpha-/\alpha-$ or $\alpha\alpha/--$)	HbA ($\alpha_2\beta_2$)	Mild anaemia
Hb H disease	1 ($\alpha-/--$)	HbA ($\alpha_2\beta_2$) and Hb H (B_4)	Moderate haemolytic anaemia
Hydrops fetalis	0 ($--/--$)	Hb Barts (γ_4) and Hb Portland ($\zeta_2\delta_2$)	Death before birth

indices (Table 8.3) and subsequently confirmed by haemoglobin electrophoresis and PCR-based molecular analysis.

> ### Key Point
>
> α-Thalassaemia is usually caused by deletion of α-globin genes. Loss of three α-globin genes causes Hb H (β_4) disease. Loss of all four α-globin genes causes death *in utero* due to hydrops fetalis with Hb Barts (γ_4).

β-Thalassaemia

The first description of severe thalassaemia is credited to Cooley *et al.* in the 1920s who reported a group of children seen at the Children's Hospital of Michigan. To this day the condition continues to be referred to as Cooley's anaemia. The various forms of β-thalassaemia (MIM 141900) now constitute a major challenge for health services in many parts of the world, not only because of their high frequency but also because of the severity of the anaemia which, in many instances, results in lifelong transfusion dependency. β-Thalassaemia is also important as it serves as a model for how a radical therapy, bone marrow transplantation, can, in some instances, be curative.

Note that the nomenclature used for different grades of severity of β-thalassaemia is potentially confusing. Severe transfusion-dependent β°-thalassaemia is described as *thalassaemia major*. Milder β⁺-thalassaemia, which is not transfusion dependent, is sometimes classified as *thalassaemia intermedia*. Heterozygous carriers, who are usually asymptomatic, are described as having *thalassaemia minor* or *thalassaemia trait*.

Genetics and epidemiology

β-Thalassaemia shows autosomal recessive inheritance with high carrier frequencies occurring in a broad belt running from the Mediterranean through North Africa and the Middle East to the Indian subcontinent and South-east Asia. As with sickle-cell disease and α-thalassaemia, the high carrier frequencies are attributed to relative resistance to falciparum malaria. Frequencies as high as 20–25% have been observed in some Mediterranean islands, such as Cyprus and Rhodes, with lower frequencies of around 3–10% in the Indian subcontinent and South-east Asia. Many other forms of abnormal haemoglobin, notably Hb S and Hb E, are also extremely common in many of these regions. This means that the overall incidence of severely affected homzygotes and compound heterozygotes is higher than would be predicted if only the carrier frequency of β-thalassaemia is taken into account.

Molecular pathogenesis

The many mutations associated with β-thalassaemia either reduce β-globin gene expression (β⁺-type) or completely suppress it (β°-type). The net effect is a reduction in, or a complete absence of, the synthesis of β-globin chains resulting in an α : β chain imbalance with excess α-globin chains. These are unstable and precipitate in the red cell precursors to form inclusion bodies. Unfortunately these interfere with red cell maturation, so erythropoiesis is impaired. Those red cells which do enter the circulation are destroyed prematurely by the spleen.

This combination of reduced erythropoiesis and increased red cell destruction results in severe haemolytic anaemia, with increased intramedullary erythropoiesis, which leads in turn to expansion of the bone marrow cavities with bone deformity and a risk of pathological fracture.

Laboratory diagnosis

In severe cases the level of haemoglobin ranges from 3 to 8 g/100 ml with hypochromic, microcytic red cells (Fig. 8.6). The bone marrow shows marked erythroid hyperplasia. In β-thalassaemia there is no Hb A, so electrophoresis shows only Hb A_2 and Hb F. Carriers of β-thalassaemia show mild anaemia, with haemoglobin levels of 9–11 g/100 ml, and slightly elevated levels of Hb F (1–3%) and Hb A_2 (4–6%).

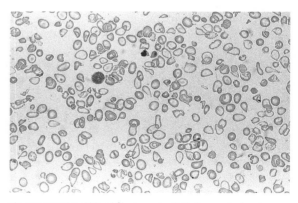

Fig. 8.6 Peripheral blood film in severe β-thalassaemia showing hypochromia, microcytosis, and nucleated red cells. Courtesy of Dr Claire Chapman, Leicester Royal Infirmary, Leicester.

Clinical features and treatment

Children with β⁰-thalassaemia (thalassaemia major) usually present at around the age of 6 months when γ-globin chains should normally be replaced by β-globin chains. Symptoms include poor feeding, failure to thrive, and recurrent infection. Without regular replacement transfusion the disease runs a downhill course with death in childhood due to general malaise and infection. Regular transfusion restores good general health, but leads to excess iron deposition in the heart, pancreas, and liver. This will prove fatal in early adult life unless prevented by adherence to a strict regime of iron chelation therapy with a drug such as desferrioxamine With a combination of regular transfusion and appropriate chelation therapy, affected individuals can survive well into adult life. In poor countries where such therapeutic regimes are not available, survival into adulthood is rare.

Bone marrow transplantation using HLA-compatible unaffected related donors is potentially curative and has been carried out in over 1000 children with success rates of over 90%. This possibility of successful therapy has prompted some parents to request *in vitro* fertilization with pre-implantation genetic diagnosis to enable an unaffected HLA-compatible 'saviour' sibling to be delivered as a potential donor. Understandably this has raised serious ethical concerns about the possible use of a 'designer baby' as an uninformed and non-consenting donor for the affected older sibling (pp. 144, 237).

Prevention

Carrier detection programmes for β-thalassaemia have been introduced in several Mediterranean regions and have met with considerable success in countries such as Cyprus where participation is strongly encouraged (p. 149). Ethical concerns about issues such as termination of pregnancy and forced participation have to be balanced against the suffering experience by affected children in countries where limited financial resources mean that there is little hope of effective treatment.

> ### Key Point
>
> β-Thalassaemia shows marked mutational heterogeneity. Homozygotes for severe β⁰-thalassaemia require life-long treatment with blood transfusion and iron chelation therapy. Alternatively, bone marrow transplantation from a histocompatible donor is potentially curative.

δβ-Thalassaemia and hereditary persistence of fetal haemoglobin

δβ-Thalassaemia and hereditary persistence of fetal haemoglobin (HPFH—MIM 142470) are rare conditions of limited clinical importance, particularly when compared to the devastating impact made by the severe forms of α- and β-thalassaemia. However, their inclusion in a discussion of the haemoglobinopathies is justified by the insight that they provide into possible approaches to therapy based on modification of gene expression.

Both conditions are characterized by reduced or absent expression of the δ- and β-globin genes. In δβ-thalassaemia this is caused by a deletion involving the contiguous (closely adjacent) δ- and β-globin genes (Fig. 8.7). Heterozygotes are asymptomatic, and homozygotes show only a mild haemolytic anaemia. This is in marked contrast to the severe transfusion-dependent anaemia seen in homozygous β⁰-thalassaemia. For reasons which are not fully understood, expression of the γ-globin genes is not suppressed in δβ-thalassaemia, so heterozygotes and homozygotes show levels of Hb F of

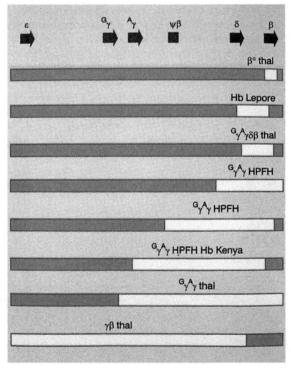

Fig. 8.7 The effects of different deletions in the β-globin gene cluster. Reproduced with permission from Lewin B (2000) *Genes VII.* Oxford University Press, New York.

4–18% and 100% respectively. This increased expression of the γ-globin genes continues throughout life and protects the individual concerned from the harmful effects of the α : β-like globin chain imbalance which impairs erythropoiesis in β-thalassaemia.

Similarly, in HPFH, which can be caused either by deletions involving the δ- and β-globin loci or by various point mutations in the β-globin gene cluster, increased expression of the γ-globin genes compensates for reduced or absent expression of the δ- and β-globin genes. Heterozygotes and homozygotes are asymptomatic, with Hb F levels of around 20% and 100% respectively. It is not known why the expression of the γ-globin genes is maintained in these conditions. Possible explanations include the prevention of normal suppression of postnatal γ-globin gene expression by point mutations in, or deletions of, γ-globin gene regulatory regions.

Whatever the correct explanation, which may well involve several different regulatory mechanisms, these 'accidents of nature' serve to emphasize the potential value of therapeutic approaches aimed at increasing γ-globin gene expression in sickle-cell disease (p. 159). This is further illustrated by the outcomes seen in individuals who inherit two different forms of haemoglobinopathy, as described in the next section.

> **Key Point**
>
> Individuals with δβ-thalassaemia and HPFH are only mildly affected because of the protective effects of sustained γ-globin gene expression.

Interaction of different haemoglobin mutations

As a result of the high gene frequencies of many different forms of haemoglobinopathy in areas where malaria is endemic, individuals are often encountered who have inherited two or more mutations in the α- and/or β-globin gene clusters. In practice it can be difficult to predict the outcome, but some consistent observations have been made.

BOX 8.3 CASE CÉLÈBRE: TIONNE 'T-BOZ' WATKINS

Tionne 'T-Boz' Watkins is the 'T' in the all-female rap group TLC. As well as receiving numerous musical accolades, including two Grammies and two American Music Awards, she has also achieved success as a songwriter, producer, actor, and poet. All of this has been achieved against a background of chronic ill-health brought on by her underlying diagnosis of sickle-cell disease, caused by compound heterozygosity for the sickle-cell trait and β-thalassaemia.

In her semi-autobiographical book, *Thoughts*, she provides a graphic description of what it is like to experience a sickle-cell crisis:

I start aching, sometimes in really excruciating pain. When it's really bad, it's as if someone is stabbing me with a butcher knife over and over where it hurts. The only parts of my body that have never hurt are my feet and my fingers. I've had pain everywhere else at one time or another. It can affect my whole body, or sometimes it can be just a leg. If it's my legs, I can't walk and I have to learn how to walk again. If it's my arms, I can't hold anything, so people have to help feed me. I feel painful and weak. I've been in so much pain that I get delirious, not knowing where I'm at. Or my face might swell up to a point where you wouldn't recognize me.

This is followed by an account of the treatment:

In the hospital, I have to drink lots of fluids and get hooked up to an IV to flush the blood. I also get hooked up to oxygen until the blood recovers. Meanwhile, my body fights against itself because the doctors prescribe drugs that constipate, then they give me other drugs that act as laxatives. They're also drugging me up to cover the pain. And these are some powerful painkillers. You go into the hospital with one problem and leave with another. Now you feel like a crackhead or a heroin addict. Once I get off the drugs I go through withdrawal, with hot and cold flashes, shaking and jumping with crazy dreams, and it's hard to breathe. In the meantime, my body is sweating out the drugs so I smell like all these chemicals.

Despite this depressing account of life as a sickle-cell patient, the underlying theme of this young woman's life, and her book, is very positive. She concludes the chapter entitled 'Monster in my veins', in which she describes her illness, with an appeal to the medical community that 'we need more research and to get more aggressive about helping the minorities out there who suffer from this disease'. The final sentence aptly sums up her attitude to life and to her illness. *'As for me, I don't intend to let it keep me from doing the things I love to do, that make me who I am. You can call that stubborn, but I just call it living'.*

Reference

Watkins, Tionne 'T-Boz' (1999) *Thoughts*. HarperCollins, New York.

α-Thalassaemia/β-thalassaemia

An individual who is a carrier of both α-thalassaemia and β-thalassaemia is referred to as a **double heterozygote** (p. 74). Such individuals are usually entirely healthy. Curiously, the presence of an α-thalassaemia mutation tends to reduce the severity of the anaemia seen in homozygotes for β-thalassaemia. This is almost certainly because the imbalance of α : β globin chains, which leads to impaired erythropoiesis, is reduced by the presence of an α-thalassaemia mutation.

β-Thalassaemia/Hb S

Individuals who are compound heterozygotes for β°-thalassaemia and Hb S have a severe disorder similar to sickle-cell disease (Box 8.3). This is predictable, given that the β°-thalassaemia gene is silenced so that the only β-globin chains manufactured are Hb S. Compound heterozygotes for HbS and milder β+-thalassaemia mutations usually have only mild anaemia or are asymptomatic.

Haemoglobin S/C disease

This results from compound heterozygosity for mutations in the sixth codon of the β-globin gene (Fig. 8.3), if an individual inherits Hb S from one parent and Hb C from the other. Clinically this results in a relatively mild form of sickle-cell disease, which presents in late childhood or adult life, but which is associated with a high incidence of vaso-occlusive complications, particularly involving the lungs and the retinal vessels.

Other β-globin variants which react with Hb S to cause sickle-cell disease include Hb D, Hb E and Hb O Arab. Hb D exists in several different forms, of which the most common, Hb D Punjab, shows a particularly high frequency in India and China. Hb E is the commonest haemoglobin variant in the world, and in the homozygous state usually presents as a mild form of β-thalassaemia. Some compound heterozygotes for β°-thalassaemia and Hb E can be so severely affected that they are transfusion dependent.

Summary

The inherited disorders of haemoglobin structure and synthesis, known collectively as the haemoglobinopathies, together constitute the most common autosomal recessive disorders in the world. The haemoglobin molecule contains four globin chains, $\alpha_2\gamma_2$ in fetal haemoglobin and $\alpha_2\beta_2$ in adult haemoglobin. These chains are encoded by genes in two globin gene clusters, an α or α-like cluster on chromosome 16 and a β or β-like cluster on chromosome 11.

Sickle-cell disease is the most common disorder of haemoglobin structure and affects approximately 1 in 625 African-Americans. It is caused by an A to T point mutation in the sixth codon of the β-globin gene, resulting in a substitution of valine for glutamic acid. The resulting globin tetramer is unstable in the deoxygenated form, causing the red blood cells to become sickle-shaped. This results in chronic haemolytic anaemia and peripheral vaso-occlusion leading to extremely painful sickle-cell crises.

The thalassaemias are disorders of haemoglobin synthesis. α-Thalassaemia results from reduced synthesis of α-globin chains, usually because of deletion of one or more of the two α-globin genes on each number 16 chromosome. Loss of three α-globin genes results in Hb H (β_4) disease characterized by moderately severe lifelong anaemia. Loss of all four α-globin genes results in severe anaemia before birth leading to hydrops fetalis and death *in utero*. The chief haemoglobin in these infants consists of four γ-chains (γ_4), and is known as Hb Barts.

β-Thalassaemia results from reduced synthesis of β-globin chains with marked underlying mutational heterogeneity in the β-globin genes on chromosome 11. The resulting α : β chain imbalance results in precipitation of α-chains and impaired erythropoiesis with severe chronic haemolytic anaemia. Severely affected children are dependent on regular blood transfusion, which conveys a risk of chronic life-threatening iron overload. Bone marrow transplantation is potentially curative.

Both α- and β-thalassaemia are common in areas where malaria is endogenous, with carrier frequencies of up to 1 in 5 in South-east Asia (α-thalassaemia) and Mediterranean regions (β-thalassaemia). Together with sickle-cell disease, these conditions are major causes of chronic morbidity and premature mortality, particularly in some of the world's poorest countries where facilities and resources for health care are limited. The management of these disorders and the suffering that they cause remains a major challenge for health care agencies throughout the world.

Further reading

Old J (2002) Hemoglobinopathies and thalassemias. In: Rimoin DL, Comra JM, Pyeritz RE, Korf BR (eds) *Principles and practice of medical genetics*, 4th edn. Churchill Livingstone, Edinburgh, pp. 1861–1898.

Rodgers GP (1998) *Sickle cell disease and thalassaemia. Baillière's Clinical Haematology*, **11**(1). Baillière Trindall, London.

Sergeant GR (1992) *Sickle cell disease*, 2nd edn. Oxford University Press, Oxford.

Weatherall DJ, Clegg JB (2001) *The thalassaemia syndromes.* Blackwell Science, Oxford.

Multiple choice questions

1 The following statements about haemoglobin are correct:

(a) fetal haemoglobin contains two α-globin and two δ-globin chains

(b) adult haemoglobin contains two α- and two β-globin chains

(c) the α- and β-globin gene clusters are arranged in tandem on chromosome 11

(d) normal humans have two α- and four β-globin genes

(e) expression of the globin genes involves chromatin remodelling

2 The following statements about abnormal forms of haemoglobin are correct:

(a) Hb H contains four α chains

(b) Hb Barts contains four β chains

(c) Hb S is caused by a single point mutation

(d) Hb S and Hb C are caused by different point mutations in the same codon

(e) Hb Lepore is caused by formation of a fusion gene

3 Healthy, unaffected parents have a 6-month-old infant who has just been diagnosed with sickle-cell disease. It would be correct to tell them that

(a) the chance that their next baby will be affected is 1 in 4

(b) babies with sickle-cell disease are anaemic from birth

(c) treatment involves regular folic acid supplementation

(d) early treatment of infection is important

(e) it is important to try to avoid dehydration

4 Healthy, unaffected parents have a 6-month-old infant who has just been diagnosed with β-thalassaemia. It would be correct to tell them that

(a) the chance that their next baby will be affected is 1 in 4

(b) their next baby is likely to have hydrops fetalis

(c) most affected babies present at birth with severe anaemia

(d) treatment involves regular iron supplementation

(e) bone marrow transplantation is potentially curative

5 The following statements about thalassaemia are correct

(a) α-thalassaemia is usually caused by a frame-shift insertion

(b) β-thalassaemia is usually caused by a complete deletion of both β-globin genes

(c) loss of three α-globin genes causes Hb H disease

(d) homozygotes for δβ-thalassaemia are usually only mildly affected

(e) someone who inherits an α-thalassaemia mutation from one parent and a β-thalassaemia mutation from their other parent will have severe anaemia

Answers

1 (a) false—fetal haemoglobin contains two α- and two γ-globin chains

(b) true

(c) false—the α-globin gene cluster is on chromosome 16

(d) false—normal humans have four α- and two β-globin genes

(e) true—this is initiated by the relevant locus control regions

2 (a) false—Hb H contains four β-globin chains

(b) false—Hb Barts contains four γ-globin chains

(c) true—a point mutation in the sixth codon of the β-globin gene

(d) true—glutamine to valine in Hb S and glutamine to lysine in Hb C

(e) true—it is the result of a small fusion gene generated by unequal crossing-over

3 (a) true—inheritance is autosomal recessive

(b) false—presentation is from 3 months onwards as β-globin chains start to replace γ-globin chains

(c) true—this is to prevent coincidental folic acid deficiency anaemia

(d) true—children with sickle-cell disease are very susceptible to bacterial infection because of autosplenectomy

(e) true—this can predispose to vaso-occlusive crises

4 (a) true—inheritance is autosomal recessive

(b) false—this is a presentation of α-thalassaemia

(c) false—as with sickle-cell disease presentation is from 3 months onwards as β-globin chains start to replace γ-globin chains

(d) false—iron overload is a major hazard in β-thalassaemia

(e) true—bone marrow transplant from a histocompatible donor is very successful

5 (a) false—the most common mutational mechanism is a deletion of the entire gene

(b) false—there is marked mutational heterogeneity which can include small intragenic deletions

(c) true—the excess β-globin chains form Hb H

(d) true—there is a compensatory increase in Hb F due to persistent γ-globin chain synthesis

(e) false—such a person is a double carrier and is usually not anaemic

Genes and development

The term **morphogenesis** refers to the extremely complex process whereby large numbers of genes interact in a carefully orchestrated manner to generate a three-dimensional embryo. Unfortunately the underlying complexity of the multitude of different developmental process involved renders the embryo very susceptible to morphogenetic errors. Approximately 20% of recognized human conceptions result in early spontaneous miscarriage. Around 50% of these are caused by gross chromosomal imbalance (p. 51) and over 80% show major structural abnormalities, ranging from complete disorganization to localized developmental disturbance. Among liveborn infants the incidence of major structural abnormalities obvious at birth is 1 in 40, with a similar proportion having an abnormality which becomes apparent later in childhood.

Developmental biology is the scientific field which relates to the study of the processes involved in embryogenesis. These include embryology, physiology, biochemistry, cell biology, and developmental genetics, the subject to which this chapter is devoted. Initially, important developmental processes that are under genetic control are considered. This is followed by a discussion of some of the gene families and pathways involved in human development, giving due consideration to the outcome when errors arise. The latter part of the chapter focuses on congenital abnormalities and specifically on how they are classified, how they can be caused by external teratogens and intrinsic gene mutations, and finally how they are studied in the medical discipline known as **dysmorphology**.

Basic developmental processes

The effects of developmental genes are mediated through a number of fundamental cellular processes integral to an understanding of developmental genetics. These are now considered.

Embryonic induction (signal transduction)

The development of a human embryo is dependent on a tightly orchestrated sequence of events whereby one group of cells modifies the behaviour of an adjacent group of cells, thereby determining their fate. This can involve growth, proliferation, differentiation, migration, or death. This process of cellular interaction is referred to as **induction**. When cell-membrane proteins on one group of cells interact directly with membrane-bound receptors on adjacent cells, this is described as **juxtacrine** induction or signalling. When proteins produced by one group of cells can diffuse over short distances to receptors on more distant cells, this is referred to as **paracrine** induction or signalling. The diffusing proteins are called **paracrine factors**. Paracrine factors include fibroblast growth factors and hedgehog proteins, as discussed later in this chapter. Proteins which diffuse outwards to form a concentration gradient which directs morphogenesis are referred to as **morphogens**.

Paracrine induction is mediated through the process of signal transduction (Fig. 9.1). This involves the attachment of a **ligand**, the paracrine factor, to a membrane-bound receptor with extracellular, transmembrane, and intracellular (cytoplasmic) components. Ligand binding induces a conformational change in the receptor which activates an intracellular kinase domain. The activated receptor initiates an intracellular signalling cascade, usually by phosphorylation, culminating in activation of a nuclear transcription factor which initiates or suppresses gene expression.

The growth factor/receptor kinase signalling pathway is one of the best understood examples of signal transduction. This is discussed further in the section on the fibroblast growth factor gene family.

Cell–cell adhesion

Cell–cell adhesion is essentially an example of juxtacrine induction, whereby cell adhesion molecules, such as cadherins, integrins, and selectins, facilitate cell binding via ligand-receptor interactions. This process enables large groups of cells to develop together into a specific organ or tissue, or to migrate as a unit as in the migration of neural crest cells from the neural tube throughout the developing embryo (p. 175).

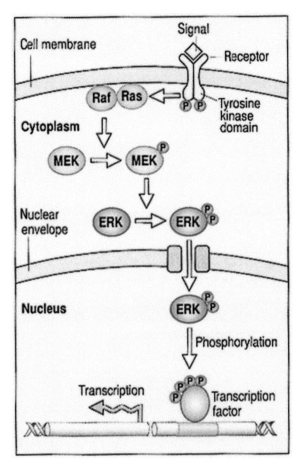

Fig. 9.1 An example of signal transduction showing a simplified outline of the Raf/Ras signalling pathway which is involved in the regulation of cell growth, proliferation and differentiation. Binding of the signal ligand to the cell surface receptor initiates an intracellular cascade of protein (Ras, Raf, MEK, and ERK) phosphorylation culminating in intranuclear activation of a transcription factor. Reproduced with permission from Wolpert L (2002) *Principles of development*, 2nd edn. Oxford University Press, Oxford.

Specific junctions form between adjacent cells either to enhance cell adhesion, e.g. adherens junctions, or to provide channels between cells, e.g. gap junctions. A mutation in a specific gap junction gene (*GJB2*), which encodes a protein known as connexin-26, is the commonest cause of sensorineural hearing loss in the European population. The mutation leads to impaired recycling of potassium ions needed for the initiation of action potentials in the cochlear hair cells.

Programmed cell death

Programmed cell death, also known as **apoptosis**, is an important part of normal development. It is responsible for the canalization of solid structures such as the urethra, for the creation of cavities such as the middle ear, and for the destruction of redundant tissue such as the webbing between embryonic digits. Failure of apoptosis can result in severe developmental abnormalities such as syndactyly (fusion of the digits) and is an important factor in the genesis of cancer (p. 194). Apoptosis is mediated by a family of genes known as BCL2 (B-cell leukaemia 2), some members of which promote apoptosis whereas others suppress it. The activation of genes that prevent apoptosis has been considered as a possible approach in gene therapy for conditions associated with premature cell death.

Developmental gene families

Recent research has shown that many gene families and developmental pathways are involved in human embryogenesis. Gene families consist of genes which show a high degree of homology, i.e. they contain similar DNA sequences. Some of the most important are described in this section.

Homeobox (HOX) gene family

In humans this consists of a total of 39 genes organized into four clusters on chromosomes 7 (*HOXA*), 17 (*HOXB*), 12 (*HOXC*), and 2 (*HOXD*). These genes were first identified in *Drosophila* in which they were named as *homeotic* or *homeobox* (hence *Hox*) genes. This was because mutations in these genes could change the identity of a body part, resulting, for example, in growth of a leg instead of an antenna. **Homeosis** is defined as the transformation of one body segment into another.

Each *HOX* gene contains a conserved sequence of 183 nucleotides, which encode a 61 amino acid DNA binding domain known as the homeodomain or homeobox. The *HOX* genes encode transcription factors, which regulate a cascading activity of downstream genes in various developmental pathways involved in morphogenesis. These gene clusters, which have been highly conserved during evolution, show sequential expression, both temporally and spatially as illustrated in Fig. 9.2. This demonstrates the pattern of expression along the anterior–posterior axis in both *Drosophila* and the human embryo. The *HOXA* and *HOXB* clusters are specifically involved in establishing the rostral–caudal (head–tail) axis, with the *HOXA* and *HOXD* clusters being important in limb development.

Given the large number of *HOX* genes in humans and their important role in embryogenesis, it is curious that little is known about the effects of *HOX* gene mutations

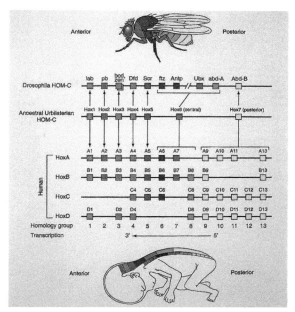

Fig. 9.2 The genomic organization and expression pattern of the *HOX* gene family in *Drosophila* and humans. The second row shows a hypothetical ancestral Hox cluster in the evolutionary pathway from *Drosophila* to mammals. Reproduced with permission from Epstein CJ, Erickson RP, Wynshaw-Boris A (eds) (2004). *Inborn errors of development*. Oxford University Press, New York.

in humans. This could be because of redundancy, in that one *HOX* gene may be able to compensate for deficiency in another. In this context the *HOX* genes are **paralogous**, in that they are very similar in structure, consistent with their origin from a common ancestral gene. The human *HOX* clusters are thought to have evolved from two successive duplications of a single primordial cluster. An alternative explanation for the lack of observed *HOX* gene mutations in humans is that their effects are so devastating that pregnancy inevitably ends in early miscarriage.

Three specific conditions have been attributed to mutations in the final genes to be expressed in the *HOXA* and *HOXD* clusters. A single nucleotide deletion in *HOXA11* causes radioulnar synostosis and thrombocytopenia due to absence of megakaryocytes (MIM 605432). Loss-of-function mutations in *HOXA13* cause the hand–foot–genital syndrome (MIM 140000), in which small thumbs and halluces are associated with uterine malformations in females and hypospadias in males. Expansions of a trinucleotide polyalanine tract in *HOXD13* cause synpolydactyly (MIM 186000), in which there is webbing (syndactyly) between the third and fourth fingers and the fourth and fifth toes with digit duplication in the syndactylous web (Fig. 9.3).

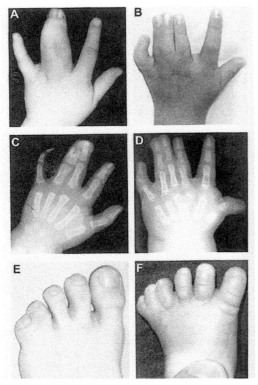

Fig. 9.3 Appearance of the hands and feet in synpolydactyly caused by a mutation in *HOXD13*. Reproduced with permission from Epstein CJ, Erickson RP, Wynshaw-Boris A (eds) (2004) *Inborn errors of development*. Oxford University Press, New York.

Unlike the dynamic triplet repeat expansions which cause Huntington disease and fragile X syndrome (p. 12), the polyalanine expansions which cause synpolydactyly are stable and do not change in size from generation to generation. A positive correlation has been noted between the size of the expansion and the severity of the limb changes.

Other homeodomain-containing genes

It is now known that many other genes encoding transcription factors also contain a homeodomain. These include several members of the paired-box (*PAX*) gene family described in the next section. Other examples of homeodomain containing genes important in human development and homeostasis are given in Table 9.1. The diversity of clinical outcomes associated with mutations in these genes emphasizes their importance. *IDX1* provides a useful example of the different effects of homozygous and heterozygous mutations. Homozygosity for a loss-of-function mutation results in pancreatic agenesis which presents at birth as diabetes mellitus and absence of pancreatic exocrine function. Heterozygotes for the same mutation develop adult onset diabetes mellitus at an average age of 35 years. *IDX1* has been designated as the gene which causes maturity-onset diabetes of the young (MODY) type 4 (p. 127).

Paired-box (*PAX*) gene family

In humans this comprises a total of nine genes which encode transcription factors involved in early embryo-

TABLE 9.1		Genes which contain a homeodomain and are associated with developmental disorders in humans		
Gene	**Locus**	**Disorder**	**MIM**	**Features**
HLXB9	7q36	Currarino syndrome	176450	Anorectal malformation, hemisacrum, presacral mass
IDX1	13q12	Maturity-onset diabetes of the young (MODY), type 4	606392	MODY in heterozygotes; pancreatic agenesis in homozygotes
LMX1B	9q34	Nail–patella syndrome	161200	Absent nails and patellae; also nephropathy and glaucoma
MSX1	4p16	Witkop syndrome	189500	Hypodontia, cleft lip/palate, nail defects
MSX2	5q34	Craniosynostosis, Boston type	604757	Variable craniosynostosis caused by gain-of-function mutations
		Parietal foramina	168500	Parietal foramina (gaps in skull ossification) caused by loss-of- function mutations
NKX2-5	5q34	Congenital heart defects	108900	Atrial septal defects and atrioventricular conduction block
PITX2	4q25	Reiger syndrome	180500	Eye, teeth and umbilical defects
SHOX	[a]Xp22/Yp11	Dyschondrosteosis	127300	Short stature and wrist deformity (p. 60)

[a] *SHOX* is located in the pseudoautosomal region shared by the X and Y chromosomes.

TABLE 9.2 *PAX* genes and developmental abnormalities

Gene	Locus	Expression pattern	MIM	Abnormality and features
PAX2	10q24	Brain, eye, ear, kidney	120330	Renal-coloboma syndrome (renal hypoplasia and ocular coloboma)
PAX3	2q35	Brain, neural tube, neural crest	193500	Waardenburg syndrome type I (deafness, dystopia canthorum and pigmentary disturbance).
PAX6	11p13	Brain, neural tube, eye and pituitary	106210	Aniridia (hypoplasia or aplasia of the iris)
PAX8	2q12-14	Neural tube and thyroid gland	167415	Congenital hypothyroidism; mutations cause thyroid dysplasia
PAX9	14q12	Teeth, ribs and vertebra	106600	Hypodontia (absence of secondary molars and premolars)

genesis, specifically in the determination of cell fate. They all share a DNA-binding domain known as a paired-box. This was first identified in a series of genes in *Drosophila* that, when mutated, cause defects in body patterning and segmentation. Many of the *PAX* genes also contain a second DNA-binding domain identical to that seen in the homeotic (*HOX*) gene family. In contrast to the *HOX* genes, the *PAX* genes are not arranged in clusters.

Loss-of-function mutations in five *PAX* genes have been shown to cause specific abnormalities consistent with the embryonic expression pattern of each of the relevant genes (Table 9.2). For example, *PAX2* and *PAX6* are both expressed in the eye.

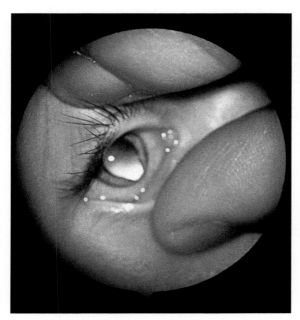

Fig. 9.4 An iris coloboma as caused by a mutation in *PAX2*. Courtesy of Mr G. Woodruff, Leicester Royal Infirmary, Leicester.

Mutations in *PAX2* lead to defective closure of the optic fissure, resulting in a defect of the optic nerve and iris known as a coloboma (Fig. 9.4). Mutations in *PAX6* result in abnormal development or complete absence of the iris. *PAX6* is one of the genes which is lost in the WAGR contiguous gene deletion syndrome (p. 64). Although *PAX3* is not expressed directly in the developing eye, mutations in *PAX3* can lead to changes in the pigmentary pattern of the irides as the pigment cells in the iris are derived from migrating neural crest cells in which *PAX3* is actively expressed.

In contrast to the effects of haploinsufficiency resulting from mutations with a loss-of-function effect, increased *PAX* transcriptional activity is implicated in oncogenesis. *PAX2* and *PAX8* are expressed in the embryonal renal Wilms tumour. Translocations resulting in the fusion of the *PAX* DNA-binding domain to activating domains in other genes have been identified in alveolar rhabdomyosarcoma and in large cell lymphoma.

Developmental gene pathways

Developmental pathways consist of genes and the proteins that they encode which are involved in processes such as signal transduction. Two of the best understood are the fibroblast growth factor/receptor and the sonic hedgehog signalling pathways.

Fibroblast growth factors and receptors

Fibroblast growth factors play an important role in the regulation of cell division, differentiation and migration through a complex series of signalling pathways. These are mediated initially by interaction between fibroblast growth factor (FGF), its receptor (FGFR), and heparin sulphate proteoglycans (HSPG). Receptor activation occurs when a trimolecular complex, consisting of FGF, FGFR, and HSPG, leads to dimerization, phosphorylation of the tyrosine kinase residues in the FGFR, and subsequent downstream signalling.

The family of genes involved in the FGF/FGFR signalling pathway consists of 23 *FGF* and 4 *FGFR* genes. To date only one of the FGF genes, *FGF23*, has been implicated as the cause of an inherited disorder, although it is known that many FGF proteins are expressed at various stages of fetal and adult life. Mutations in *FGF23* cause a rare form of vitamin D resistant rickets (MIM 193100), which, unlike the more common X-linked form (p. 80), shows autosomal dominant inheritance.

The four known *FGFR* genes share a highly conserved amino acid sequence but differ in their ligand affinities and tissue distribution. They encode a subset of receptor tyrosine kinases each of which consists of an extracellular region with ligand-binding domains, a single membrane-spanning domain, and an intracellular bipartite kinase domain. Mutations in *FGFR1*, *FGFR2*, and *FGFR3* cause 12 different conditions, which can be grouped under the headings of craniosynostosis syndromes and short-limb skeletal dysplasias (Table 9.3). Craniosynostosis refers to premature fusion of the cranial sutures. This often leads to severe disturbance of normal facial development. The relevant short-limb skeletal dysplasias show variable outcomes ranging from death before or shortly after birth, as in thanatophoric dysplasia (Fig. 9.5), to mild–moderate short stature with normal intelligence and life expectancy, as in achondroplasia (Fig. 9.6).

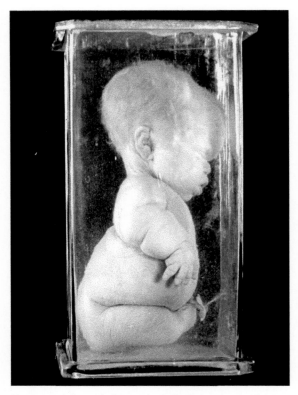

Fig. 9.5 An infant with thanatophoric dysplasia. Courtesy of Dr R J Oostra, Museum Vrolik, Amsterdam, The Netherlands; photograph prepared by C J Hersbach.

TABLE 9.3 Disorders caused by mutations in fibroblast growth factor receptor (FGFR) genes

Gene	Locus	Disorder	MIM	Features
Craniosynostosis syndromes				
FGFR1	8p11	Pfeiffer syndrome	101600	Craniosynostosis with broad thumbs and big toes
FGFR2	10q26	Apert syndrome	101200	Craniosynostosis plus hand and foot syndactyly
		Beare–Stevenson syndrome	123790	Craniosynostosis plus cutis gyrata
		Crouzon syndrome	123500	Craniosynostosis with ocular proptosis
		Pfeiffer syndrome	101600	Craniosynostosis with broad thumbs and big toes
FGFR3	4p16.3	Crouzon syndrome	123500	Craniosynostosis plus acanthosis nigricans
		Muenke syndrome	602849	Isolated coronal suture synostosis
Short-limb skeletal dysplasias				
FGFR3	4p16.3	Achondroplasia	100800	Macrocephaly, short trunk and limbs
		Hypochondroplasia	146000	Short trunk and limbs
		Thanatophoric dysplasia		
		Type I	187600	Macrocephaly, severe short stature, small chest
		Type II	187601	As type I; also craniosynostosis

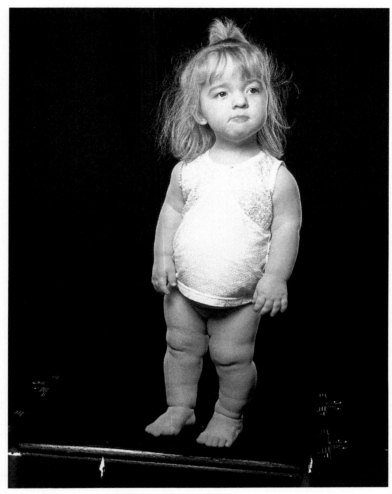

Fig. 9.6 A young girl with achondroplasia.

All of these conditions are caused by mutations which exert a gain-of-function effect, mediated by a number of different mechanisms. The common mutation which causes achondroplasia, g.1138G→A or p.Gly380Arg in *FGFR3*, increases receptor kinase activity by around 20% in a ligand-independent manner. The two mutations which cause most cases of Apert syndrome, p.Ser252Trp and pPro253Arg in *FGFR2*, cause increased receptor signalling by increasing affinity for the ligand FGF2. Finally, the mutation which causes a particularly severe form of thanatophoric dysplasia (which translates from the Greek as 'death-bearing') acts by removing intrinsic inhibitors of the kinase domain resulting in a 100-fold increase in kinase activity.

Sonic hedgehog signalling pathway

The hedgehog signalling pathway is made up of a complex network of genes and gene products involved in various sequential interactions in early embryonic signalling. The term 'hedgehog' was used originally when a mutation in one member of the gene family was shown to cause duplication of the bristles in *Drosophila* larvae. Three hedgehog genes have been identified in humans. These are sonic hedgehog (*SHH*) on chromosome 7, Indian hedgehog (*IHH*) on chromosome 2, and desert hedgehog (*DHH*) on chromosome 12. *SHH* appears to be the most important of these, with a wide pattern of expression including the brain, heart, limb buds, and axial skeleton.

To simplify understanding, the sonic hedgehog pathway can be considered as consisting of three key steps (Fig. 9.7). The first is the modification of the *SHH* protein product by the addition of a cholesterol moiety. The second involves short- and long-range *SHH* signalling by binding of SHH as a ligand to its receptor, PTCH, encoded by *Patched*, (or *PTCH*) on chromosome 9.

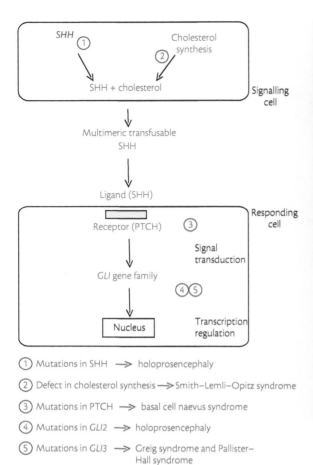

① Mutations in SHH ⟶ holoprosencephaly

② Defect in cholesterol synthesis ⟶ Smith–Lemli–Opitz syndrome

③ Mutations in PTCH ⟶ basal cell naevus syndrome

④ Mutations in GLI2 ⟶ holoprosencephaly

⑤ Mutations in GLI3 ⟶ Greig syndrome and Pallister–Hall syndrome

Fig. 9.7 Simplified representation of the Sonic hedgehog signalling pathway.

In the third step, PTCH activates intracellular signalling of a family of *GLI* genes, originally so named because the first was identified in a tumour known as a glioblastoma. The *GLI* genes act as transcription regulators.

The importance of the hedgehog signalling pathway is illustrated by the diverse conditions caused by mutations in it (Fig. 9.7). Mutations in *SHH* resulting in haploinsufficiency cause holoprosencephaly (MIM 142945) (Fig. 9.8). This is a severe and often lethal malformation, which is also common in trisomy 13 (p. 57) and in which there is failure of normal cleavage of the prosencephalon into the cerebral hemispheres. This is usually associated with midline facial developmental abnormalities including hypotelorism (closely spaced eyes), flat nose, and median cleft lip (Fig. 9.8). Mutations in *PTCH* with a loss-of-function effect cause the basal cell naevus (Gorlin—MIM 109400) syndrome in which numerous basal cell carcinomas occur in association with

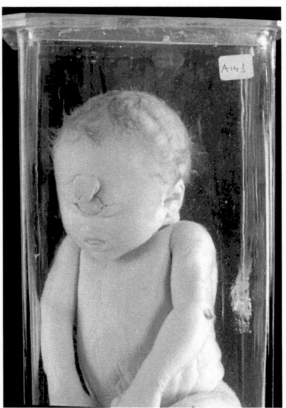

Fig. 9.8 Facial appearance in holoprosencephaly. Courtesy of Dr R J Oostra, Museum Vrolik, Amsterdam, The Netherlands; photograph prepared by C J Hersbach.

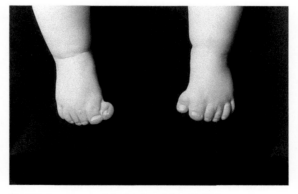

Fig. 9.9 Appearance of the feet in a child with Greig cephalopolysyndactyly syndrome.

jaw cysts and rib abnormalities. Mutations in *GLI2* cause holoprosencephaly. Mutations in *GLI3* cause two conditions, the Greig cephalopolysyndactyly syndrome (MIM 175700) and the Pallister–Hall syndrome (MIM

146510). In the Greig cephalopolysyndactyly syndrome the forehead is broad and prominent and the limbs show both extra digits (polydactyly) and webbing between the digits (syndactyly) (Fig. 9.9). Characteristic findings in the Pallister–Hall syndrome are polydactyly, imperforate anus, and a hypothalamic hamartoma.

Errors in cholesterol synthesis can also have an adverse effect on *SHH* signalling because of impaired cholesterol modification. This is illustrated by the Smith–Lemli–Opitz syndrome (MIM 270400), which is caused by inability to convert 7-dehydrocholesterol to cholesterol. Many affected patients have polydactyly, and holoprosencephaly occurs in 4–5%.

Although the *IHH* and *DHH* signalling pathways are not as well characterized as the *SHH* pathway, it has recently been shown that loss-of-function mutations in *IHH* cause type A-1 brachydactyly (MIM 112500), in which the middle phalanges of the fingers and toes are short (Fig. 9.10). This harmless condition was the first human characteristic recognized to show autosomal dominant inheritance, by Farabee in 1903.

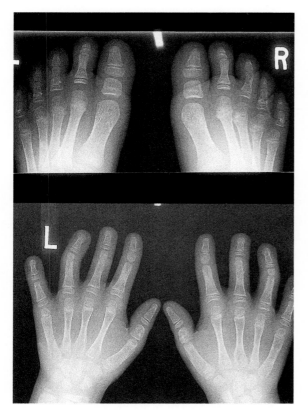

Fig. 9.10 Radiograph showing shortening of the middle phalanges of the hands and feet in type A-1 brachydactyly.

> **Key Point**
>
> Large numbers of gene families and developmental pathways are involved in human embryogenesis. Mutations in these genes cause various abnormalities and developmental syndromes.

Integrated developmental pathways

One of the goals of developmental biology is to understand how gene families and signalling pathways interact to determine how organs and tissues are formed. At present this integration of molecular signalling and tissue maturation is not well understood. However, some insight is provided by the study of developmental pathways involved in various disease processes. Two such examples are considered here: the neural crest migratory pathway and the process of sex determination.

Neural crest cell migration

Neural crest cells are derived from the neuroectoderm lying along the crest of the neural fold. As the neural tube separates from the surface ectoderm, the neural crest cells start to migrate to form various body tissues including the muscles of the head and neck, the C cells of the thyroid gland, the adrenal medulla, the peripheral nervous system, and the pigment-producing cells known as melanocytes. Many gene families have been implicated in the induction and subsequent migration of neural crest cells to form these various tissues. Evidence from the study of Waardenburg syndrome and Hirschsprung disease indicates that two of these pathways are closely integrated (Fig. 9.11).

Four types of Waardenburg syndrome (MIM 193500) are recognized:

♦ *Type I*: the characteristic features are hearing loss, pigmentary abnormalities of the hair, skin and eyes, and dystopia canthorum (lateral displacement of the inner canthi of the eyes). It is caused by heterozygous *PAX3* mutations which cause haploinsufficiency.

♦ *Type II*: the clinical features are very similar to type I, with the exception that the distance between the inner canthi is normal. It is caused by mutations in *MITF* (microphthalmia associated transcription factor) which is activated by *PAX3* and *SOX10*.

♦ *Type III*: a severe form of type I in which affected individuals also have limb abnormalities.

♦ *Type IV*: patients have the typical features of both Waardenburg syndrome *and* Hirschsprung disease. It is caused by mutations in either *SOX10* or the

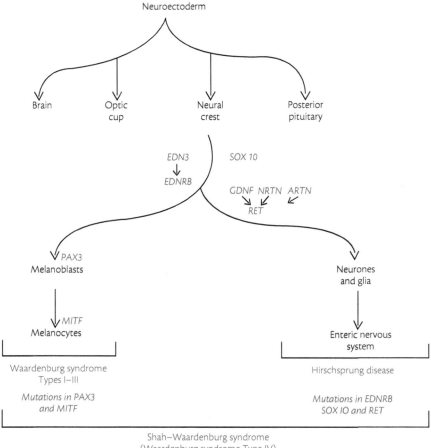

Fig. 9.11 Simplified diagrammatic representation of genes involved in neural crest migration.

Endothelin signalling pathway. Also known as the Shah–Waardenburg syndrome, this type is of particular relevance to the present discussion.

The genetic contribution to Hirschsprung disease is complex and has been considered already in Chapter 6 (p. 127). Clinically, Hirschsprung disease presents as severe constipation or abdominal obstruction due to failure of normal neural crest cell migration resulting in aganglionosis of the colon and rectum. Several genes, including the endothelin and RET signalling pathways, contribute to what has been described as the **oligogenic** aetiology of this condition. Although it is not known precisely how these genes interact, the prevailing view is that Hirschsprung disease is a disorder of abnormal neural crest migration, i.e. a *neurocristopathy*, which results from mutations in a small number of important cell-signalling genes.

In view of the integration of the signalling pathways involved in neural crest cell migration (Fig. 9.11) it is not difficult to see how mutations in some of the relevant genes can cause both Waardenburg syndrome and Hirschsprung disease, i.e. the Shah–Waardenburg syn-

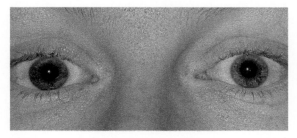

Fig. 9.12 View of the eyes in a woman with Waardenburg syndrome showing iris heterochromism (= different coloured irides).

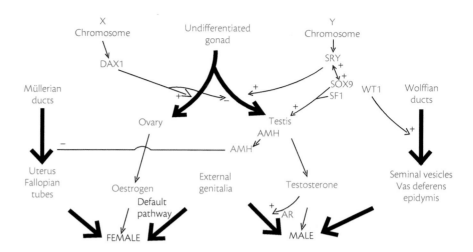

Fig. 9.13 Diagrammatic representation of the sex determination pathway.

drome. The pigmentary abnormalities are caused by absence of melanocytes in skin, hair, and the irides (Fig. 9.12). The hearing loss results from absence of melanocytes in the stria vascularis of the inner ear. The functional bowel obstruction reflects abnormal innervation of the colon and rectum. Shah–Waardenburg syndrome represents an excellent example of how molecular biology has provided an explanation for the way in which mutations in a single gene can cause diverse and apparently unrelated clinical effects, a phenomenon referred to a **pleiotropy**.

Sex determination pathway

The determination of sex in the human embryo is remarkably complex and is still not fully understood. A simplified and somewhat hypothetical schematic representation is shown in Fig. 9.13.

Until 5–6 weeks after conception the embryonic gonads are undifferentiated. Their subsequent development into testes or ovaries is determined by several genes, most notably *SRY* (sex determining region of the Y chromosome) on the Y chromosome (Box 9.1) and *DAX1* on the X chromosome. (*DAX1* stands for dosage-sensitive sex reversal, adrenal hypoplasia congenita critical region on the X chromosome, gene 1.) *SRY* interacts with various other transcription regulatory genes, including *SF1*, *SOX9*, and *WT1*, to promote the testis to produce testosterone and anti-Müllerian hormone (AMH). Testosterone acts through the androgen receptor (AR) to masculinize the external genitalia and promote the development of the Wolffian ducts into male internal genitalia. AMH inhibits the development of the Müllerian duct system.

In the absence of *SRY*, a double dose of *DAX1* (i.e. two X chromosomes) supports normal development of the undifferentiated gonads into ovaries. The Müllerian ducts develop into the uterus, Fallopian tubes, and proximal vagina. The external genitalia proceed along a default pathway to develop into the clitoris, labia, and distal vagina.

Chromosomal causes of abnormal sex development

Structural or numerical abnormalities of the X or Y chromosome often influence the normal sex determination pathway. The presence of a single X chromosome, as in Turner syndrome (45,X—p. 59), leads to failure of sustained ovarian development ('gonadal dysgenesis') possibly due at least in part to loss of one copy of *DAX1*. *DAX1* encodes a nuclear receptor which is necessary for normal development of the adrenal glands and ovaries. *DAX1* also suppresses testicular development. This may partly explain why men with Klinefelter syndrome (47,XXY—p. 60) have small testes. Duplication of Xp21.3, which includes the *DAX1* locus, in a 46,XY embryo can lead to gonadal dysgenesis resulting in a female phenotype. This is referred to as XY sex reversal or male **pseudohermaphroditism** (Table 9.4).

Translocation of Yp11.2, which contains the *SRY* locus, to the short arm of the X chromosome can occasionally occur as a result of *illegitimate crossing-over* between the X and Y chromosomes in meiosis I (Fig. 9.14). This can result in an XX karyotype with a male phenotype, or an XY karyotype with a female phenotype. The presence of two Y chromosomes has no obvious effect on gonadal development, as illustrated by normal sex determination in men with the XYY syndrome (p. 61).

BOX 9.1 LANDMARK PUBLICATION: THE DISCOVERY OF *SRY*

After the development of a reliable technique for the study of human chromosomes in 1956, it was soon established that the presence of a Y chromosome leads to maleness regardless of the number of X chromosomes present in each cell. Absence of a Y chromosome results in female development. The hypothetical Y chromosome gene responsible for determining maleness was given the name testis determining factor or *TDF*.

Clues to the location of *TDF* were provided by the study of XX males and XY females who were found to have small translocations involving the short arms of the X and Y chromosomes. Thus 46,XX males were found to have the end of the short arm of a Y chromosome attached to the end of the short arm of an X chromosome. Similarly some 46,XY females were identified in whom the end of the short arm of the Y chromosome was replaced by the end of the short arm of an X chromosome.

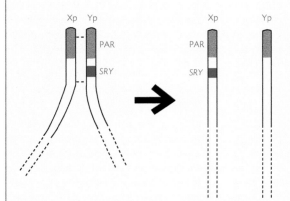

Fig. 9.14 Shows illegitimate crossing over at the ends of the X and Y chromosome short arms in male meiosis I. PAR = pseudoautosomal region; *SRY* = sex-determining region of the Y chromosome.

Molecular analysis of the different portions of Y chromosome present in 46,XX males helped establish that *TDF* was almost certainly located very close to the pseudoautosomal region (PAR) of Yp. This is the portion of the Y chromosome which pairs with the corresponding region of the X chromosome in meiosis I in males. It was plausible that rarely an 'illegitimate' crossover could occur just outside the PAR, in which case the *TDF* would be transferred to the short arm of the X chromosome (Fig. 9.14).

The search for *TDF* was concentrated on the region of the Y chromosome just proximal to the PAR. Subclones from this region were hybridized to DNA from males of different mammals. One subclone was found to contain male-specific sequences with a transcript which was expressed only in testes. This male determining gene was given the name *SRY*. Subsequently mutations in *SRY* were reported in a small number of 46,XY females.

SRY is now accepted to be the conductor of the orchestra of genes which determine male development (Fig. 9.13). It consists of a single exon which encodes a 204-amino-acid protein containing a high mobility group (HMG) domain. This binds to DNA thereby regulating the transcription of other genes involved in the sex-determination pathway. A subgroup of genes containing a similar HMG domain were designated as <u>S</u>RY-related b<u>ox</u> or *SOX* genes. *SOX9* has been shown to play a major role in male determination (p. 179).

Reference

Sinclair AH, Berta P, Palmer MS *et al.* (1990) A gene from the human sex-determining region encodes a protein with homology to a conserved DNA-binding motif. *Nature*, **346**, 240–244.

TABLE 9.4   Causes of sex reversal

Male pseudohermaphroditism	Female pseudohermaphroditism
(46,XY with ambiguous or female genitalia)	(46,XX with ambiguous or male genitalia)
Androgen insensitivity (due to mutation in *AR*)	Congenital adrenal hyperplasia due to 21 hydroxylase deficiency
Loss of function mutation/deletion in *SRY*	Exogenous androgen ingestion in pregnancy
Duplication of *DAX1* on Xp21.3	Translocation of *SRY* from Y to X chromosome
Campomelic dysplasia (caused by mutation in *SOX9*)	
Denys–Drash syndrome (caused by mutation in *WT1*)	
Adrenal failure (caused by mutation in *SF1*)	

Single-gene causes of abnormal sex development
Male pseudohermaphroditism

Loss-of-function mutations in *SRY, SF1, SOX9*, or *WT1* can all cause ambiguous genitalia or complete sex reversal. *SF1* (steroidogenic factor 1) encodes a nuclear receptor that is a key regulator of steroid synthesis in the adrenal glands and developing testes. Loss-of-function mutations have been identified in 46,XY females with adrenal failure. *SOX9* encodes a transcription factor with an HMG domain, which shows strong homology with the HMG domain in *SRY*. Loss-of-function mutations in *SOX9* cause campomelic dysplasia (MIM 114290) in which lethal skeletal abnormalities are associated with variable 46,XY sex reversal. The skeletal abnormalities are explained by the fact that *SOX9* also regulates the expression of the collagen gene, *COL2A1*, which is expressed in cartilage. Finally, *WT1* (Wilms tumour suppressor) encodes a DNA-binding zinc finger transcription factor which is actively expressed in the developing gonads and Wolffian ducts. Loss-of-function mutations in *WT1* cause the Denys–Drash syndrome (MIM 194080), in which 46,XY sex reversal occurs in association with a nephropathy of early childhood onset and/or Wilms tumour. *WT1* is one of the genes deleted in the WAGR contiguous gene deletion syndrome, in which the W stands for Wilms tumour and the G for genitourinary abnormalities (p. 64).

Mutations in *AR*, the gene which encodes the androgen receptor, result in a condition that used to be called testicular feminization and is now referred to as androgen insensitivity (MIM 300068). Affected individuals are chromosomal males with testes which produce testosterone and AMH. The absence of functional androgen receptors means that the embryo cannot respond to testosterone, so development of the external genitalia proceeds along the default female pathway. A normal response to AMH results in regression of the Müllerian ducts so that the phenotypic female, with a 46,XY karyotype, has no uterus or Fallopian tubes. The testes are usually present in the inguinal canal.

Mutations in *AMH* in males result in the persistent Müllerian duct syndrome. These individuals are phenotypically male but with undescended testes which are attached to the uterus in the pelvis. Mutations in *AMH* have no effect in females as *AMH* is normally only expressed in the testes.

Female pseudohermaphroditism

Any factor which raises the levels of circulating androgens in a female embryo or fetus can lead to masculinization of the external genitalia. This can be exogenous, as in maternal androgen ingestion, or endogenous, as

in congenital adrenal hyperplasia. The commonest cause of congenital adrenal hyperplasia is 21-hydroxylase deficiency (MIM 201910), which shows autosomal recessive inheritance and is caused by mutations in *CYP21* which lies within the HLA gene cluster on chromosome 6 (p. 144). 21-Hydroxylase is responsible for the conversion of progesterone to deoxycorticosterone and of 17-OH progesterone to 11-deoxycortisol in the biosynthetic pathways of aldosterone and cortisol. Failure of normal synthesis leads to increased secretion of pituitary adrenocorticotrophic hormone (ACTH). This in turn drives the adrenal gland to produce cortisol precursors and androgens which have a virilizing effect on the developing female embryo. Attempts have been made to prevent this virilization by giving the mother dexamethasone from around weeks 7–8 of gestation onwards, in the hope that this will suppress the embryo's pituitary–adrenal axis. Early prenatal diagnosis is carried out by chorionic villus sampling (p. 274). If the fetus is found to be both female and affected, the maternal dexamethasone is continued to term.

It is very important that congenital adrenal hyperplasia is diagnosed soon after birth, as affected babies need urgent replacement therapy with aldosterone and cortisol. Without adequate treatment there is a high risk of death from adrenal insufficiency in early infancy. Because of this high risk, congenital adrenal hyperplasia is now included in many newborn screening programmes (p. 219).

> **Key Point**
>
> Human sex development is determined by a complex series of sequential events under the control of a large number of regulatory genes. In male pseudohermaphroditism, as caused by mutations in the androgen receptor, the karyotype is male and the external genitalia are female. In female pseudohermaphroditism, as caused by congenital adrenal hyperplasia, the karyotype is female and the external genitalia are virilized.

Birth defects

Approximately 1 in 40 of all infants has a developmental abnormality which is apparent at birth, and a similar proportion have a more subtle internal defect which becomes apparent later in childhood or adult life. These abnormalities are said to be **congenital** because they are present at, or date from, birth. Some of these abnormalities are treatable, e.g. cleft lip or palate, whereas others, such as bilateral renal agenesis, are invariably lethal. A third category consists of abnormalities such as severe brain defects, which are compatible with long-term survival but result in severe disability.

The term *birth defects* is used in a generic sense to embrace all types of abnormalities, regardless of their nature or cause. As perinatal mortality and morbidity started to decline due to improvements in obstetrics and public health, the relative impact of birth defects steadily became more important, leading to the realization that a more precise description and classification of birth defects was required. This led to the establishment of an international working group which reported its findings in 1982.

Classification of birth defects

The working group devised a practical set of guidelines for describing and classifying *errors of morphogenesis* (Table 9.5). This was based around the underlying pathogenesis, relating both to the nature of the abnormality and to its underlying cause. The classification is accepted universally and provides the starting point for the study of abnormal morphogenesis in the medical specialty now referred to as **dysmorphology**.

Malformation

A **malformation** is defined as a morphological defect of an organ which results from an 'intrinsically abnormal developmental process'. Intrinsic implies that the developmental potential of the organ was abnormal from conception or, as the working group neatly summarized, 'the organ never had a chance' to develop normally. Malformations can be divided into major or

minor depending upon whether or not they are of either medical or cosmetic significance.

Disruption

A **disruption** is defined as a morphological defect which results from the 'extrinsic breakdown of, or interference with, an originally normal developmental process'. This implies that an external factor, such as trauma or a teratogen, has interfered with what would otherwise have been a normal developmental outcome.

Deformation

A **deformation** describes a developmental defect which results from an abnormal 'mechanical force'. Thus the intrinsic developmental programming of the embryo/fetus is normal, but abnormal pressure in the uterine cavity results in a positional defect.

Dysplasia

The term **dysplasia** refers to the 'abnormal organization of cells into tissue'. In contrast to a malformation which involves a single organ, a dysplasia usually affects a tissue or structure, such as bone, which is present in different parts of the body.

Sequence

A **sequence** refers to the cascade of defects which arise from a single known 'prior anomaly or mechanical factor'. In older literature this was sometimes referred to as an 'anomalad' or a 'complex'. The concept of a sequence was developed to account for the developmental consequences of defects such as urethral agenesis or severe spina bifida. Urethral agenesis results in

TABLE 9.5 Classification of birth defects

Defect	Examples	Causes and genetic implications
Malformation	Cleft lip/palate	Most isolated malformations show multifactorial inheritance
	Cardiac defects	
	Neural tube defects	
Disruption	Cataracts caused by congenital rubella	Caused by environmental factors. Recurrence risk is usually very low
	Limb defects caused by amniotic bands	
Deformation	Congenital hip dislocation	Caused by mechanical compression. Recurrence risk depends on cause
	Talipes (club foot)	
Dysplasia	Skeletal dysplasias, e.g. achondroplasia	Often caused by single-gene defects
Sequence	Potter (oligodramnios) sequence	Usually sporadic with low recurrence risk
Syndrome	Apert syndrome, Down syndrome, fetal alcohol syndrome	Can be chromosomal, single gene or non-genetic
Association	VATER assocation (vertebral, anal, tracheo-esophageal and renal abnormalities)	Not genetic, although cause is not known

oligohydromnios, which in turn leads to pulmonary hypoplasia and deformations such as talipes. This is referred to as the oligohydramnios or Potter sequence. A lumbar meningomyelocoele causes wasting of the lower limbs with paralysis and neurological talipes.

Syndrome

A **syndrome** describes a pattern of abnormalities which are 'thought to be pathogenetically related' and do not represent a sequence. Generally a syndrome implies that the pattern of abnormalities is consistent and recognizable. Use of the term also implies that there is a known cause, e.g. Down syndrome, although in practice this does not always apply.

Association

An **association** refers to the non-random occurrence of multiple abnormalities not known to represent a sequence or syndrome. Essentially this is a statistical concept rather than a factual explanation for why certain malformations occur together more often than would be expected by chance. Nevertheless, certain abnormalities do show unexplained but statistically significant associations.

Cause of birth defects

Despite the enormous progress which has been made in molecular genetics and developmental biology over the last 20 years, it is a sad fact that around 50% of all congenital abnormalities remain unexplained (Table 9.6). A review of the results of large numbers of birth surveys indicates that around 5–6% of all defects are caused by chromosome abnormalities, with a similar proportion being attributed to single-gene inheritance. Multifactorial inheritance constitutes the major genetic contribution and accounts for approximately 30% including most isolated 'non-syndromal' cardiac defects, cleft lip or palate, and neural tube defects (Table 6.3, p. 124). Around 5–10% are caused by environmental factors, as discussed in the following section.

The causes of the remaining 50% of congenital abnormalities remain elusive. This unknown category includes many brain abnormalities such as agenesis of the corpus callosum, limb reduction defects such as missing fingers or hands, and internal structural defects such as oesophageal atresia and diaphragmatic hernia. It is obviously very frustrating for parents of a child with one of these conditions that they cannot be given a satisfactory explanation for why it has occurred. However, they can at least be reassured that there is no suggestion of blame on anyone's part and that the sporadic nature of the abnormality means that it is unlikely to recur in a future pregnancy.

Teratogenesis

Teratogenesis can be defined as the process whereby an environmental factor interferes with the normal development of an embryo or fetus. Such a factor is referred to as a **teratogen**. These terms are derived from the Greek 'teratos', which means monster. Sadly, it is still a popular misconception amongst the general public that many abnormal babies have a monster-like appearance. Nothing could be further from the truth. It is also a popular misconception that an agent is only a teratogen if it results in a disturbance of morphogenesis. This is certainly one possible outcome but there are many others including early pregnancy loss, intra-uterine growth retardation, and long-term intellectual impairment (Table 9.7).

Basic principles of teratogenesis

Several factors can influence the potential teratogenic effect of an agent such as a drug or intrauterine infection.

Time of exposure

The most sensitive period for organogenesis is from 4 to 8 weeks after conception, although exposure up to

TABLE 9.6 Causes of birth defects	
Cause	**Proportion of all defects (%)**
Unknown	50
Genetic factors	
Chromosomal	5–6
Single gene	5–6
Multifactorial	30
Environmental factors	
Intra-uterine infection	
Teratogenic drugs	5–10
Maternal illness	

TABLE 9.7 Examples of teratogenic effects
Death of the embryo leading to early miscarriage
Defects of morphogenesis including characteristic facial features and structural abnormalities
Intra-uterine growth retardation and premature delivery
Abnormal brain development resulting in variable microcephaly, learning disability, and/or behavioural disturbance

TABLE 9.8	Critical periods for teratogenesis
Neural tube	2–5 weeks
Heart defects	3–6 weeks
Limb defects	4–8 weeks
Cleft lip/palate	4–10 weeks
Facies	4–8 weeks
Eyes	3–8 weeks
External genitalia	8–16 weeks
Brain	4 weeks–term

12 weeks or later can affect development of the central nervous system, eyes, teeth, and external genitalia (Table 9.8). Exposure within 2 weeks of conception results either in death of the embryo and early miscarriage or in survival of a normal embryo because of the totipotential nature of cells at that stage.

The concept of a critical period for teratogenicity was emphasized by experience with the drug thalidomide. This powerful sedative was prescribed to a large number of women in Europe until in 1961 an association with limb abnormalities was identified. By this time over 10 000 babies had been affected. The critical period of exposure was between 25 and 40 days after conception. The precise mechanism whereby thalidomide exerts a teratogenic effect is unknown. Possible explanations include neurotoxicity resulting in loss of segmental innervation or a destructive effect on blood vessel formation. Although the abnormality most commonly associated with thalidomide embryopathy is absence of the middle portion of the limbs, known as phocomelia, many other organs and structures were often also affected including the ears, eyes, heart, and kidneys.

Dose and duration

Most teratogenic drugs show a dose–response relationship, with teratogenic effects occurring only beyond a specific threshold dose. This may differ from pregnancy to pregnancy, depending upon the genotypes of the mother and embryo and the duration of the exposure. For example, sustained heavy consumption of alcohol during pregnancy is likely to cause the fetal alcohol syndrome, in which the affected child has a characteristic facies with short palpebral fissures and a smooth philtrum in association with microcephaly and occasional cardiac defects. More moderate alcohol ingestion, or even occasional binge drinking, while unlikely to result in the full syndrome, is believed to convey a high risk for learning difficulties and behavioural disturbance.

Known teratogens

Environmental factors with proven teratogenic effects can be considered under the headings of infections, drugs, and physical agents. Maternal illness constitutes a fourth category (Table 9.9).

Infectious agents illustrate a number of the basic principles of teratogenesis, most notably in relation to the critical time of exposure. The rubella virus readily crosses the placenta into the embryo or fetus and conveys a risk of up to 80% for causing abnormalities if the mother is infected during the first trimester. Thereafter

TABLE 9.9	Examples of known teratogens
Agent	**Teratogenic effects**
Drugs	
ACE inhibitors	Renal dysplasia
Alcohol	Fetal alcohol syndrome (see text)
Androgens	Female pseudohermaphroditism
Diethylstilboestrol	Uterine and vaginal defects
Folic acid antagonists	Facial and skeletal abnormalities
Phenytoin	Facial and digital abnormalities
Streptomycin	Deafness
Tetracycline	Dental staining
Thalidomide	Abnormalities of the limbs, face, heart and kidneys
Valproate	Fetal valproate syndrome—characteristic facies and neural tube defects
Infections	
Cytomegalovirus	Microcephaly, mental retardation and deafness
Herpes simplex	Skin vesicles and retinopathy
HIV	Immune deficiency; also growth retardation and microcephaly
Parvovirus 19	Anaemia and hydrops fetalis
Rubella	Mental retardation, deafness, retinopathy, and heart defects
Syphilis	Mental retardation, osteitis, and rhinitis
Toxoplasmosis	Mental retardation, microcephaly, and retinopathy
Varicella	Skin vesicles, microcephaly, and retinopathy
Physical	
Mechanical constraint	Deformations (see text)
Radiation	Microcephaly and mental retardation

the risk for fetal abnormality becomes much lower, although hearing loss and learning disability can result from second- or even third-trimester infection. Mass immunization has greatly reduced the incidence of rubella embryopathy.

The list of drugs with a known teratogenic effect is long, and is even longer if drugs with a suspected but unproven teratogenic effect are included. This illustrates the difficulty of confirming that a suspected low-grade teratogenic agent is indeed a teratogen, particularly if the adverse effects are non-specific. Anecdotal case reports are meaningless and large-scale prospective studies, even if they are acceptable ethically, are difficult to organize and interpret. This means that information can only be acquired by epidemiological surveys, usually retrospective, looking for recognizable patterns of dysmorphology or statistically significant increases in the incidence of specific birth defects. A teratogenic effect has been proposed for many drugs such as Debendox and caffeine, and for physical agents such as electromagnetic fields and ultrasound, but definitive proof has not been forthcoming.

Maternal illnesses compromise a small but important category of agents which can have serious adverse effects on the embryo and/or fetus. High circulating levels of phenylalanine associated with untreated or poorly treated maternal phenylketonuria (p. 211) convey a very high risk for mental retardation and structural abnormalities, including heart defects, in the fetus. Poorly controlled maternal diabetes mellitus in pregnancy is associated with a 2–3-fold increase in congenital abnormalities such as holoprosencephaly, heart defects, and sacral agenesis. The incidence of congenital abnormalities shows a close correlation with maternal levels of glucose as monitored by regular assay of the level of glycosylated haemoglobin (HbA_{1c}). Maternal systemic lupus erythematosus can cause congenital cardiac conduction defects in offspring and is also an established cause of recurrent early pregnancy loss. This is thought to be the explanation for the dreadful obstetric history of Queen Anne, who ruled England from 1702 to 1714. During her lifetime she had 17 pregnancies, with 3 stillbirths/neonatal deaths and 11 miscarriages. None of her liveborn children survived to adulthood.

Syndromes and dysmorphology

As attention has started to focus on children who are born with congenital abnormalities, it has become increasingly apparent that a significant proportion

have multiple defects. Often these are associated with unusual facial features. Such children are described as being **dysmorphic**. In addition, many have some neurodevelopmental disability manifesting as mental retardation, convulsions, movement disorders, or abnormalities of tone such as hypotonia or spasticity. Collectively these children are described as having a **syndrome**, or as being *syndromal*, although strictly speaking many actually have another pattern of birth defects such as a dysplasia or a sequence.

Syndromes can be classified on the basis of their aetiology (Table 9.10), on the body system which is predominantly involved (e.g. craniosynostosis syndromes, skeletal syndromes), or on their main clinical consequences (mental retardation syndromes, overgrowth syndromes). The investigation of children with multiple abnormalities now constitutes a significant proportion of the work of clinical geneticists and of paediatricians who care for children with disability or other long-standing medical problems. The establishment of a diagnosis is not merely an interesting, and sometimes very challenging, academic exercise. An accurate diagnosis can provide parents and carers with information about possible complications and the

TABLE 9.10	An aetiological classification of syndromes

Cause	Examples
Chromosomal (Chapter 3)	
Microscopic	Down syndrome (+21)
	Patau syndrome (+13)
	Edwards syndrome (+18)
	Turner syndrome (45,X)
	Cri-du-chat syndrome (deletion 5p)
Sub-microscopic	Williams syndrome (7q microdeletion)
	Angelman and Prader–Willi syndromes (15q microdeletions)
Single gene	Apert syndrome (p 172)
	Fragile X syndrome (p 112)
	Greig syndrome (p 174)
	Smith-Lemli-Opitz syndrome (p 175)
Teratogenic	Fetal alcohol syndrome (p 182)
	Fetal valproate syndrome (p 182)
Unknown	Russell–Silver syndrome[a]
	Proteus syndrome (p. 260)

[a] Features of Russell–Silver syndrome are short stature and limb asymmetry. Cause of most cases is unknown.

BOX 9.2 CASE HISTORY: EEC SYNDROME

A 30-year-old woman was referred to the genetics clinic before starting a family because she wished to know whether she might have children with congenital abnormalities. Her concerns were based on the fact that she had been born with abnormal hands and a cleft lip, for which she had undergone successful surgical correction in early infancy. This woman was otherwise healthy, although she had noticed that her hair and her skin tended to be very dry.

On examination, the most striking finding was that both of the woman's hands were very abnormal with absence of the middle digits (Fig. 9.15). This is known as ectrodactyly or a split hand/foot malformation. Other findings were that her hair was dry and coarse and that some parts of her skin had a scaly appearance. It was also noted that several of her nails were small and pitted, and that some of her secondary dentition had not erupted.

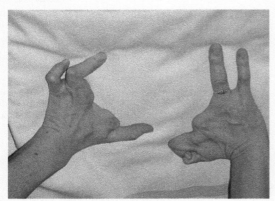

Fig. 9.15 Appearance of the hands showing ectrodactyly in a woman with EEC syndrome.

On the basis of these clinical features a diagnosis of the ectrodactyly, ectodermal dysplasia, cleft lip/palate (EEC—MIM 60429) syndrome was made. This condition shows autosomal dominant inheritance with both variable expression and reduced penetrance (p. 72). Thus it is variable in its manifestations and on occasions can actually appear to skip a generation. It is now known to be caused by mutations in a gene *TP63*, which shows close homology to the tumour suppressor gene *TP53* (p. 194), although there is no evidence for an increased risk of cancer in the EEC syndrome. The precise role of *TP63* is not clear, but it is thought to be involved in the maintenance and regeneration of basal stem cells in epithelium.

The woman was informed of her diagnosis and given full details of the spectrum of its clinical effects and its underlying pattern of inheritance. She was also offered the option of having a sample of her blood sent for mutation testing. After lengthy consideration she decided that she did not wish either to take up this offer or consider prenatal diagnostic tests in pregnancy as she had coped well with the condition and felt that her children, if affected, would do likewise. When seen again several years later, she had given birth to two children, one of whom appeared completely unaffected while the other had only dry skin and small pitted nails.

Reference

Buss PW, Hughes HE, Clarke A (1995) Twenty-four cases of the EEC syndrome: clinical presentation and management. *Journal of Medical Genetics*, **32**, 716–723.

long-term prognosis. It also ensures that accurate genetic counselling can be provided (Box 9.2). Finally, and perhaps of greatest importance, it enables parents and carers to make contact with parents of other affected children through national networks of well organized and very responsible support groups.

An outline approach to the investigation of children with multiple abnormalities can be found in Chapter 13 (p. 257).

Summary

The development of a human embryo is controlled by the complex interaction of large numbers of genes in processes such as signal transduction, cell–cell adhesion, and programmed cell death. Several gene families and gene pathways have been identified. Notable among these is the *HOX* gene family consisting of 39 genes which sequentially encode transcription factors responsible for the regulation of a cascade of downstream genes involved in various aspects of morphogenesis.

Important gene pathways include the interaction of fibroblast growth factors with receptors and sonic hedgehog signalling. Mutations in genes in these pathways result in recognized developmental abnormalities including craniosynostosis syndromes, short-limb skeletal dysplasias, digital anomalies, and holoprosencephaly. Neural crest migration and sex determination provide examples of embryogenic developmental pathways in which there is now a basic understanding of how molecular and cellular processes interact. Abnormalities of sex determination are considered under the headings of male pseudohermaphroditism (46,XY with female genitalia) and female pseudohermaphroditism (46,XX with male genitalia).

Approximately 1 in 40 of all infants has a congenital abnormality which is apparent at birth. A classification of birth defects has been proposed based on whether abnormalities are single or multiple and the underlying causal mechanism. Genetic factors contribute to 40–50% of all congenital abnormalities. The cause of most of the remaining 50% is unknown. Environmental factors (teratogens) account for less than 10%. These include drugs and chemicals, intrauterine infection, physical agents and intrinsic environmental factors associated with maternal illness.

Further reading

Epstein CJ, Erickson RP, Wynshaw-Boris A (eds) (2004) *Inborn errors of development. The molecular basis of clinical disorders of morphogenesis.* Oxford University Press, New York.

Moore KL, Persaud TVN (2003) *Before we are born. Essentials of embryology and birth defects*, 6th edn. Saunders, Philadelphia.

Spranger J, Benirschke K, Hall JG *et al.* (1982) Errors of morphogenesis: concepts and terms. Recommendations of an international working group. *Journal of Pediatrics*, **100**, 160–165.

Strachan T, Lindsay S, Wilson DI (1997) *Molecular genetics of early development.* Bios Scientific Publishers, Oxford.

Wolpert L (2002) *Principles of development*, 2nd edn. Oxford University Press, Oxford.

Multiple choice questions

1 Which of the following associations between a developmental gene and a clinical condition are valid?

 (a) a *HOX* mutation and digital abnormalities

 (b) a *PAX* mutation and ocular abnormalities

 (c) an *FGFR* mutation and short stature

 (d) an *SHH* mutation and Hirschsprung disease

 (e) a *SOX9* mutation and sex reversal

2 A baby born with a squashed face and talipes because of large fibroids in its mother's uterus would be classified as having

 (a) a malformation

 (b) a deformation

 (c) a dysplasia

 (d) a sequence

 (e) a syndrome

3 Proven teratogenic agents include

 (a) anticonvulsants such as phenytoin and valproate

 (b) folic acid

 (c) alcohol

 (d) ultrasound

 (e) toxoplasmosis

4 Sex determination in the human

 (a) is complete by 6 weeks after conception

 (b) involves genes on the autosomes and the sex chromosomes

 (c) can be influenced by maternal ingestion of drugs

 (d) is abnormal in congenital adrenal hyperplasia because of lack of testosterone

 (e) is abnormal in Turner syndrome because of masculinization of the external genitalia

5 Examination of a newborn infant reveals that the genitalia are ambiguous (pseudohermaphroditism). It would be correct to tell the parents that

 (a) this could be due to the XYY syndrome

 (b) urgent chromosome analysis is indicated

 (c) their baby can go home with an out-patient appointment for 6 weeks

 (d) the recurrence risk for future pregnancies is negligible

 (e) this abnormality would be classified as an association

Answers

1 (a) true—mutations in *HOXA13* and *HOXD13* cause digital abnormalities

(b) true—mutations in *PAX2* and *PAX6* cause ocular abnormalities

(c) true—mutations in *FGFR3* cause achondroplasia

(d) false—mutations in *SSH* cause holoprosencephaly

(e) true mutations in *SOX9* cause campomelic dysplasia in which males can show sex reversal

2 (b) is correct—the features described are caused by an abnormal mechanical force

3 (a) true—most anticonvulsants have teratogenic effects

(b) false—folic acid *antagonists* are teratogens whereas folic acid is recommended in pregnancy to reduce the risk of a neural tube defect

(c) true—alcohol, particularly when ingested in high quantities, adversely affects brain development in the fetus

(d) false—diagnostic ultrasound has not been proven to be teratogenic

(e) true—toxoplasmosis conveys a high risk for teratogenesis

4 (a) false—sex determination does not commence until around 6 weeks gestation

(b) true—many genes are involved on both the autosomes and the sex chromosomes

(c) true—maternal ingestion of androgens can have a virilizing effect on a female fetus

(d) false—the circulating levels of androgens are *raised* in congenital adrenal hyperplasia

(e) false—the external genitalia are normal in Turner syndrome

5 (a) false—the external genitalia are normal in the XYY syndrome

(b) true—chromosome analysis should be undertaken as soon as possible

(c) false—the baby should not be discharged until a diagnosis of congenital adrenal hyperplasia has been ruled out as this could lead to metabolic collapse in the first weeks of life

(d) false—many forms of pseudohermaphroditism have a genetic basis and could recur

(e) false—an association is the non-random occurrence of two or more abnormalities. Ambiguous genitalia could represent a malformation or a dysplasia

Genes and cancer

Cancer is common, with a lifetime general population incidence in western countries of approximately 1 in 3. Between 20% and 25% of all deaths in these countries are caused by cancer, and the contribution of cancer to mortality in the developed nations is likely to increase as methods are found for preventing and treating other common causes of death such as cardiovascular disease and dementia. Most of us can identify at least one close relative who died as a result of cancer, so it is not surprising that around one third of all referrals to genetics clinics are prompted by a family history of cancer.

Less than 5% of cancer is inherited in a single-gene fashion, but, paradoxically, most forms of cancer have an underlying genetic basis as they result from the accumulation of a small number of somatic mutations in a single cell. These mutations occur in various genes involved in the regulation of the cell cycle and cell growth. The factors that induce these changes include mutagenic agents such as ionizing radiation, carcinogens such as cigarette smoke and asbestos, and, rarely, viruses such as HTLV-1, the retrovirus which causes T-cell leukaemia in adults.

The genes which, when mutated, cause cancer can be considered under the headings of oncogenes, tumour suppressor genes, and DNA repair genes. When an individual inherits a mutation in one of these genes in the germline, then that individual is likely to have an increased susceptibility to develop cancer as a consequence of an inherited *cancer-predisposing syndrome*. As already indicated, these account for less than 5% of all cancers. Most malignant tumours result from the accumulation of several mutations in a cell through a process referred to as the *multistage* or *multi-step* genesis of cancer.

Oncogenes

Genes which are involved in the promotion of cell division and proliferation are known as **proto-oncogenes**. An **oncogene** is a mutant form of a proto-oncogene, which results in increased stimulation of these normal cellular processes. Mutations which cause oncogenes exert a gain-of-function effect (p. 13), which is usually apparent at a cellular level in the heterozygous state. Thus at a cellular level these mutations act in a dominant fashion.

Oncogenes were first identified through the study of RNA tumour viruses, also known as *retroviral* tumour viruses as they contain a gene which encodes a reverse transcriptase that converts RNA to DNA. A subgroup of RNA tumour viruses, known as *acute transforming viruses*, can cause rapid tumour development in animals. This is achieved by the acquisition of normal host genes into the viral genome through recombination. This process of viral-mediated gene transfer is known as **transduction** (p. 233). When incorporated into the viral genome, the host gene shows increased activity which results in increased cell growth and tumour formation. The infected cells are said to have undergone **transformation**. The gene derived from the host cell is referred to as a viral oncogene, or *v-onc*. The version present in the host genome is known as a cellular oncogene or *c-onc*.

Over 100 cellular oncogenes which can cause tumours in animals and/or humans have now been identified. These can be classified according to the type of tumour that they cause, or on the basis of their role in the regulation of cell proliferation. As shown in Fig. 10.1, oncogenes encode four main groups of regulatory proteins: growth factors, cell-surface growth factor receptors, intracellular mediators of signal transduction, and nuclear proteins including transcription factors. Activating gain-of-function mutations in any of these factors can lead to an increase in the rate of cell proliferation. Most known proto-oncogenes are components of signal transduction pathways (p. 168). Activation results in deregulation of the relevant pathway, leading to autonomous cell proliferation and survival.

In addition, proto-oncogenes have been shown to be involved in other processes including regulation of the cell cycle and programmed cell death (apoptosis—p. 168). The *BCL2* gene family (p. 169), which regulates apoptosis, was first identified at a translocation breakpoint in follicular lymphomas. Proto-oncogenes are also involved in the regulation of angiogenesis and cell adhesion, so that oncogenic mutations in these regulatory genes can promote tumour growth, migration, and metastasis.

Activating mutational mechanisms in oncogenesis

Although oncogenes were first identified as mutated proto-oncogenes in retroviruses, most oncogenic mutations occur spontaneously in somatic cells and very few human tumours are caused by viral transformation. Three common activating mutational mechanisms are recognized: point mutations, amplification, and translocation (Table 10.1).

Activation by point mutation

A family of genes known as RAS, because they were first identified as a cause of <u>ra</u>t <u>s</u>arcomas, play a key role in G-protein receptor mediated signal transduction (Fig. 10.2). Activating point mutations have been identified in *RAS* genes in several tumours, including pancreatic, colorectal, lung, and bladder cancer. The first oncogene identified in human cancer was a mutated *H-RAS* allele detected in a bladder cancer cell line by DNA transfection studies. **Transfection** is the term

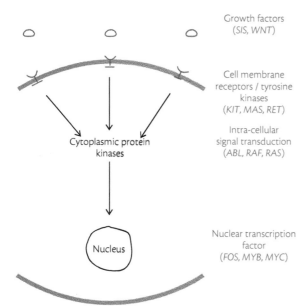

Growth factors
(*SIS, WNT*)

Cell membrane receptors / tyrosine kinases
(*KIT, MAS, RET*)

Intra-cellular signal transduction
(*ABL, RAF, RAS*)

Cytoplasmic protein kinases

Nucleus

Nuclear transcription factor
(*FOS, MYB, MYC*)

Fig. 10.1 Examples of known oncogenes and their roles in cell signalling.

TABLE 10.1 Examples of oncogenes involved in human cancer

Activating mechanism	Oncogene	Protein function	Cancer
Point mutations	K-RAS	GTPase	Pancreatic, colorectal, lung
	H-RAS	GTPase	Bladder
	N-RAS	GTPase	Myeloid leukaemia
Amplification	NMYC	Transcription factor	Neuroblastoma
	ERBB2	Growth factor receptor	Breast and ovarian
	TERC	Telomerase	Many tumours
Translocation	BCR-ABL	Tyrosine kinase	Acute lymphocytic and chronic myeloid leukaemia
	RET	Tyrosine kinase	Papillary thyroid cancer
	PAX3	Transcription regulator	Alveolar rhabdomyosarcoma
	CMYC	Transcription regulator	Burkitt's lymphoma

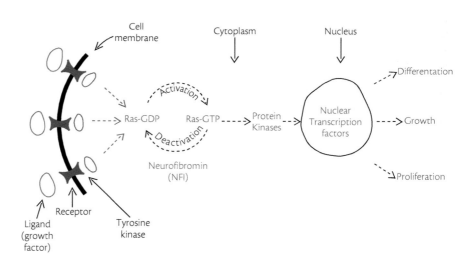

Fig. 10.2 The G-protein receptor *RAS* signalling pathway.

used to describe the transfer of DNA from one cell to another by non-viral methods.

Normally in G-protein receptor signalling, GTP binds to the RAS protein forming a RAS–GTP complex. This is converted to an inactive RAS–GDP by intrinsic GTPase. The mutant RAS proteins found in several tumours have reduced intrinsic GTPase activity leading to increased cell signalling.

Activation by gene amplification

Although the underlying mechanism is not fully understood, many cancer cells contain multiple copies of structurally normal proto-oncogenes. Several hundred copies of the gene may be present, either as insertions within a chromosome known as **homogeneously staining regions** or as very small extra chromosomes known as **double minutes** (Fig. 10.3). Examples of common gene amplifications associated with tumour development include *NMYC* in neuroblastoma and *ERBB2* in breast cancer (see Table 10.1). *NMYC* encodes a transcription regulator and *ERBB2* encodes a growth factor receptor. The gene which encodes telomerase often shows an increase in copy number in various human tumours. Telomerase normally maintains DNA synthesis at the ends of each chromosome, and increased telomerase activity is associated with cellular immor-

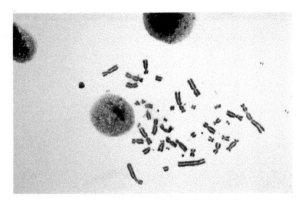

Fig. 10.3 Metaphase spread from a child with neuroblastoma showing several additional tiny extra chromosomes known as double minutes. Courtesy of Mrs Jean Sadler, Department of Cytogenetics, Leicester Royal Infirmary, Leicester.

tality and tumour progression. Telomerase can be activated by other mutational mechanisms including increased expression of other oncogenic transcription factors. The level of telomerase activity in cells is sometimes used as an indication of their malignant potential.

Oncogene amplification in tumours can provide useful prognostic information. For example, *NMYC* amplification is usually found only in advanced tumours and conveys a poor prognosis. Amplified segments of DNA are generally detected by comparative genomic hybridization (CGH—p. 30) using either conventional analysis or newer microarray-based methods (p. 21).

Activation by translocation

Chromosome abnormalities including reciprocal translocations and aneuploidy are often present in leukaemias and lymphomas. They also occur, but less frequently, in solid tumours. Initially it was thought that these might be secondary to the underlying onco-genic process, but it is now known that they play an important causal role.

Translocations can activate a proto-oncogene in two ways. The first involves the creation of a novel chimaeric (fusion) gene. The second involves the relocation of a relatively inert gene into a region of the genome where it is much more actively expressed.

Creation of a novel chimaeric gene

Chromosome analysis in patients with chronic myeloid leukaemia reveals that 90% have a balanced reciprocal translocation involving chromosomes 9 and 22 (Fig. 10.4). The small derivative number 22 chromosome is known as the *Philadelphia* chromosome. The breakpoint on chromosome 9 involves a proto-oncogene known as *ABL*. As a result of the translocation the 3′ region of *ABL* is fused to the 5′ region of a gene on chromosome 22 known as *BCR* (breakpoint cluster region) (Fig. 10.5). The new *BCR–ABL* fusion gene, which can be identified by FISH (Fig. 10.6), encodes an activated tyrosine kinase with the ability to transform white blood cell precursors in the bone marrow.

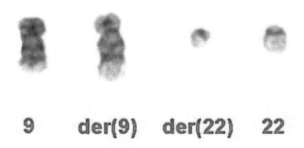

9   der(9)   der(22)   22

Fig. 10.4 Partial karyotype showing a derivative number 22 (Philadelphia) chromosome. Courtesy of Mrs Karen Marshall, Department of Cytogenetics, Leicester Royal Infirmary, Leicester.

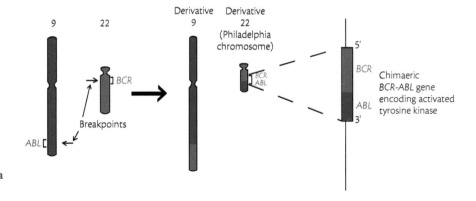

Fig. 10.5 Diagrammatic representation of the 9:22 translocation seen in chronic myeloid leukaemia and how this leads to the formation of a chimaeric *BCR-ABL* oncogene.

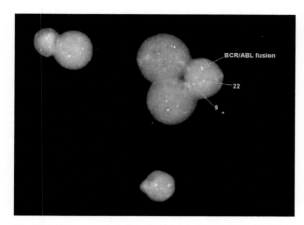

Fig. 10.6 Demonstration of the *BCR-ABL* fusion oncogene by FISH. Courtesy of Mrs Karen Marshall, Department of Cytogenetics, Leicester Royal Infirmary, Leicester.

Activation by relocation

Approximately 80% of children with a tumour known as Burkitt's lymphoma, which is common in parts of Africa, have a reciprocal translocation involving chromosome 8 and either chromosome 2, 14 or 22 (Fig. 10.7). In each of these translocations the *CMYC* proto-oncogene is moved to a region of the genome where immunoglobulin genes are actively transcribed. In this new setting *CMYC* is overexpressed. This is a key step in the development of the tumour. The initial step is thought to be malaria-induced B-lymphocyte hyperplasia, which increases the target cell population for a translocation. Infection with the Epstein–Barr virus results in further B-cell proliferation leading to tumour formation.

> **Key Point**
>
> Oncogenes are mutated proto-oncogenes which exert a dominant gain-of-function effect leading to promotion of cell division.

Tumour suppressor genes

At a simplistic level, cell division and proliferation can be viewed as representing a balance between the actions of proto-oncogenes, which exert a positive effect, and a group of genes which exert a negative regulatory effect, collectively known as **tumour suppressor** genes. The existence of tumour suppressor genes was first suspected on the basis of cell hybrid studies, when it was noted that if malignant cells were fused with normal cells they lost their malignant properties. These studies, combined with Knudson's pioneering observations on the genetics of retinoblastoma (Box 10.1), supported the concept that some genes or chromosome regions can normally prevent tumour development.

Genetic basis of retinoblastoma

Retinoblastoma is a rare malignant embryonal tumour which arises in the retina before the age of 5 years in approximately 1 in 20 000 children. In around 60% of cases the tumour arises as a sporadic event with no relevant previous family history. In these sporadic, or isolated, cases only one eye is affected and the mean age at presentation is 24 months. In familial cases, or in cases with tumours affecting both eyes, the mean age at presentation is 8 months. Early treatment is curative, although survivors of familial cases face a risk of up to 25% for developing another malignant tumour, most commonly an osteosarcoma, in later life (Table 10.2).

Research following on from the proposal by Knudson of the 'two hit' hypothesis (see Box 10.1) has confirmed the validity of his observations. The gene which, when mutated, causes retinoblastoma is known as *RB1*. It was localized to chromosome 13q14 through the discovery that in about 5% of affected children this chromosome region is deleted. Polymorphic molecular markers from this region were shown to segregate with the disease in familial cases. In all such families the disorder showed

Fig. 10.7 Diagrammatic representation of the 8;14 translocation seen in many cases of Burkitt lymphoma. *CMYC* is the proto-oncogene which is activated by translocation to a region of increased transcriptional activity, in this situation brought about by the regulatory activity of the immunoglobulin heavy chain gene *IGH*.

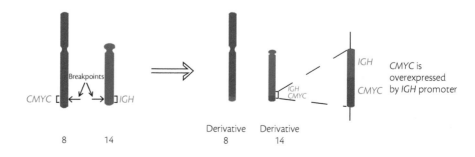

BOX 10.1 LANDMARK PUBLICATIONS: RETINOBLASTOMA AND THE TWO-HIT HYPOTHESIS

Retinoblastoma is generally regarded as the classic example of a tumour caused by a two-mutation mechanism. The possibility that it could be caused by a two-stage process such as initiation and promotion had been considered for several years, but it was not until Alfred Knudson's publication in 1971 that formal proof for a two-hit mechanism was forthcoming.

Knudson's proposal was based on a detailed statistical analysis of 48 affected children treated at the University of Texas, together with a series of 52 published cases from the UK. He observed that the proportions of unaffected, unilateral, and bilateral cases amongst probable carriers of a germline mutation were consistent with a Poisson distribution and calculated that the mean number of tumours formed by a carrier is three. This enabled him to estimate that germline and somatic mutation rates were approximately equal and on this basis he proposed that a retinoblastoma is caused by two mutational events, one germline and one somatic, or, in the non-hereditary form, two somatic events.

Confirmation of the validity of Knudson's hypothesis was provided 12 years later when Cavenee et al. demonstrated loss of heterozygosity for DNA markers closely linked to the (then unidentified) retinoblastoma locus in several retinoblastoma tumours. This loss of heterozygosity was brought about by various mechanisms including non-disjunction with reduplication and mitotic recombination (Fig. 10.8). They proposed that retinoblastoma results from the 'development of homozygosity for recessive mutant alleles at the retinoblastoma locus'. An alternative possibility is that the second locus is deleted through a non-disjunction or deletional event leading to hemizygosity for the mutant allele. In subsequent studies it was shown that in tumour tissue it is always the normal allele that is lost.

As well as providing an explanation for how many tumours develop, these observations indicated that a search for loss of heterozygosity in tumour tissue would be a very effective way to identify the loci of tumour suppressor genes. Conversely, several tumour suppressor genes identified through the study of rare familial cancer syndromes have proved to be important factors in the multistep generation of sporadic cancers.

References

Knudson AG (1971) Mutations and cancer: statistical study of retinoblastoma. *Proceedings of the National Academy of Sciences of the USA*, **68**, 820–823.
Cavenee WK, Dryja TP, Phillips RA (1983) Expression of recessive alleles by chromosomal mechanisms in retinoblastoma. *Nature*, **305**, 779–784.

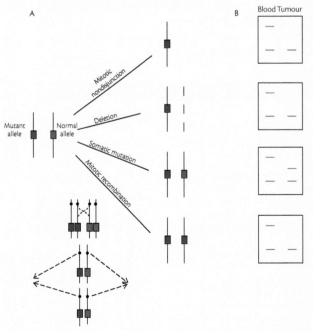

Fig. 10.8 (A) Diagrammatic illustration of mechanisms which can cause a second 'hit' resulting in development of a retinoblastoma in a child who inherits a germline mutation (shown in red). (B) The appearance on a gel of markers for the normal and mutant alleles in blood and tumour tissue.

TABLE 10.2 Characteristics of inherited (germline) and sporadic retinoblastoma

	Inherited	Sporadic
Family history	Often positive, but affected child may represent a new mutation	No affected relatives
Average age at presentation	8 months	24 months
Tumour distribution	Multifocal and bilateral	Single and unilateral
Increased risk for other primary tumours	Osteosarcoma, soft tissue sarcoma, malignant melanoma, bladder cancer	None
Mutational origin	One germline and one somatic	Both somatic

autosomal dominant inheritance with reduced penetrance of 80–90%.

After the isolation of *RB1*, it was shown by comparing DNA from blood and from the tumours of affected familial cases that the normal copy of *RB1* was often missing from tumour tissue. This is consistent with the first hit being provided by the inherited germline mutation, followed by loss of the normal *RB1* allele providing the second hit necessary for tumour development (see Fig. 10.8). This loss of the normal allele is the event which triggers abnormal proliferation in a cell which is already susceptible due to the presence of the inherited mutation. Loss of the normal allele in a tumour is referred to a **loss of heterozygosity** (LOH). A search for LOH in tumour tissue as compared with blood is a widely used approach for identifying tumour suppression genes. This is usually undertaken by comparative genomic hybridization (p. 30).

In sporadic retinoblastoma both of the mutations occur as somatic events. These occur only rarely, which explains why usually only one eye is affected and the age of onset is later than in familial cases (see Table 10.2). Reduced penetrance in familial retinoblastoma is explained by some heterozygotes simply being fortunate in that no somatic mutational event occurs in a retinal cell during the critical first few years of life when the retinal cells are proliferating.

Causal mechanisms for the second hit

Possible explanations for the second hit are illustrated in Fig. 10.8. All of these exert a loss-of-function effect. LOH can be generated by a deletion which removes the normal allele or by mitotic non-disjunction with failure of normal separation of the sister chromatids in anaphase.

Possible mutational mechanisms which would not be discernible as LOH but which would generate a second hit include simple point mutations, mitotic recombination, and gene conversion. A **mitotic recombination** is a rare event in which non-sister homologous chromosomes exchange material before chromatid separation so that there is a theoretical possibility that a daughter cell will become homozygous for the mutant allele (Fig. 10.8). A **gene conversion** is an equally rare event in which one member of a pair of heterozygous alleles becomes identical to its homologue, possibly because of formation of a heteroduplex between the two alleles with subsequent 'repair' of the mismatched bases.

Recently it has been recognized that the second hit can also be caused by a non-structural change involving silencing of expression by DNA methylation. This is associated with a closed chromatin configuration which prevents binding by transcription factors. Hypermethylation of CpG dinucleotides in the promoter regions of tumour suppressor genes is a common finding in many tumours. This is referred to as an **epigenetic** change because although gene function is altered there is no structural change within the gene.

Dominant inheritance but a recessive mechanism

The issue of whether familial retinoblastoma shows dominant or recessive inheritance can cause considerable confusion. In families the disorder shows vertical transmission from generation to generation, which is typical of autosomal dominant inheritance. Yet at a cellular level both alleles have to be abnormal for a tumour to develop, so at this level the mechanism is recessive. This apparent contradiction in terms applies to all of the inherited conditions caused by mutations in tumour suppressor genes, as discussed later in this chapter. For the purposes of genetic counselling, and answering multiple-choice questions in examinations, the mode of inheritance is dominant even though oncogenesis at a cellular level involves a recessive mechanism. This represents a fundamental difference between oncogenes and tumour suppressor genes (see Table 10.3).

TABLE 10.3 Distinguishing features of oncogenes and tumour suppressor genes

	Oncogenes	Tumour suppressor genes
Mutational effect	Gain-of-function	Loss-of-function
Cellular genetic mechanism	Dominant	Recessive
Normal biological function	Promotion of cell division and proliferation	Suppression of cell division; arrest of cell cycle; promotion of apoptosis

Key Point

Tumour suppressor genes normally exert a negative regulatory effect on cell division and proliferation. Loss-of-function mutations in both alleles result in uncontrolled cell division and tumour formation

Cellular role of tumour suppressor genes

Tumour suppressor genes are generally involved in the same processes as proto-oncogenes but with opposing effects. Because of their roles in arresting the cell cycle to allow time for repair of damaged DNA many of these genes are described as *gatekeepers*. The genes that are actively involved in DNA repair, as discussed in the next section, are in turn referred to as *caretakers*. Mutations in the same tumour suppressor gene are often found in different sporadic tumours, indicating that these genes play important general roles in the regulation of cell proliferation. Inherited germline mutations in several tumour suppressor genes result in specific inherited cancer syndromes, as outlined later in this chapter.

Retinoblastoma (RB1) gene

RB1 is expressed in many tissues, and mutations in *RB1* have been identified not just in retinoblastomas, but also in osteosarcomas, soft tissue sarcomas, small-cell carcinomas of the lung, and breast, bladder, pancreatic and prostate cancers. As already indicated, patients with germline mutations in *RB1* who survive after treatment for a retinoblastoma in childhood are known to have a significant risk for developing osteosarcoma in later life. For reasons which are not understood, these individuals do not seem to be at increased risk for developing some of the other types of cancer in which *RB1* mutations are found.

RB1 encodes a protein which, depending upon its level of phosphorylation, either blocks or allows cell cycle progression from G_1 into the S phase. This is achieved by binding to transcription factors which promote DNA synthesis. Several rare DNA tumour viruses which can cause cancer, such as the human papilloma virus, which causes warts and urogenital cancer, act by encoding an oncogenic protein which binds to the *RB1*-encoded protein, thereby preventing its normal gatekeeping role.

TP53 tumour suppressor gene

TP53 on chromosome 17 encodes a protein p53, so named because it has a molecular weight of 53 000. Loss of function mutations in *TP53* are the most common genetic changes observed in human cancer, with both alleles inactivated in over 50% of all tumours. This key role in preventing oncogenesis has led to *TP53* being labelled as the 'guardian of the genome'. Germline mutations in *TP53* cause the inherited cancer-predisposing condition known as the Li–Fraumeni syndrome (p. 198).

The encoded protein, p53, normally plays two key roles in preventing oncogenesis (Fig. 10.9). The first involves the activation by transcription regulation of genes which arrest the cell cycle at the G_1–S checkpoint to allow repair of DNA damage. The second is to trigger apoptosis, by promoting the *BCL-2*-associated pathway (p. 168), when extensive DNA damage has occurred. The importance of p53 in cancer has prompted a search for drugs that could be used to increase *TP53* expression in malignant cells.

NF1 tumour suppressor gene

NF1 is a large gene on chromosome 17 which encodes a 300-kDa protein known as neurofibromin. Somatic loss-of-function mutations have been found in *NF1* in several tumours including colorectal cancer, melanoma, and neuroblastoma. Germline mutations in *NF1* cause type 1 neurofibromatosis (p. 198).

Neurofibromin shows close homology with proteins which activate GTPase in the RAS proteins. As already discussed (p. 189), oncogenic mutant RAS proteins have reduced intrinsic GTPase activity. Normal levels of neurofibromin stimulate GTPase activity and thereby prevent over-expression of RAS signalling proteins. Loss-of-function mutations in *NF1* probably contribute to tumour formation by increasing RAS signalling and promoting cell proliferation (see Fig. 10.2).

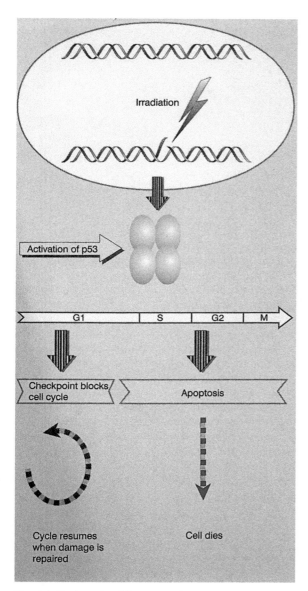

Fig. 10.9 The key roles of the p53 protein in preventing oncogenesis. Reproduced with permission from Lewin B (2000) *Genes VII*. Oxford University Press, New York.

APC tumour suppressor gene

The *APC* gene on chromosome 5 encodes a 300-kDa protein which is localized in the cell cytoplasm. Somatic mutations in *APC* are found in over 80% of colorectal cancers. Germline mutations in *APC* cause the inherited condition known as adenomatous polyposis coli or familial adenomatous polyposis (p. 198).

Normally the *APC*-encoded protein regulates the intracellular level of β-catenin, which, when present in the free state, binds with intranuclear transcription factors to activate transcription. Loss-of-function mutations in *APC* lead to increased levels of free β-catenin, which in turn leads to increased transcription of growth-promoting genes including *CMYC*, the gene which is activated by translocation in Burkitt's lymphoma.

DNA repair genes

As discussed in Chapter 3 (p. 66), several complex systems exist to repair damage to DNA sustained either by extrinsic mutagenic agents, such as ionizing radiation or ultraviolet light, or by intrinsic errors in DNA replication. These systems utilize several families of enzymes to unwind the double helix (helicases), remove the abnormal DNA (endonucleases and exonucleases), synthesize new DNA (polymerases), and join the new strands together (ligases). Errors in any of these systems can lead to the accumulation of mutations in oncogenes and tumour suppressor genes, leading in turn to tumour formation. Relevant processes known to be associated with specific cancer-predisposing syndromes include nucleotide excision, the repair of double strand breaks, and the repair of mismatched base pairs. In some respects DNA repair genes can be viewed as a subgroup of tumour suppressor genes as they share the same loss-of-function two-hit mechanism.

Impaired nucleotide excision

Nucleotide excision repair involves a complex pathway commencing with chromatin unwinding, followed by excision of nucleotides in the damaged strand, and completed by synthesis and integration of new DNA by DNA polymerase and DNA ligase. Errors in this system cause xeroderma pigmentosum, which is associated with the development of skin tumours following exposure to ultraviolet light (p. 66). Several different forms exist. All show autosomal recessive inheritance.

Impaired double-strand breakage repair

The repair of double-strand breaks, as can be caused by ionizing radiation, involves a complicated and incompletely understood process, whereby the cell cycle is arrested to allow a multiprotein complex to initiate and complete the repair process. The multiprotein complex includes several proteins, including ATM, BRCA1, and BRCA2, which are associated with known cancer-predisposing syndromes. ATM is the protein encoded by the gene, which, when mutated, causes ataxia telangiectasia (p. 66) in homozygotes. Heterozygotes have a small increased risk for developing breast cancer. ATM plays an important role in arresting the cell cycle and in activating the double-strand repair complex.

BRCA1 and BRCA2 are integral components of the multiprotein complex which repairs double-strand breaks by various processes including homologous recombination and binding to the ends of the broken DNA fragments. Heterozygous mutations in the genes which encode BRCA1 and BRCA2 cause hereditary breast and ovarian cancer as discussed later in this chapter. Homozygous mutations in *BRCA2* cause a rare form of Fanconi anaemia (p. 66).

Impaired mismatched base-pair repair

Mismatched base pairs arise as a result of errors in DNA replication. A multimeric protein complex consisting of proteins encoded by up to six different genes interacts with the strand in which the mismatched base is detected. A segment of DNA surrounding the mismatch is excised and the excised tract is then repaired.

Errors in the mismatch repair system lead to a marked increase of up to 1000-fold in the number of mutations within a cell. Microsatellite repeats are particularly susceptible to mutation through mispairing or slippage (p. 11). Thus errors in the mismatch repair system lead to a dramatic increase in the number of new microsatellites present, in addition to mutations in oncogenes and tumour suppressor genes. A marked increase in the number of microsatellites present within a tumour is referred to as **microsatellite instability** and the tumour is said to be **replication error positive**.

Many tumours show a mild increase in the number of microsatellites because of hypermethylation of one or more of the genes encoding the multimeric repair complex. However, a germline mutation in one of these genes causes a marked increase in the number of microsatellites. Heterozygous mutations in these genes cause hereditary non-polyposis colon cancer (HNPCC). Although this accounts for only 5% of all colorectal cancers it is an important cause of hereditary colon cancer, as discussed later in this chapter.

> **Key Point**
>
> Mutations in DNA repair genes (caretakers) predispose to tumour formation by increasing the rate of mutations in genes which are normally involved in the regulation of cell division.

Sporadic cancer—a multistep genetic process

The prevailing view is that most sporadic tumours arise as the result of a series of mutations in proto-oncogenes and tumour suppressor genes which influence the normal cell proliferation regulatory processes. As a result, cell division proceeds autonomously. Other characteristics of tumour cells are that they achieve immortality, possibly through amplification of the gene which encodes telomerase, and avoid apoptosis as a result of mutations in genes such as *TP53*. Malignancy is determined by the ability to invade adjacent tissue and metastasize. This is mediated by mutations in genes such as *NF2* and *VHL* which regulate processes such as cell–cell adhesion and angiogenesis.

This underlying theme has been best elucidated for colon cancer. A hyperproliferative focus develops in normal colonic epithelium as a result of a mutation in a proto-oncogene such as *K-RAS* or homozygous mutations in the *APC* tumour suppressor gene. Progression to a benign tumour known as an adenoma occurs as a result of accumulating mutations in other tumour suppressor genes and through alterations in the methylation status of regulatory proto-oncogenes. Subsequent mutations and/or loss of heterozygosity in *TP53* convey malignant potential. The whole process takes place over several years, which explains why screening on an annual basis is viewed as effective prophylaxis for colorectal cancer (p. 204).

Familial cancer syndromes

Reference has already been made to several inherited conditions caused by germline mutations in oncogenes, tumour suppressor genes and DNA repair genes. Although most of these conditions are rare, accounting for less than 5% of all cancer cases, their recognition is important, because of the high risks they convey for causing multiple tumours in affected individuals and also because of the importance of offering appropriate screening to other family members.

The 10 most commonly encountered familial cancer syndromes are listed in Table 10.4. Surprisingly few of these are caused by mutations in proto-oncogenes. The explanation for this is unknown. One possibility is that the effects of gain-of-function mutations in proto-oncogenes on embryonic development are so severe that early pregnancy loss usually ensues. It is now recognized that mutations in many cancer-related genes can also cause developmental syndromes, as discussed later in the context of *RET* (p. 198).

TABLE 10.4 Familial cancer syndromes

Syndrome	MIM	Gene	Locus	Tumours
Activated proto-oncogene				
Multiple endocrine adenomatosis 2	171400, 162300, 155240	RET	10q11	Medullary thyroid carcinoma, parathyroid adenomas, phaeochromocytoma
Tumour suppressors				
Familial adenomatous polyposis	175100	APC	5q21	Colorectal cancer, duodenal and gastric cancer, desmoid tumours, osteomas of the jaw
Li–Fraumeni syndrome	151623	TP53	17p13	Brain tumours, breast cancer, leukaemia, sarcomas
Neurofibromatosis 1	162200	NF1	17q11	Neurofibromas and CNS tumours
Neurofibromatosis 2	101000	NF2	22q12	CNS tumours and vestibular schwannomas
Retinoblastoma	180200	RB1	13q14	Retinoblastoma and osteosarcoma
Von Hippel–Lindau syndrome	193300	VHL	3p25	Retinal angiomas, CNS haemangioblastomas, phaeochromocytoma
DNA repair genes				
Familial breast cancer 1	113705	BRCA1	17q21	Breast and ovarian cancer
Familial breast cancer 2	600185	BRCA2	13q12	Breast and ovarian cancer; also breast cancer in men
Hereditary non-polyposis colon cancer	114500			Colorectal, endometrial, gastric, biliary and urinary tract cancer
	Common	MLH1	3p21	
		MSH2	2p16	
	Rare	MSH3	5q11	
		MSH6	2p16	
		PMS2	7p22	

Note that all of these conditions show autosomal dominant inheritance. Cancer-predisposing syndromes which are caused by mutations in DNA repair genes and which are associated with chromosome breakage are indicated in Table 3.7 (p. 66).

The familial cancer syndromes share a number of characteristics (Table 10.5), which should raise suspicion of an underlying inherited cancer susceptibility syndrome. Enquiry should always be made about factors such as early age of onset, bilateral involvement, and tumours in different systems when assessing a family history for inherited cancer susceptibility.

TABLE 10.5 Characteristics of a familial cancer syndrome

Early age of onset
Bilateral involvement or multifocal tumours
Two rare tumours in one person
Two or more relatives with the same rare tumour
Several relatives with a common cancer (e.g. breast or colon) or with related cancers (e.g. breast and ovary, colon and endometrial)

Multiple endocrine adenomatosis type 2

Multiple endocrine adenomatosis type 2 (MEN2) refers to three rare conditions which all show autosomal dominant inheritance and are caused by mutations in the *RET* oncogene.

◆ In MEN2A presentation is usually in early adult life with medullary thyroid carcinoma, phaeochromocytoma and/or hyperparathyroidism due to a parathyroid adenoma.

◆ MEN2B often presents in late childhood with a similar spectrum of tumours, in association with a tall thin body habitus similar to that seen in Marfan syndrome (p. 106).

- In the third type of MEN2, involvement is limited to medullary thyroid carcinoma. This is sometimes referred to as familial medullary thyroid carcinoma.

Specific gain-of-function mutations in *RET* are associated with each of these presentations. *RET* (<u>re</u>arranged during <u>t</u>ransfection) encodes a membrane-bound receptor tyrosine kinase, which on binding to a ligand, undergoes a conformational change that activates several intracellular downstream signalling pathways in neuro-endocrine tissues.

Note that *RET* is an example of a gene in which a wide spectrum of mutations can cause diverse phenotypes (Table 10.6). Somatic activation through the formation of a chimaeric gene, made up of the intracellular tyrosine kinase domain of *RET* and other genes normally expressed in the thyroid gland, can result in papillary thyroid carcinoma. This accounts for a proportion of the increased incidence of malignancy seen in children who were exposed to radiation following the accident at Chernobyl in 1986. Germline mutations with a loss-of-function effect are found in a large proportion of children with Hirschsprung disease (p. 127).

Thus *RET* is an example of a gene in which different mutations can cause either developmental birth defects or malignancy in later life. It is somewhat ironic that the genes which contribute to the development of a human embryo can mutate to bring that embryo's life to a premature conclusion.

TABLE 10.6 Disorders caused by mutations in the *RET* proto-oncogene

Type of mutation	Disorder caused
Gain-of-function	
Point mutation	Multiple endocrine adenomatosis types 2A, 2B and familial medullary thyroid carcinoma
Rearrangement	Papillary thyroid carcinoma
Loss-of-function	
Point mutations and deletions	Hirschsprung disease
	Central hypoventilation syndrome[a]

[a] In this condition there is a tendency to hypoventilate when asleep. This is also referred to as Ondine's curse. Ondine was a mythical sea nymph who sacrificed her immortality for a lover who subsequently proved to be unfaithful. In her anger she imposed a curse upon him that he would stop breathing and die as soon as he fell asleep.

Familial adenomatous polyposis

Familial adenomatous polyposis, also known as *adenomatous polyposis coli*, is characterized by the development of hundreds of small adenomatous polyps in the colon and rectum with onset in early adult life. The overall incidence is approximately 1 in 10 000. In almost all cases at least one of the polyps will become malignant, so surgical removal of the colon is recommended for all confirmed cases. In a variant form of familial adenomatous polyposis, known as Gardner syndrome, other manifestations include osteomas of the mandible, multiple sebaceous cysts, and desmoid tumours which arise from the small bowel mesentery or abdominal wall.

Familial adenomatous polyposis and Gardner syndrome are caused by germline inactivating mutations in *APC* (p. 195). The adenomas contain a second hit involving the other *APC* allele. Mutations in the 5′ end of the gene often cause a mild attenuated form of familial adenomatous polyposis with fewer polyps and later age of onset. Gardner syndrome shows an association with mutations in the 3′ end of *APC*. Mutations elsewhere in *APC* result in the typical features of familial adenomatous polyposis.

Li–Fraumeni syndrome

In this rare condition there is a predisposition to develop a wide range of tumours and malignancy including sarcomas, breast cancer, brain tumours, leukaemia, and adrenocortical carcinoma. Approximately 50% of affected individuals develop a tumour by the age of 30 years and the lifetime risk for developing cancer is more than 90%. Multiple tumours are common and treatment with radiotherapy predisposes to the formation of other tumours in the radiation field.

Li–Fraumeni syndrome is caused by germline mutations in the tumour suppressor gene *TP53* (p. 194). The widespread distribution and high incidence of tumours in this condition is in accord with the key roles normally played by *TP53* in arresting the cell cycle and promoting apoptosis.

Neurofibromatosis type 1

Neurofibromatosis type 1 is one of the most common autosomal dominant disorders, with an incidence of 1 in 3000. Approximately 50% of cases arise as a result of new mutations. The main clinical features are smooth round areas of increased pigmentation, known as café-au-lait patches, freckling in the axillary and inguinal regions, and multiple small cutaneous tumours known as neurofibromata (Fig. 10.10). Up to 50% of affected individuals develop larger locally invasive tumours known as plexiform neurofibromas (Fig. 10.11). Overall, neuro-

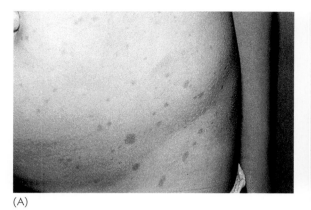

(A)

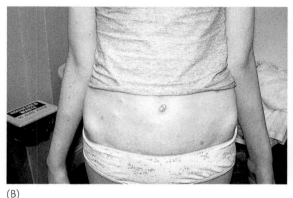

(B)

Fig. 10.10 Café-au-lait patches (A) and neurofibromata (B) in neurofibromatosis type 1. Courtesy of Dr William Reardon, National Centre for Medical Genetics, Dublin.

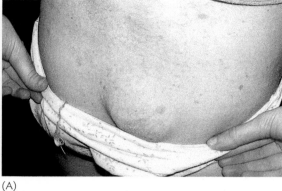

(A)

(B)

Fig. 10.11 (A) Sacral plexiform neurofibroma in a girl with neurofibromatosis type 1. (B) Displacement of the orbit due to a periorbital plexiform neurofibroma in a child with neurofibromatosis type 1. Courtesy of Dr William Reardon, National Centre for Medical Genetics, Dublin.

fibromatosis type 1 conveys a risk of approximately 5% for the development of malignant or central nervous system tumours. Other possible features include mild learning difficulties, hypertension and a spinal curvature (scoliosis).

The disease is caused by mutations in *NF1* which normally suppresses the *RAS* signalling pathway (p. 194). Most mutations result in a truncated protein. Approximately 5% of cases are caused by deletions involving the entire gene. These individuals are more severely affected with significant learning difficulties and numerous neurofibromata. Other genes including *TP53* are also involved in tumour formation in neurofibromatosis type 1.

Neurofibromatosis type 2

Unlike type 1, neurofibromatosis type 2 is a rare condition with an incidence of approximately 1 in 40 000. The usual presentation is with bilateral eighth cranial nerve vestibular tumours known as schwannomas in early adulthood (Fig. 10.12). Other central nervous

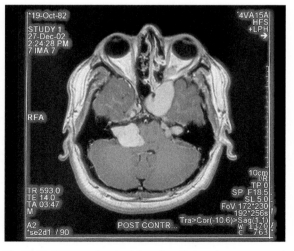

Fig. 10.12 MRI scan showing bilateral vestibular schwannomas and a retro-orbital tumour in a patient with neurofibromatosis type 2.

system tumours can also occur. Small numbers of café-au-lait patches and peripheral neurofibromas may be present. The mean age at diagnosis is 28 years and penetrance is over 95% by age 50 years.

Neurofibromatosis type 2 is caused by loss-of-function mutations in *NF2*, which encodes a cytoskeletal protein known as **merlin**. This is thought to play an important role in stabilizing cell–cell interactions and may act as a tumour suppressor by influencing cell adhesion. Somatic mutations in *NF2* have been identified in breast cancer, colorectal cancer, malignant melanoma, and mesothelioma.

Retinoblastoma

This condition has been discussed at length in the section on tumour suppressor genes. The worldwide incidence is approximately 1 in 20 000 live births. Approximately 10% of affected children have a positive family history and in 30% both eyes are affected. These children have germline mutations. Most of the remaining 60% of cases with no family history and only unilateral involvement result from two somatic mutations (hits) occurring in a single retinal cell. However, a small proportion (around 10%) have a new germline mutation, as revealed by an empirical risk of 2–5% to offspring of these apparent sporadic cases.

Children with germline mutations are at increased risk of developing other tumours, particularly osteosarcomas, in later life. This risk is increased in children who were treated in the past with radiotherapy, in contrast to the newer techniques of cryotherapy and photocoagulation. Mutations in the causal gene, *RB1*, are a common finding in sporadic osteosarcoma, breast cancer, and small-cell lung carcinoma.

Von Hippel–Lindau disease

This rare condition, with an incidence of 1 in 35 000, usually presents in the third decade with retinal angiomas, cerebellar and spinal haemanigoblastomas, renal cancer, and phaeochromocytoma. Penetrance is almost 100% by age 60–65 years. Because of the anatomical location of the tumours, treatment is difficult and morbidity is high.

The disease is caused by mutations in *VHL*, which encodes a protein given the unimaginative name of pVHL. This is normally involved in the stabilisation of an hypoxia-inducible factor, which, when activated, promotes expression of genes such as vascular endothelial growth factor. This is consistent with the observation that tumours with loss-of-function mutations in both *VHL* alleles are often very vascular. Somatic mutations in *VHL* and/or loss of heterozygosity for the *VHL*

locus are common in renal, lung, breast, ovarian and testicular tumours.

Familial breast cancer

Approximately 5% of all cases of breast cancer are caused by germline mutations in one of two genes, *BRCA1* on chromosome 17q and *BRCA2* on chromosome 13q. As already indicated, these genes encode proteins which are components of the multimeric protein complex involved in the repair of radiation-induced DNA double-strand breaks. Familial breast cancer caused by mutations in one or other of these genes is typical of an inherited cancer susceptibility syndrome in that it shows early age of onset, bilateral involvement, and tumours in multiple organs.

Mutations in *BRCA1* account for 50% of autosomal dominant breast cancer families, with the lifetime risk for a heterozygous woman developing breast cancer being around 50% by 50 years and 85% by age 70 years. These women also face a risk of 40–60% for developing ovarian cancer by age 70 years. Different mutations convey slightly different risks. For example the common 185 del AG mutation found in the Ashkenazi Jewish population (p. 140) conveys risks for developing breast and ovarian cancer by age 70 years of 56% and 16% respectively.

Mutations in *BRCA2* account for 30% of all autosomal dominant breast cancer families. Risks for women developing breast and ovarian cancer are similar to those for women with mutations in *BRCA1*. Males with a germline mutation in *BRCA2* have a lifetime risk for developing breast cancer of around 5%. Men with Klinefelter syndrome (p. 60) have a similar risk. The comparable risk for males with a mutation in *BRCA1* is much lower. Males with mutations in *BRCA1* or *BRCA2* also have an increased risk of 6–10% for developing prostate cancer by age 75 years.

In keeping with the two-hit hypothesis, tumours in *BRCA1* and *BRCA2* heterozygotes usually show a mutation in the other allele or loss of heterozygosity for the relevant locus. In contrast, mutations in *BRCA1* and *BRCA2* are found in very few sporadic tumours. However, hypermethylation of the *BRCA1* promoter occurs in many sporadic breast cancers and this is thought to be a key event in the multistep development of many of these tumours.

Hereditary non-polyposis colon cancer

This has an incidence of 1 in 3000 and accounts for approximately 5% of all cases of colorectal cancer. It used to be known as the Lynch family cancer syndrome. Characteristic features are a relatively early

---

BOX 10.2 CASE HISTORY: GASTRIC CANCER AND NAPOLEON BONAPARTE

Napoleon Bonaparte was born in 1769 in Corsica. After training at French military schools he rose to become commander of the French army in 1796 and Emperor of France in 1804. For several years he waged successful conquests against his European neighbours until his territorial aspirations were thwarted first by a disastrous attempt to invade Russia and finally by the Duke of Wellington at Waterloo in 1815. Thereafter he was exiled to St Helena, an island off the west coast of Africa.

Shortly after his arrival on St Helena, Napoleon developed scurvy followed by a chest infection and hepatitis. In 1820 he developed severe sharp pain in the right hypochondrium (the upper right abdomen), which heralded a rapid decline in general health with persistent vomiting culminating in his death 9 months later. A detailed autopsy was performed, on the basis of which it was concluded that the primary cause of death was cancer in the pyloric region of the stomach.

Over the years there has been much debate and speculation about the precise cause of Napoleon's death. However, review of his family history provides strong support for a diagnosis of familial stomach cancer. His father and paternal grandfather died at the ages of 39 and 40 years respectively, both with a probable diagnosis of stomach cancer. This diagnosis is also thought to account for the death of four of his seven siblings at the ages of 44, 49, 56, and 65 years. Napoleon's only son died at the age of 21 years as a result of tuberculosis. Napoleon himself is said to have been concerned by his family history of cancer and to have anticipated this as the cause of an early death.

Most stomach cancer is thought to have a multifactorial aetiology, with environmental factors such as diet and infection with *Helicobacter pylori* making a major contribution. In around 10% of cases there is evidence of familial clustering. Rarely, this can be attributed to a diagnosis of familial adenomatous polyposis or hereditary non-polyposis colon cancer. More often it is caused by a germline mutation in *CDH1*, the gene which encodes E-cadherin, a calcium-dependent cell adhesion protein. As with most familial forms of cancer the age of onset in familial stomach cancer is earlier than in sporadic cases, in which onset is usually after the age of 60 years. Napoleon's family history is typical of that seen in families with a confirmed *CDH1* mutation.

Reference

Sokoloff B (1938) Predisposition to cancer in the Bonaparte family. *American Journal of Surgery*, **40**, 673–679.

---

onset of colon cancer at mean age of 42 years, an excess of right-sided colonic tumours, and multiple primary tumours. Affected individuals also face an increased risk for developing other tumours, particularly gastric (Box 10.2), biliary, and urinary tract cancers. The risk for developing colorectal cancer by age 70 years is 90% for men and 60% for women. However, women also have a 40% chance of developing endometrial cancer.

Most tumours in patients with hereditary non-polyposis colon cancer show marked microsatellite instability, and germline mutations have been identified in five of the known mismatched base-pair repair genes (see Table 10.4). Somatic mutations in one or more of these genes are often found in sporadic colorectal tumours.

Assessment and management of familial breast and colorectal cancer

Both breast and colorectal cancer are common. The lifetime risk for a woman in the developed world to develop breast cancer is 1 in 11. After lung cancer, breast cancer is the commonest cause of death due to malignancy in women. The lifetime risk for a man or woman to develop colorectal cancer is approximately 1 in 50, and after lung cancer colorectal cancer is the commonest cause of death due to malignancy in men.

Given that breast and colorectal cancer are common, it is not surprising that a family history of one or other of these conditions is also common. Nor is it surprising that, in view of the associated high mortality figures, a family history of breast or colorectal cancer now prompts around one third of all referrals to genetic clinics. Recognition of the contribution of genetic susceptibility to both of these conditions has led to the development of guidelines and protocols for their assessment and management in families.

Family history of breast cancer

A positive family history of breast cancer is known to convey an increased risk for female relatives. Approximately 5% of all cases are caused by germline mutations in genes such as *BRCA1* or *BRCA2*. The mechanisms conveying familial susceptibility in other cases are not known. Multifactorial inheritance implying the

TABLE 10.7 Guidelines for referral and screening mammography with one, two, or three relatives affected with breast cancer at various ages

Family history of breast cancer	Expected breast cancer cases between 40–50 y	Lifetime risk (population risk is 1 in 11)	Risk group	Early mammography	Specialist genetics clinic
1 relative[a]					
1 relative > 40 y	Maximum 1 in 50	Maximum 1 in 8	Low	No	No
1 relative < 40 y	1 in 30–1 in 50	1 in 12–1 in 6	Low/ moderate	Yes	No[b]
Female < 30 or male affected at any age	Maximum 1 in 25	Maximum 1 in 6			
2 relatives[a]					
2 relatives 50–60 y	1 in 40	1 in 8	Low	No	No
2 relatives average age 40–49 y	1 in 25	1 in 6–1 in 4	Moderate	Yes	No[b]
2 relatives average age 30–39 y	1 in 14	1 in 4 – 1 in 3	High	Yes	Yes
3 relatives[a]					
3 relatives average age 50–60 y	1 in 15	1 in 4	Moderate	Yes	Yes
3 relatives average age 40–50 y	1 in 11	1 in 3	High	Yes	Yes
Breast and other cancers					
1 or more relative with breast cancer ≤ 50 y + ≥ 1 relative with ovarian cancer at any age *or* one relative with both	Usually more than 1 in 25	Usually more than 1 in 6	Moderate/ high	Yes	Yes
1 or more relative with breast cancer < 40 y plus relative with childhood malignancy			May be high	Avoid mammograms pending genetics review	Yes

[a] 'Relative' includes first degree relative and their first degree relatives (first-degree relatives = mother, father, brother, sister, child). A relative with clearly bilateral breast cancer can be counted as two relatives for simplicity. A male relative with breast cancer counts as a young female (<40).

[b] Ethnic origin may make mutation searching easier, for example Ashkenazi Jewish ancestry might mean genetic testing would be more helpful even with a less striking family history.

Adapted with permission from Eccles DM, Evans DGR, Mackay J (2000) Guidelines for a genetic risk based approach to advising women with a family history of breast cancer. *Journal of Medical Genetics,* **37**, 204–209.

interaction of shared environment with genes conveying low-grade susceptibility may account for around 20–25% of all cases. Alternatively, other as yet unidentified autosomal dominant genes with low penetrance may be implicated. The remaining 70–75% of cases probably occur as a consequence of random somatic mutations.

The challenge in a clinical setting is to try to determine, on the basis of family history, firstly whether a woman should be referred for detailed assessment and, if so, whether her risk should be categorized as low, moderate, or high. Towards this end various guidelines have been drawn up by groups such as the UK Cancer Family Study Group (Table 10.7). These guidelines

BOX 10.3 CASE CÉLÈBRE: GENE PATENTS

A patent is a formal legal entitlement to exploit an invention or process for a fixed period of time, usually 20 years. The right of exploitation is granted exclusively to the inventor, who can, within the terms of patent law, license others to use the invention. The concept of patenting was developed to encourage and reward research and development.

This all seems entirely reasonable when viewed in the context of a new type of vacuum cleaner or, of much greater relevance to medicine, a new drug. However, the issues become much more contentious when they relate to the patenting of a genetic sequence. Does the successful sequencing of a new fragment of DNA represent an invention or a discovery? Is it acceptable to patent a scientific fact or a piece of genetic information simply because a company or institution is the first to identify it?

In the hard-nosed world of business, profit-driven biotechnology companies and cash-strapped universities can cogently argue that their investments in research and development should be protected through patenting if their discoveries should prove to have commercial applications. For example, the identification of the DNA sequence of a gene encoding a cell receptor might ultimately lead to the discovery of a new drug for its activation or suppression. The counter argument is that patenting a DNA sequence is a 'misappropriation of the human heritage' which could stifle future research and delay the development of clinical applications.

Initial rumblings of concern over these issues rose to a crescendo when, in the mid 1990s, the American company Myriad Genetics successfully sought and won patenting rights over the DNA sequences and diagnostic uses of *BRCA1* and *BRCA2* in the USA. This included the exclusive right to full sequencing, which could only be undertaken at the company's laboratories in Salt Lake City for a fee in excess of $2000. In 2001 Myriad won a European right on the genes, which was fiercely opposed by several leading European institutions, notably the Curie Institute in Paris which was backed by the French government. To date the issue remains unresolved, with most European laboratories continuing to offer *BRCA1* and *BRCA2* mutation testing with no payment to Myriad Genetics.

The philosophical, ethical, and legal arguments are complex, with a need for clarification which will become more acute when successful gene therapy strategies are implemented. Ultimately there will have to be an international consensus which acknowledges both the financial rights of the sequencers and the therapeutic rights of the patients.

Reference

Westphal SP (2002) Your money or your life. *New Scientist*, **175**, 29–33.

---

require that a careful family history is taken and that all diagnoses are confirmed with accurate details of age at onset.

Generally the screening and management offered to a woman will depend on the category into which her risk is assessed:

- A 'low' risk of 1 in 8 or less would lead to reassurance and general guidance on breast awareness and self-examination.

- A 'moderate' risk, of from 1 in 8 to 1 in 4, would be managed by careful clinical examination, guidance on regular self-examination, and an offer of annual mammography from 40 years onwards.

- In the 'high' risk category, with a risk of 1 in 4 or greater, consideration is given to *BRCA1/BRCA2* mutation testing if there is a surviving affected relative who can be tested. Screening is offered on the basis of annual examination, regular self-examination, and annual mammography from 35 years onwards. Annual transvaginal ultrasound is offered from 30 years onwards to try to detect early ovarian cancer. Some women in the high-risk category opt for prophylactic bilateral mastectomy and oophorectomy. Ideally it would be desirable to offer mutation testing to all women with a positive family history, but *BRCA1* and *BRCA2* are large genes, so mutation analysis is time consuming and expensive (Box 10.3).

Several studies are being carried out in various parts of the world to try to determine the most effective methods for surveillance and intervention. The role of MRI is being evaluated and the potential prophylactic benefits of anti-oestrogen agents such as tamoxifen are also being assessed.

Family history of colorectal cancer

As with breast cancer, a positive family history of colorectal cancer conveys an increased risk for relatives.

TABLE 10.8   Example of suggested screening for colon cancer

1st degree relative age <40 y	5 yearly colonoscopy	Begin 5 y before earliest colorectal cancer diagnosis but usually at 35 y, not <25 y
Two 1st degree relatives both > age 70 y	Reassure. No colonoscopy	
Two 1st degree relatives average age 60–70 y	Single colonoscopy	Age 55 y
Two 1st degree relatives average age 50–60 y	5 yearly colonoscopy	Age 35–65 y
Two 1st degree relatives average age <50 y	3–5 yearly colonoscopy	Refer to genetics centre Begin age 30–35 y or 5 y before earliest colon cancer diagnosis in family
Three close relatives Amsterdam criteria positive	1–2 yearly colonoscopy	Refer to genetics centre Begin age 25 y
Three close relatives not Amsterdam criteria positive	3–5 yearly colonoscopy	Refer to genetics centre Begin age 30–40 y (5 y before earliest cancer diagnosis in family)
FAP	Annual sigmoidoscopy/colonoscopy	Refer to genetics centre

Adapted with permission from Hodgson SV, Maher ER (1999) *A practical guide to human cancer genetics*. Cambridge University Press, Cambridge.

This risk is influenced both by the age of onset in the family and by the number of affected relatives. Just as with breast cancer, around 5% of colorectal cancer is caused by single-gene familial cancer syndromes such as familial adenomatous polyposis and hereditary non-polyposis colon cancer. In 20–25% of cases genetic susceptibility is thought to play a role, but how this is conveyed is not known. Environmental factors, specifically diet, are also important.

An individual's risk can be categorized as low, moderate, or high on the basis of the number of affected relatives and their age of onset (Table 10.8). The diagnosis of hereditary non-polyposis colon cancer is based on specific patterns of family involvement known as the Amsterdam or modified Amsterdam criteria. When familial adenomatous polyposis or hereditary non-polyposis colon cancer is suspected, specific mutation analysis is undertaken.

Individuals deemed to fall within a low-risk category are reassured and offered a single colonoscopy at age 55 years. Those with a moderate risk are offered colonoscopy on a 3–5-yearly basis from the age of 35 years onwards. Individuals at high risk, including proven mutation carriers for hereditary non-polyposis colon cancer, are offered annual colonoscopy from 25 years onwards together with endometrial and ovarian screening by pelvic examination and transvaginal ultrasound in women. In some centres screening for relatives at moderate or high risk by colonoscopy is offered from 5 years before the earliest cancer diagnosis in the family. Colonectomy is recommended for those known to have a mutation causing familial adenomatous polyposis.

The prevailing evidence indicates that screening protocols of this nature do lead to early detection of colorectal tumours and significant reduction in mortality. These protocols are flexible and still being evaluated. Research is also being carried out into the long-term beneficial effects of high-starch diets and regular intake of the non-steroidal anti-inflammatory drugs aspirin and sulindac.

It is important to emphasize that screening for cancer almost invariably generates anxiety, so counselling and long term support are vital.

Key Point

Protocols have been established for the assessment and management of individuals who present with a family history of breast or colon cancer.

Summary

Most human cancers are caused by the accumulation of mutations in proto-oncogenes, tumour suppressor genes, and DNA repair genes. These genes are normally involved in processes which regulate important cellular functions such as division, proliferation, and apoptosis.

Mutations in proto-oncogenes to form oncogenes exert a gain-of-function effect and act in a dominant manner at the cellular level. These activating mutations are caused by various mechanisms including point mutations, amplification and the formation of new chimaeric genes by translocations.

Mutations in tumour suppressor genes and DNA repair genes exert a loss-of-function effect and act in a recessive fashion at cellular level. These mutations occur as a result of cellular events such as chromosome non-disjunction in mitosis. This can be identified in tumour tissue by looking for loss of heterozygosity. Mutations in specific mismatched base-pair repair genes result in microsatellite instability and lead to cancer by failure to repair acquired mutations in proto-oncogenes and tumour suppressor genes.

It has been shown that several familial cancer syndromes are caused by mutations in proto-oncogenes, tumour suppressor genes, and DNA repair genes. Most of these show autosomal dominant inheritance. Some cancers such as retinoblastoma are caused by two allelic mutations ('two hits') in a tumour suppressor gene. One of these mutations can be inherited in the germline and the other somatic, or both can be somatic. Characteristics of the familial cancer syndromes are that tumours develop at an earlier age than in acquired cancer, they are often bilateral and/or multifocal, and they can affect different organs.

In many families there is a history of breast or colon cancer. Guidelines have been drawn up to assist in the counselling and management of relatives who are at increased risk of developing one of these common cancers. Screening programmes based on annual mammography or colonoscopy have been developed for relatives deemed to be at high risk.

Further reading

Cowell JK (ed.) (2001) *Molecular genetics of cancer*, 2nd edn. Bios Scientific Publishers, Oxford.

Eeles RA, Ponder BAJ, Easton DF, Horwich A (ed.) (1996) *Genetic predisposition to cancer*. Chapman & Hall Medical, London.

Henderson BE, Ponder B, Ross RK (2003) *Hormones, genes and cancer*. Oxford University Press, Oxford.

Hodgson SV, Maher ER (1999) *A practical guide to human cancer genetics*, 2nd edn. Cambridge University Press, Cambridge.

Multiple choice questions

1 Mutational mechanisms which can result in activation of an oncogene include

(a) point mutations

(b) amplification

(c) formation of a chimaeric gene

(d) loss of heterozygosity

(e) deletion

2 Factors which point to familial rather than non-familial retinoblastoma include

(a) unaffected first-cousin parents

(b) presentation soon after birth

(c) multifocal tumours

(d) tumours in other organs

(e) a family history of trisomy 13

3 The following tumours or cancer syndromes are caused by mutations in tumour suppressor genes:

(a) retinoblastoma

(b) multiple endocrine adenomatosis type 2

(c) Li–Fraumeni syndrome

(d) Burkitt lymphoma

(e) familial adenomatous polyposis

4 A woman presents with concern about her family history of cancer. Which of the following would raise suspicion of a familial cancer syndrome?

(a) breast cancer in her mother at age 70 years

(b) colon cancer in her father at age 70 years

(c) breast and ovarian cancer in her sister at age 40 years

(d) a sister and a brother with breast cancer

(e) multiple sarcomas in her father

5 Mutations in DNA repair genes can cause

(a) ataxia telangiectasia

(b) breast cancer

(c) colon cancer

(d) neurofibromatosis type 1

(e) retinoblastoma

Answers

1 (a) true—these account for many oncogene mutations, e.g. *RET* in multiple endocrine adeno matosis type 2

(b) true—amplified genes can sometimes be seen as double minute additional chromosomes

(c) true—this is seen in chronic myeloid leukaemia

(d) false—this describes the loss of a tumour suppressor gene

(e) false—oncogenes exert a gain-of-function effect, whereas deletions lead to haploinsufficiency i.e. loss-of-function

2 (a) false—this would only be relevant if at least one of the parents was affected

(b) true—sporadic retinoblastoma has an average age at presentation of 24 months

(c) true—multifocal tumours would be consistent with one of the two hits being inherited

(d) true—as with multifocal tumours, tumours in different organs are consistent with one of the two hits being inherited

(e) false—retinoblastoma can be associated with a deletion of the *RB1* locus on chromosome 13 but is not a feature of trisomy 13

3 (a) true—retinoblastoma is caused by mutations in *RB1*

(b) false—multiple endocrine adenomatosis type 2 is caused by mutations in the *RET* proto-oncogene

(c) true—Li–Fraumeni syndrome is caused by mutations in *TP53*

(d) false—Burkitt lymphoma is associated with activating mutations involving the *CMYC* proto-oncogene

(e) true—familial adenomatous polyposis is caused by mutations in *APC*

4 (a) false—this is more likely to be a sporadic event

(b) false—as with late-onset breast cancer, this late age of onset points to a sporadic form of cancer

(c) true—this is consistent with a mutation in *BRCA1* or *BRCA2*

(d) true—breast cancer in a male is a feature of mutations in *BRCA2*

(e) true—this would point to a diagnosis of Li–Fraumeni syndrome

5 (a) true—ataxia telangiectasia is caused by homozygous mutations in a gene involved in the repair of double-strand breaks

(b) true—both *BRCA1* and *BRCA2* are involved in the repair of double-strand breaks

(c) true—hereditary non-polyposis colon cancer is caused by mutations in the DNA mismatch repair genes

(d) false—*NF1* is a tumour suppressor gene

(e) false—*RB1* is a tumour suppressor gene

CHAPTER 11

Genes and biochemistry

Human metabolism involves the coordinated interaction of large numbers of biochemical pathways, each of which consists of sequential reactions mediated by enzymes. Intracellular organelles, cell surface receptors, and intercellular transport systems also play important roles. It is impossible to estimate the number of genes involved in the regulation and maintenance of the body's metabolic processes, but the fact that over 500 inborn errors of metabolism have been described gives an indication of the underlying complexity.

The term *inborn error of metabolism* was coined by the London physician, Archibald Garrod, who is generally acknowledged to be the founding father of biochemical genetics (Box 11.1). Over 100 years ago he proposed that individuals 'differ in their chemistry, as they do in their structure'. It is now recognized that there is enormous diversity in humans at the biochemical level and that variation in metabolic pathways and receptor activity is likely to play key roles in conveying susceptibility for common multifactorial disorders such as coronary artery disease, diabetes mellitus, and hypertension.

This chapter is concerned primarily with biochemical disorders that are determined by single-gene inheritance. Most enzyme deficiencies show recessive inheritance, either autosomal or X-linked, indicating that the underlying mutations usually exert loss-of-function effects. Thus most metabolic pathways operate satisfactorily when enzyme activity is 50% of normal, but beware of multiple-choice questions which state that all enzyme deficiencies show recessive

BOX 11.1 LANDMARK PUBLICATION: THE GENETIC BASIS OF ALKAPTONURIA

Alkaptonuria (MIM 203500) is a rare and relatively harmless condition characterized by the presence of large quantities of homogentisic acid in urine, which turns dark in the presence of alkali or on exposure to air. The diagnosis is often first suspected because of staining of the nappies (diapers) and underlying skin in young infants. In older children ear wax shows increased pigmentation. In adults, the deposition of pigment in cartilage and connective tissue (ochronosis) can cause progressive degenerative arthritis. This results from the binding of homogentisic acid and its polymers to collagen.

Around 1900 Archibald Garrod, a London physician who was based at St Bartholomew's Hospital and the Great Ormond Street Children's Hospital, gathered information about all of the cases known locally, together with those previously published in what was then a meagre medical literature. He observed that the condition showed 'a peculiar mode of inheritance' in that it would appear in two or more siblings whose parents were unaffected, as illustrated by the fact that 19 of the 32 cases known to him occurred in 7 sibships. In addition Garrod noted that the parents of many of these children were first cousins, a finding which he confirmed in the parents of previously published cases by contacting their authors. The overall incidence of consanguinity amongst the parents of alkaptonuric children was 60%, in contrast to figures of 4% for the English aristocracy and 1.5% for the general population of London.

In a much-quoted example of intellectual collaboration, Garrod's observations were interpreted by his contemporary, William Bateson, as being consistent with autosomal recessive inheritance as 'discovered by Mendel'. Garrod published his observations in a modestly entitled paper in 1902, in which he speculated that albinism and cystinuria might share the same inheritance pattern (cystinuria is a rare condition in which increased levels of cystine in urine predispose to the formation of renal calculi). He developed his theme of biochemical individuality in a series of lectures delivered in 1908 and in his book entitled *Inborn Errors of Metabolism* published in 1909.

Together, Garrod and Bateson laid the foundation stone for the concepts of biochemical genetics by this first assignment of autosomal recessive inheritance to a human disorder. In 1958, 22 years after Garrod's death, his prediction of an 'alternative course of metabolism' was confirmed by the discovery that alkaptonuria is caused by a deficiency of homogentisic acid oxidase (Fig. 11.1). The gene was mapped to chromosome 3q by autozygosity mapping in 1993 and finally cloned in 1996. Readers interested in learning more about Garrod's work and life should consult his 1909 textbook, which was republished in 1963, and his biography by Bearn, published in 1993.

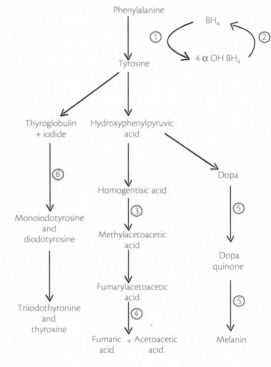

Enzyme	Disorder
① Phenylalanine hydroxylase	Phenylketonuria
② Dihydropteridine reductase	Biopterin deficiency
③ Homogentisic acid oxidase	Alkaptonuria
④ Fumarylacetoacetic acid hydrolase	Tyrosinaemia type I
⑤ Tyrosinase	Oculocutaneous albinism
⑥ Thyroid peroxidase	Hypothyroidism

Fig. 11.1 Simplified diagram of the phenylalanine and tyrosine metabolic pathway.

References

Bearn AG (1993) *Archibald Garrod and the individuality of men.* Clarendon Press, Oxford.

Garrod AE (1902) The incidence of alkaptonuria: a study in chemical individuality. *Lancet*, **ii**, 1616–1620.

Harris H (1963) *Garrod's inborn errors of metabolism.* Oxford University Press, London.

inheritance. A few, notably those which cause the various forms of porphyria, are rate-limiting and manifest clinically in heterozygotes, so that the inheritance pattern is dominant.

It is clearly not possible for students, or indeed for anyone, to have a working knowledge of all known metabolic disorders. The latest (eighth) edition of the definitive textbook on metabolic disease now consists of 4 large volumes and extends to over 7000 pages! However, students are expected to be familiar with the more common disorders which illustrate basic principles, as discussed in the following pages. They are also required to be aware of the range of disorders for which neonatal screening is available and the criteria that screening programmes should fulfil.

Disorders of carbohydrate metabolism

Carbohydrates provide the main source of energy in the human diet and are metabolized into the three main monosaccharides—fructose, galactose, and glucose. Fructose and galactose are in turn converted into glucose, which is then converted to pyruvate and ATP through the process of glycolysis. Pyruvate is then metabolized through the Krebs citric acid cycle to provide energy.

Over 40 disorders of carbohydrate metabolism have been identified. Several of these involve the generation of energy in mitochondria as discussed later in this chapter. In this section two specific conditions are considered. The first, galactosaemia, was one of the first disorders to be included in newborn screening programmes. The second, von Gierke disease, is the best-known example of a group of disorders characterized by the storage of glycogen.

Galactosaemia (MIM 230400)

The incidence of galactosaemia, based on evidence from newborn screening programmes, is approximately 1 in 55 000. Inheritance is autosomal recessive. Affected infants present soon after the institution of milk feeds with vomiting, failure to thrive, and infection which can prove fatal. Surviving infants who are untreated develop cataracts, liver failure, and developmental delay. The diagnosis can be suspected by the presence of galactose in urine and is confirmed by specific enzyme assay.

Classical galactosaemia is caused by mutations in the gene which encodes galactose-1-phosphate uridyl transferase; affected children cannot convert galactose to glucose (Fig. 11.2). Over 150 mutations have been

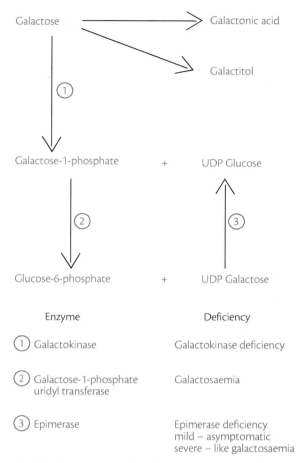

Fig. 11.2 Galactose metabolic pathway showing enzyme defects which can cause raised levels of galactose in blood. UPD = uridine diphosphate.

reported. Two common mutations (p.Gln188Arg and p.Lys258Asn) account for 70% of cases in the white population and one (p.Ser135Leu) accounts for 60% in the black population. Homozygous mutations in the gene which encodes galactokinase cause a much milder form of galactosaemia in which clinical manifestations are limited to cataracts.

Classical galactosaemia is treated by the complete removal of milk and milk-containing products from the diet. This prevents the physical complications, but long-term intellectual development is often impaired with verbal dyspraxia being a common finding. Many affected women develop ovarian failure due to galactose toxicity. Galactosaemia is one of the conditions which is now screened for routinely in the neonatal period.

Glycogen storage disease type Ia (von Gierke disease—MIM 232200)

The glycogen storage disorders are caused by deficiencies of enzymes involved in the synthesis or breakdown of glycogen. Carbohydrates are normally stored in the form of glycogen in the liver to maintain the blood glucose level and in muscles for energy requirements. Thus defects in glycogen synthesis or degradation present with features such as hypoglycaemia, hepatomegaly, muscle weakness, and, if cardiac muscle is involved, heart failure. At least 8 different types have been described; they are numbered on the basis of the order in which they were reported.

In glycogen storage disease type Ia, affected children present with marked hepatomegaly and hypoglycaemia precipitated by short periods of fasting. The blood levels of uric acid, cholesterol, and triglycerides

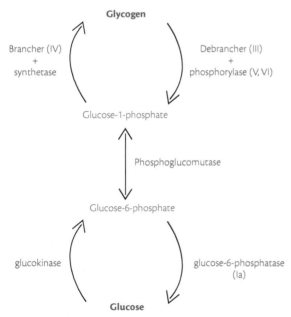

Fig. 11.3 Simplified diagram of the glycogen metabolic pathway. Roman numerals indicate the site of the defect in various glycogen storage disorders. Ia = von Gierke disease; III = Cori disease; IV = Anderson disease; V = McCardle disease; VI = Hers disease.

are elevated. Without treatment affected children show poor growth and a high incidence of hepatic adenoma.

The basic defect lies in the gene which encodes glucose-6-phosphatase, the enzyme responsible for the conversion of glucose-6-phosphate to glucose (Fig. 11.3). (Note that this is not the same as glucose-6-phosphate dehydrogenase, deficiency of which causes haemolytic anaemia—p. 227.) Inheritance is autosomal recessive. The enzyme is only expressed in liver, so specific enzyme diagnosis can only be achieved by liver biopsy. In practice the diagnosis is based on the combination of the clinical and biochemical findings supplemented by specific mutation analysis. Treatment involves maintaining the normal level of blood glucose by frequent feeding during the day and the infusion of glucose overnight. Successful treatment helps maintain normal growth and reduces the incidence of hepatic adenoma.

Disorders of amino acid metabolism

The disorders of amino acid metabolism hold an honoured place in the history of both biochemical genetics and medical genetics as one of these conditions, alkaptonuria, was the first disorder recognized to show autosomal recessive inheritance (Box 11.1). The recognition of another of these conditions, phenylketonuria, led to the establishment of neonatal screening programmes.

Phenylketonuria (MIM 261600)

Phenylketonuria (PKU) is one of a group of disorders, collectively known as the hyperphenylalaninaemias, which result from defects in the metabolism of phenylalanine (Fig. 11.1). Classical PKU is caused by defects in the gene which encodes phenylalanine hydroxylase (PAH), the enzyme which converts phenylalanine to tyrosine. Inheritance is autosomal recessive. The incidence in western Europeans is approximately 1 in 10 000. Over 400 mutations have been reported in the gene which encodes PAH. Prenatal diagnosis can only be carried out by molecular analysis (linkage or mutation detection), as PAH is only expressed in the liver.

Without treatment, children with classical PKU develop severe mental retardation in association with eczema, fair skin, blonde hair, and neurological abnormalities such as tremor and hypertonicity. The developmental and neurological abnormalities are caused by the toxic effects of high levels of phenylalanine on myelination and brain development. Hypopigmentation results from reduced synthesis of melanin due to competitive inhibition of tyrosine hydroxylase by the high levels of phenylalanine.

The discovery that the early introduction of a low-phenylalanine diet prevents mental retardation in classical PKU led to the establishment of neonatal screening in many parts of the world. Dietary restriction should begin as soon after birth as possible and should be maintained throughout childhood and ideally throughout adult life. Maintenance of low phenylalanine levels is particularly important in pregnant women, who should ideally go on a low-phenylalanine diet in advance of pregnancy, as a high circulating level of maternal phenylalanine is neurotoxic to the developing embryonic and fetal brain. The incidence of mental retardation in the offspring of mothers with untreated classical PKU is close to 100%. These children also show a high incidence of microcephaly and structural cardiac defects.

Key Point

Dietary restriction of phenyalanine prevents the development of mental retardation in children with phenylketonuria. The blood levels of phenylalanine should be monitored carefully in affected women who are pregnant, to prevent damage to the brain of the unborn infant.

Other causes of hyperphenylalaninaemia

Classical PKU is the commonest cause of hyperphenylalaninaemia. Other causes can be subdivided on the basis of their clinical effects into mild and severe.

Mild hyperphenylalaninaemia is caused by mutations in the gene encoding PAH which result in a degree of residual enzyme activity. These mild forms of *variant PKU* can be managed by a less rigorous low-phenylalanine diet. Severe hyperphenylalaninaemia is due to classical PKU in 99% of cases. In the remaining 1% it results from defects in the formation or recycling of tetrahydrobiopterin (BH_4), the cofactor for PAH in the conversion of phenylalanine to tyrosine (Fig. 11.3). BH_4 is also required for the maintenance of normal levels of dopamine and serotonin in the brain. For this reason, children with defects in BH_4 metabolism usually develop mental retardation despite careful dietary restriction of phenylalanine.

Other disorders of phenylalanine and tyrosine metabolism

Normally phenylalanine is converted to tyrosine, which is required for the synthesis of catecholamines, melanin, and thyroid hormones (Fig. 11.1). Enzyme defects in these various pathways account for several well-characterized inborn errors. Deficiency of homogentisic acid oxidase causes alkaptonuria (MIM 203500), which presents with darkening of the urine in children and arthritis in adults (Box 11.1). Tyrosinaemia type I (MIM 276700) presents with either acute or chronic liver failure, renal tubular disease, and hepatic carcinoma. Treatment is by liver transplantation or with a drug, NTBC, which prevents the synthesis of hepatotoxic fumarylacetoacetic acid. Tyrosinase-negative oculocutaneous albinism (MIM 203100) presents with complete lack of pigment in the skin and hair, together with nystagmus and photophobia. Finally, tyrosine residues in thyroglobulin normally undergo iodination to form triiodothyronine and thyroxine. Defects in thyroid peroxidase result in a goitre with hypothyroidism. All of these disorders show autosomal recessive inheritance.

Maple syrup urine disease (MIM 248600)

This rare condition is caused by a defect in a multimeric enzyme complex known as branched chain α-ketoacid decarboxylase (BCKAD). This is involved in the metabolism of the branched-chain amino acids valine, leucine, and isoleucine. Any defect in BCKAD, which is encoded by at least four different genes, results in elevated levels of these branched-chain amino acids in blood and urine, in which their presence gives rise to the characteristic smell of maple syrup. Inheritance in all forms is autosomal recessive.

Clinically, maple syrup urine disease usually presents in early infancy with anorexia, apathy, and dehydration. If untreated, there is rapid progression with convulsions and spasticity to coma and death. The institution of early treatment based on dietary restriction of branched-chain amino acids and, in some cases, thiamine has allowed some children detected through newborn screening to survive into adult life with relatively normal intellectual development. Unfortunately some children show episodic deterioration regardless of treatment, particularly during intercurrent illness.

Disorders of the urea cycle

The pathways considered so far have involved the generation of energy or the manufacture of proteins. In contrast, the urea cycle is concerned with the removal of waste nitrogen formed from protein ingestion or turnover. This takes place in the liver and involves five major biochemical reactions (Fig. 11.4). Defects in any of these result in high levels of ammonia, with age of onset varying from the neonatal period to adulthood depending on the degree of enzyme deficiency.

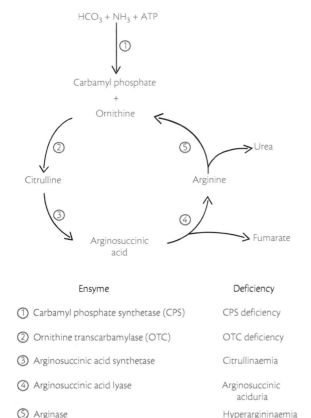

Ensyme | Deficiency

① Carbamyl phosphate synthetase (CPS) | CPS deficiency

② Ornithine transcarbamylase (OTC) | OTC deficiency

③ Arginosuccinic acid synthetase | Citrullinaemia

④ Arginosuccinic acid lyase | Arginosuccinic aciduria

⑤ Arginase | Hyperargininaemia

Fig. 11.4 The urea cycle metabolic pathway.

Ornithine transcarbamylase deficiency (MIM 311250)

This condition illustrates the spectrum of clinical presentations which can occur in the urea cycle disorders. Ornithine transcarbamylase (OTC) is responsible for the condensation of carbamyl phosphate and ornithine to form citrilline. Unlike the other enzymes involved in the urea cycle, it is encoded by a gene on the X chromosome, *OTC*. Thus transmission is X-linked and, as some females show mild manifestations, strictly speaking it should be classified as showing X-linked dominant inheritance. In the classic severe form, affected boys present soon after birth with severe hyperammonaemia, lethargy, convulsions, and coma leading to a rapidly fatal outcome. These boys have mutations which result in less than 1% of normal enzyme activity. Mutations with greater residual activity are associated with a milder presentation in childhood or adult life. This can take the form of behavioural disturbance or altered levels of consciousness after a heavy protein load. These men, and some heterozygous women, tend to show an aversion to protein meals.

The diagnosis is suspected on the basis of the clinical findings, high blood levels of ammonia, and increased levels of orotic acid in urine. The enzyme is expressed only in liver, so definitive diagnosis can only be achieved by mutation analysis or linkage analysis in informative families. Outcome in the severe form is poor. Drugs such as sodium benzoate and phenylacetate can be used to reduce the circulating level of ammonia. Liver transplantation offers a potential cure. OTC deficiency and the other urea cycle disorders are excellent candidates for gene therapy when this becomes safe and reliable.

Disorders of lipid metabolism and transport

Lipids and lipoproteins are believed to play a major role in conveying susceptibility to cardiovascular disease. Genetic and environmental factors contribute to an individual's susceptibility through a complex network of lipoprotein metabolic pathways. Physiologically, fatty acids stored in triglycerides of adipose tissue are an important source of energy, particularly during fasting or when the body's glycogen stores are depleted. Thus disorders of lipid metabolism and transport can present in one of two ways: either as an acute illness precipitated by an energy crisis, or as a chronic disorder manifesting with coronary artery disease or a cerebrovascular accident (stroke).

Medium chain acyl-CoA dehydrogenase deficiency (MIM 201450)

Medium chain acyl-CoA dehydrogenase (MCAD) deficiency is a disorder of fatty acid oxidation, which has been the subject of increasing publicity over the last few years because the clinical presentation can closely mimic that of sudden infant death syndrome (cot death).

Normally, fatty acids are released during fasting and taken up by mitochondria in liver and muscle. There they are oxidized to produce acetyl-CoA, which in muscle is converted to energy through the Krebs citric acid cycle, and in liver is converted to ketone bodies to provide energy elsewhere. MCAD is a mitochondrial matrix enzyme which catalyses the initial reaction in the β-oxidation of C4–C12 straight-chain acyl-CoAs.

Children with MCAD deficiency present with intermittent hypoglycaemia and vomiting precipitated by transient fasting, which is often associated with an intercurrent minor infection. If untreated, this can lead rapidly to coma and death. These episodes usually occur between the ages of 3 months and 2 years. MCAD deficiency affects approximately 1 in 20 000 children of western European origin, and 90% of mutations involve a single transition resulting in the substitution of lysine by glutamic acid.

This dramatic presentation has sometimes led to the erroneous diagnosis of cot death. The finding of fatty infiltration in the liver at postmortem examination should prompt consideration of a disorder of fatty acid oxidation. Further investigations can then be undertaken on stored samples from the child, including the blood spot collected soon after birth as part of the neonatal screening programme.

> **Key Point**
>
> The diagnosis of MCAD deficiency should be suspected in any child who presents with unexplained hypoglycaemia or sudden death.

Familial hypercholesterolaemia (MIM 143890)

Familial hypercholesterolaemia (FH) is caused by mutations in the gene that encodes the low density lipoprotein (LDL) receptor on chromosome 19. Biochemically it results in elevated levels of plasma cholesterol bound to LDL. Clinically it presents with premature atherosclerosis

TABLE 11.1 Classes of LDL receptor mutations in familial hypercholesterolaemia

Class	Type	Effect
1	Null alleles	No receptors produced
2	Transport deficient	Receptors cannot leave the endoplasmic reticulum
3	Binding deficient	Receptors cannot bind LDL at the cell membrane
4	Localization deficient	Receptors cannot migrate to coated pit on membrane for internalization
5	Recycling deficient	Receptors cannot release bound LDL after internalization

and deposits of cholesterol in skin and tendons known as xanthomata. Inheritance is autosomal dominant with an incidence of 1 in 500 in most populations. Heterozygotes show an average age of onset of 45 years in men and 53 years in women. The homozygous state affects approximately 1 in 1 million individuals (i.e. $1/500 \times 1/500 \times 1/4$) who have usually developed symptomatic coronary artery disease by the age of 20 years.

Over 400 mutations have been identified in the LDL receptor gene. These have been divided into five different classes based on their effect on receptor function (Table 11.1). The net effect of all these mutations is

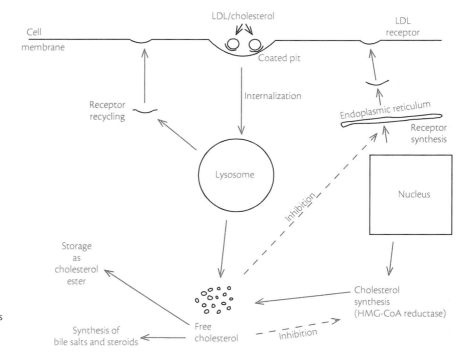

Fig. 11.5 Diagrammatic representation of the role of the LDL receptor and the intracellular processing of cholesterol. A reduction in the number of functional receptors leads to an increase in the plasma level of cholesterol.

that the blood level of LDL cholesterol rises and the synthesis of new cholesterol is not inhibited (Fig. 11.5).

FH is one of the most common autosomal dominant disorders known in humans and accounts for approximately 5% of all myocardial infarctions occurring before the age of 60 years. This high incidence has prompted the UK government to propose in a White Paper published in 2003 (entitled *Our Inheritance, Our Future*) that pilot studies be carried out in several centres with a view to establishing how many heterozygotes can be identified by screening the families of known cases. This approach is known as *cascade screening*. It has been estimated that a similar undertaking in the Netherlands identified around 30% of all predicted heterozygotes in that country. Through this type of approach it is hoped that large numbers of young heterozygotes can be informed of the benefits of a low cholesterol diet and offered prophylactic treatment with 3-hydroxy-3-methylglutaryl-CoA (HMG-CoA) reductase inhibitors, also known as statins.

> **Key Point**
>
> Familial hypercholesterolaemia affects 1 in 500 individuals and accounts for 5% of all myocardial infarctions before the age of 60 years. Effective prevention consists of a low-cholesterol diet and the use of HMG-CoA reductase inhibitors.

Lysosomal disorders

Lysosomes are tiny structures located in the cytoplasm which contain a large number of enzymes responsible for the degradation of specific molecules. Defective enzyme activity leads to accumulation of the relevant substrate which is stored in various organs and tissues. This inexorably leads to progressive physical abnormalities such as coarsening of facial features, hepatosplenomegaly, and heart failure. If the brain is involved then there is progressive loss of intellectual and other neurological skills, a process referred to as neurodegeneration.

Most lysosomal storage disorders are caused by enzyme deficiencies, but a few result from a defect in the targeting of enzymes to their location in the lysosomes. All of the lysosomal storage disorders are serious and those which involve the central nervous system are usually devastating. This is illustrated by consideration of two conditions, Hurler syndrome and Tay–Sachs disease.

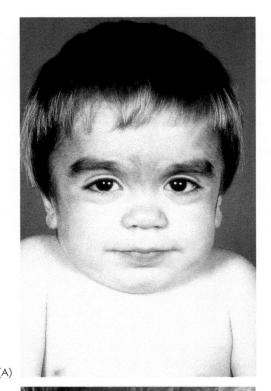

(A)

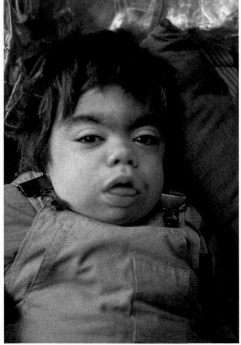

(B)

Fig. 11.6 Facial views of a child with Hurler syndrome at ages (A) 3 years and (B) 9 years.

TABLE 11.2 Classification of the mucopolysaccharidoses

Type	Name	Enzyme deficiency	Features
I	Hurler	α-L-iduronidase	Coarse facies, hepatosplenomegaly, joint contractures, mental retardation
	Scheie	α-L-iduronidsae	Mild facial coarsening, normal intelligence
II	Hunter	Iduronate sulphatase	Severe form—as for Hurler Mild form—severe physical features but normal intelligence
III	Sanfilippo A–D	4 different enzyme defects	Mild coarsening, severe behaviour problems, mental retardation
IV	Morquio A Morquio B	Galactosamine-6-sulphatase β-galactosidase	Coarse facies, severe skeletal involvement with short stature and kyphosis
V	Scheie	α-L-iduronidase	Now classified as Type I
VI	Maroteaux-Lamy	Aryl-sulphatase B	Coarse facies, short stature and joint contractures
VII	Sly	β-glucuronidase	Vary from mild to severe with normal intelligence

Hurler syndrome (MIM 607014)

Hurler syndrome (MPS I-H) is caused by a severe reduction in the level of activity of a lysosomal enzyme, α-iduronidase, which contributes to the degradation of dermatan sulphate and heparan sulphate. These complex carbohydrates, known as mucopolysaccharides or glycosasminoglycans, are major constituents of the ground substance of connective tissue. In affected children the diagnosis can usually be suspected by the age of 1 year because of facial coarsening (Fig. 11.6), hepatosplenomegaly, and developmental delay. Other features are thickening of the skin, corneal clouding, and joint contractures. Without treatment the disorder shows relentless progression, resulting in death by the age of 10–14 years. Treatment with enzyme infusion and bone marrow transplantation partially modifies the disease course (p. 232).

Mutations which result in greater residual levels of enzyme activity cause a much milder condition known as Scheie syndrome (MPS I-S). Affected individuals are of normal intelligence and have a relatively normal life expectancy but show mild facial coarsening, corneal clouding, and joint contractures. A spectrum of intermediate phenotypes exists. These are caused by compound heterozygosity for a severe MPS I-H mutation and a mild MPS I-S mutation. It was the study of conditions such as the mucopolysaccharidoses which led to the realization that many genetic disorders show both allelic and locus heterogeneity (Table 11.2).

Tay–Sachs disease (MIM 272800)

This condition has already been considered in the context of population screening (p. 148). It is caused by severely reduced activity of hexosaminidase A, one of several enzymes necessary for the degradation of a sphingolipid, GM2 ganglioside. Hexosaminidase A is a complex enzyme consisting of α and β subunits and an activator protein. Tay–Sachs disease is caused by mutations in the gene *HEXA*, which encodes the α subunit. Mutations in the gene *HEXB*, which encodes the β subunit, produce a similar condition known as Sandhoff disease.

Children with the classic, also known as infantile, form of Tay–Sachs disease appear normal for the first few

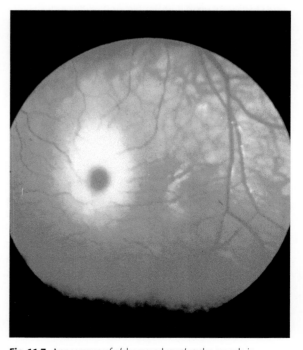

Fig. 11.7 Appearance of a 'cherry red spot' at the macula in Tay–Sachs disease.

months of life until the effects of the storage of GM2 ganglioside in the brain become apparent. From around 6 months onwards there is progressive neurodegeneration with convulsions and spasticity preceding a vegetative state and death by age 4 years. Progressive visual loss occurs from 1 year onwards because of storage in the retinal ganglion cells. This manifests as a cherry-red spot at the macula (Fig. 11.7). At present there is no effective treatment for this devastating condition.

As with many other lysosomal storage disorders, less severe degrees of enzyme deficiency result in milder clinical presentations. In the juvenile-onset form of Tay–Sachs disease, deterioration in neurological function begins in mid-childhood and the rate of progression is much slower than in the severe form. The chronic late-onset form is characterized by slowly progressive localized neurological disability which can present as ataxia, spinal muscular atrophy, or psychiatric illness such as depression or schizophrenia.

Mitochondrial disorders

Each mitochondrion contains a small circular chromosome which encodes 2 ribosomal RNAs, 22 tranfer RNAs, and 13 polypeptides involved in the production of ATP through the process of oxidative phosphorylation. This process also involves another 80–90 polypeptides that are encoded by nuclear genes and then transported into mitochondria. These organelles provide the main source of energy for the cell. Thus mitochondrial defects particularly affect organs, such as brain and muscle, that have a high requirement for ATP.

The interaction of mitochondrial and nuclear encoded polypeptides in the process of oxidative phosphorylation means that in practice it can be very difficult to counsel the parents of a child with a disorder of mitochondrial function. Many disorders of oxidative phosphorylation show autosomal recessive inheritance but some result from mutations in mitochondrial DNA so that the pattern of inheritance is mitochondrial (see p. 86 and Box 4.3). Interpretation is further complicated by the fact that some nuclear mutations can disrupt the integrity of the mitochondrial genome by causing either mitochondrial DNA deletions or mitochondrial DNA depletion. The clinical consequences of changes in mitochondrial DNA show a close correlation with the nature of the underlying mutation.

Mitochondrial DNA deletions and duplications

These are associated with two specific conditions. In the Kearns–Sayre syndrome (MIM 530000) there is progressive ophthalmoplegia with ptosis, retinal pigmentation, and cardiac conduction defects. Muscle biopsy shows the presence of ragged red fibres due to clusters of abnormal mitochondria. Pearson's syndrome (MIM 260560) is characterized by pancreatic insufficiency, pancytopaenia, and lactic acidosis.

Point mutations in transfer and ribosomal RNA

These impair mitochondrial protein synthesis. Two specific mutations in one of the genes encoding for mitochondrial rRNA cause sensorineural hearing loss. One of these mutations causes aminoglycoside-induced hearing loss (MIM 580000), in which affected individuals are particularly at risk of developing hearing loss when treated with antibiotics such as gentamicin, neomycin, and streptomycin.

Mutations in the mitochondrial genes encoding tRNA cause MELAS (mitochondrial encephalopathy, lactic acidosis and stroke-like episodes—MIM 540000) and MERRF (myoclonic epilepsy and ragged red fibres in muscle—MIM 545000).

Mutations in genes involved in oxidative phosphorylation

These can have diverse and sometimes unpredictable consequences, ranging from the relatively mild optic atrophy seen in Leber's hereditary optic atrophy (Box 4.3, MIM 535000) to the devastating and potentially lethal metabolic collapse seen in conditions such as Leigh's disease (MIM 256000). This type of presentation is associated with severe hypotonia, convulsions, altered levels of consciousness, and profound lactic acidosis.

> **Key Point**
>
> Mitochondrial disorders can present in many different ways. Organs with a high requirement for ATP are particularly susceptible to defects in mitochondrial metabolism.

Peroxisomal disorders

Peroxisomes are tiny organelles located in the cell cytoplasm. They are bounded by a single lipid membrane surrounding a granular matrix containing enzymes and other proteins which are imported from the ribosomes by two peroxisomal assembly proteins. Peroxisomes have several important metabolic functions. These include:

◆ oxidation of fatty acids

◆ synthesis of cholesterol and bile acids

- synthesis of plasmalogens which are an important constituent of myelin

- degradation of pipecolic acid.

Over 15 peroxisomal disorders have been identified. In some of these, such as Zellweger syndrome, which is discussed below, there is a defect in peroxisomal biogenesis and all aspects of peroxisomal function are disturbed. In others, such as X-linked adrenoleukodystrophy, there is a defect in a single peroxisomal enzyme or protein.

Zellweger syndrome (MIM 214100)

Infants with Zellweger syndrome, which is also known as the cerebrohepatorenal syndrome, usually come to attention soon after birth with hypotonia, poor feeding, failure to thrive, and convulsions. On examination they have a high forehead with a large anterior fontanelle (soft spot in the skull). Other findings can include cataracts, hepatomegaly, renal cysts, and stippling of the epiphyses of the long bones. The long-term outlook is poor and most of these infants die in early childhood.

The biochemical abnormalities in Zellweger syndrome are consistent with failure of peroxisomal biogenesis. The blood shows elevated levels of very long chain fatty acids and pipecolic acid, with low levels of plasmalogens. Zellweger syndrome can be caused by mutations in at least six different genes involved in peroxisomal synthesis. Inheritance is autosomal recessive, and there is no effective treatment.

X-linked adrenoleukodystrophy (MIM 300100)

X-linked adrenoleukodystrophy (ALD) usually presents in boys between the ages of 5 and 10 years with an insidious decline in intellectual skills, behavioural changes, and progressive neurological disability including ataxia, convulsions, and spasticity. Approximately 50% of these boys also develop adrenocortical insufficiency (Addison's disease). Most affected boys die within 2–3 years. Long-term survivors show profound disability, with loss of all skills. This condition was the subject of the film *Lorenzo's Oil*. Treatment with Lorenzo's oil, a mixture of glycerol trioleate and glycerol trierucate, lowers the levels of very long chain fatty acids in blood, but this has not proved to be effective in symptomatic patients.

X-linked adrenoleukodystrophy can present in a milder form, which mimics the clinical features of multiple sclerosis. This is known as adrenomyeloneuropathy. Approximately 20% of female heterozygotes develop these features. Both X-linked adrenoleukodystrophy and adrenomyeloneuropathy are caused by mutations in the gene that encodes a factor involved in the transfer of very long chain fatty acids across the peroxisomal membrane. The clinical problems seen in affected individuals result from the accumulation of very long chain fatty acids, which causes progressive damage to the brain and adrenal glands.

Disorders of porphyrin metabolism

The porphyrias comprise a group of disorders caused by reduced activity in one of the eight enzymes involved in the synthetic pathway of haem (Fig. 11.8). Some show autosomal dominant inheritance, others autosomal recessive. One particular form, porphyria cutanea tarda, can be caused either by inherited defects in the enzyme URO-decarboxylase or by exposure to toxins such as hexachlorobenzene. All forms of porphyria can be precipitated by drugs, such as barbitu-

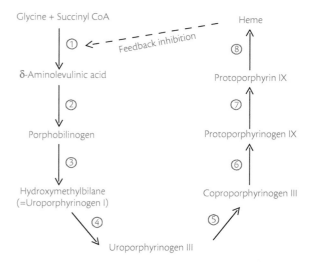

Enzyme	Associated deficiency disorder	
① ALA - synthase		
② ALA - dehydratase	ALA dehydratase deficiency (rare)	(AR)
③ HMB - synthase	Acute intermittent porphyria	(AD)
④ URO - synthase	Congenital erythropoietic porphyria	(AR)
⑤ URO - decarboxylase	Porphyria cutanea tarda	(AD)
⑥ COPRO - oxidase	Hereditary coproporphyria	(AD)
⑦ PROTO - oxidase	Variegate porphyria	(AD)
⑧ Ferrochelatase	Erythropoietic protoporphyria	(AD)

Fig. 11.8 The haem synthetic pathway showing the site of the enzyme defects in different forms of porphyria.

rates and progesterone, which increase the production of porphyrins.

The disorders of porphyrin metabolism are classified as either hepatic or erythropoietic, depending on the main site of porphyrin accumulation. The clinical features are variable and include abdominal pain, peripheral neuropathy, psychiatric disturbance, and cutaneous photosensitivity, which presents as blistering in sun-exposed skin. The diagnosis of the specific porphyrias is made by assay of porphyrins in red blood cells and urine, with confirmation by specific mutation analysis.

Variegate prophyria (MIM 176200)

Variegate porphyria is so named because the clinical features are variable. They can include abdominal pain, constipation, hypertension, fever, psychiatric disturbance, and skin photosensitivity. Presentation can be at any time from puberty onwards. Variegate porphyria is caused by reduced activity of protoporphyrinogen oxidase (PROTO-oxidase). It is particularly common amongst the South African Afrikaner population, 8000 of whom are affected as a result of transmission from a single Dutch settler. This represents a rare example of an autosomal dominant disorder achieving a relatively high frequency as a result of a founder effect in a genetic isolate (p. 140).

Variegate porphyria has also achieved fame as the proposed cause of the 'madness' of King George III (Box 11.2).

BOX 11.2 CASE CÉLÈBRE: GEORGE III

George III reigned as King of Great Britain for 60 years, from 1760 to 1820. This was a particularly eventful period in European and North American history, during which Britain managed to lose the American colonies but succeeded in curbing the territorial aspirations of the French Emperor, Napoleon Bonaparte (see Box 10.2). George is generally acknowledged to have been a sincere and well-intentioned man, but during the second half of his reign his ability to carry out his role was seriously impaired by intermittent ill-health. This has resulted in him being remembered, rather unfairly, as 'mad King George'. The film *The Madness of King George* gives a more sympathetic portrayal. This film also illustrates that one of the best ways to stay healthy at that time was to avoid the bizarre remedies prescribed by the King's physicians!

Beginning in 1788, George experienced a series of episodic illnesses in which he developed abdominal pain, constipation, a high temperature, muscle pain, and blistering on his arms. During these episodes his behaviour became unpredictable and irrational to the extent that he had to withdraw, or be withdrawn, from public life. One of his physicians noted that during these attacks, George's urine became red 'like Alicante wine'. A severe attack in 1810 coincided with the onset of dementia, following which George was deemed incapable of carrying out his duties and his role as sovereign was effectively taken over by his eldest son, the then Prince of Wales.

In hindsight it has been suggested that George III suffered from porphyria, probably of the variegate form. This would certainly explain the episodic nature of his illness and his unusual urine. It has also been suggested that several of his ancestors, including Mary Queen of Scots and her son James I, were also affected. Both of these individuals experienced severe recurrent abdominal colic, and Mary's intermittent odd behaviour led to a diagnosis of hysteria. Porphyria has also been suspected either posthumously or contemporaneously in some of George's descendants, notably Princess Charlotte, the sister of the German Emperor, Kaiser Wilhelm II, and the late Prince William of Gloucester, a first cousin of the present sovereign, Elizabeth II. Analysis of DNA extracted from bone samples taken from Princess Charlotte's coffin identified a point mutation in one of her PROTO-oxidase genes. However, this mutation was in an intron so its significance remains uncertain.

Without conclusive biochemical or molecular confirmation, the diagnosis of porphyria in George III remains somewhat speculative, although the accumulated circumstantial evidence is impressive. His medical history illustrates the diverse and unpredictable nature of the clinical features in this group of conditions. It also illustrates the enormous influence which a single point mutation may have exerted on the course of American and European history.

References

Macalpine I, Hunter R, Rimington C (1968) Porphyria in the royal houses of Stuart, Hanover and Prussia. *British Medical Journal*, **i**, 7–18.

Röhl JCG, Warren M, Hunt D (1998) *Purple secret. Genes, 'madness' and the Royal Houses of Europe*. Bantam Press, London.

Newborn screening

Neonatal screening programmes were first introduced in the early 1960s after it became apparent that careful dietary restriction could prevent the development of mental retardation in phenylketonuria. The screening procedure, known as the *Guthrie test*, was based on

---

TABLE 11.3 Criteria for neonatal screening

The disorder has potentially serious consequences

There is effective treatment or prevention

The screening test is reliable with high sensitivity and specificity (p. 148)

The disorder is sufficiently common to justify the cost of screening

A system is in place to ensure that screening is offered to all infants and that results are communicated and acted upon

For further details see Wilson JMG, Jungner G (1968) *Principles and practice of screening for disease.* World Health Organization, Geneva.

TABLE 11.4 Examples of disorders for which newborn screening is undertaken

Disorder	Treatment
Metabolic	
Biotinidase defieiency	Biotin (5–10 mg/day)
Galactosaemia	Restriction of galactose
Maple syrup urine disease	Restriction of branched-chain amino acids
MCAD deficiency	Avoidance of hypoglycaemia
Phenylketonuria	Low-phenlyalanine diet
Tyrosinaemia	Dietary restriction and NTBC to prevent liver damage
Urea cycle disorders	Protein restriction and sodium benzoate
Endocrine	
Congenital adrenal hyperplasia	Steroid replacement
Hypothyroidism	Thyroid hormone replacement
Other common inherited disorders	
α_1-Antitrypsin deficiency	Avoidance of cigarette smoke
Cystic fibrosis	Antibiotics, pancreatic supplements and physiotherapy (p. 107)
Duchenne muscular dystrophy	Physiotherapy—no effective cure
Sickle-cell anaemia	Folic acid, penicillin, and immunization (p. 159)

the observation that a high level of phenylalanine in a spot of blood could overcome the inhibition of the growth of a bacterium, *Bacillus subtilis*. Accordingly it was proposed that a drop of blood be collected from all newborn infants at age 3–7 days, after the institution of milk feeds. This is then sent to a central laboratory for analysis, and infants with a positive Guthrie test are investigated further by specific assay of the plasma level of phenylalanine.

The success of initial pilot studies in the USA led to the introduction of neonatal screening in almost all developed countries, with a general consensus, supported by the World Health Organization, that programmes should fulfil certain basic criteria (Table 11.3). Improvements in technology, particularly the recent introduction of tandem mass spectrometry, now permit screening for a large number of disorders (Table 11.4). These include a wide range of metabolic conditions, two hormonal disorders—congenital adrenal hyperplasia and hypothyroidism, and common single-gene conditions such as cystic fibrosis and sickle-cell anaemia.

Almost all programmes now include screening for phenylketonuria, hypothyroidism, and galactosaemia. Screening for other conditions, particularly the rare metabolic disorders, varies widely. For example, in the USA in 2003 screening was offered for over 30 conditions in Massachusetts, but for less than 10 in the neighbouring states of New Hampshire and Rhode Island. In most parts of the world newborn screening is required by law, but parents can refuse testing on religious grounds.

In all the conditions considered so far, some form of effective treatment exists which modifies the natural disease course. Pilot schemes have been undertaken to screen for Duchenne muscular dystrophy (p. 109), for which at present there is no effective treatment. One of the proposed benefits of these studies is that parents of affected infants can be offered genetic counselling so that, in theory, the birth of a second affected child can be prevented. The results of these studies indicate that in general parents welcome this opportunity for reproductive choice and that newborn screening for Duchenne muscular dystrophy is acceptable as long as ample provision exists for genetic counselling and long-term support.

Key Point

Newborn screening programmes should fulfil certain basic criteria. Specifically, there should be effective prevention or treatment for those disorders for which screening is offered.

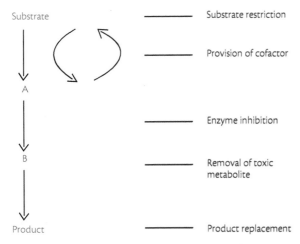

Substrate ———— Substrate restriction

———— Provision of cofactor

A

———— Enzyme inhibition

B

———— Removal of toxic metabolite

Product ———— Product replacement

Fig. 11.9 Conventional approaches to the treatment of metabolic disorders.

The treatment of biochemical disorders

Most genetic disorders cannot be cured, and until effective methods are developed for delivering gene therapy this situation is unlikely to change. However, progress in the treatment of many of the inborn errors of metabolism has been encouraging, as illustrated by the normal intellectual development of most children with phenylketonuria in the developed world. Although these conditions cannot be cured, there are several therapeutic approaches which can be utilized to control and prevent their more serious consequences (Fig. 11.9). The use of recombinant enzyme therapy is considered in Chapter 12 (p. 230).

Substrate restriction

If the accumulation of a substrate that has not been metabolized has adverse effects, these can be regulated by dietary restriction. The example of phenylketonuria indicates that this can be very successful. Phenylalanine is an essential amino acid, so care has to be taken to ensure that it is not totally excluded from the diet. Care also has to be taken to ensure that special diets contain adequate quantities of essential factors such as trace elements and vitamins. Galactosaemia provides another example of a disorder in which dietary restriction of a potentially toxic substrate has proved to be effective.

Replacement of a missing product

This approach can be used when clinical problems ensue from failure of the body to produce an important metabolite. Inborn errors involving the synthesis of hormones provide good examples. Defects in the thyroid hormone synthesis pathway can be treated very simply and effectively with thyroxine. Congenital adrenal hyperplasia, due to 21-hydroxylase deficiency (p. 179), responds well to treatment with cortisol. This inhibits the production of adrenal androgens via the hypothalamic–pituitary axis. Children with glycogen storage disease type Ia respond well to treatment by frequent feeding, which compensates for their innate inability to convert glycogen into glucose.

Removal of toxic metabolites

In a small number of conditions effective methods have been devised for the removal of toxic metabolites, which accumulate because of a downstream defect or the use of an alternative metabolic pathway. The example of the urea cycle disorders has already been cited. These conditions can be treated by a combination of dietary restriction and the use of drugs, such as sodium benzoate, which lower the level of ammonia. This is achieved by conjugation with glycine and glutamine to form products with a high nitrogen content which are excreted in urine. Disorders of copper and iron storage provide other examples. In Wilson's disease (MIM 277900) excess levels of copper can be removed using the chelating agent penicillamine. In haemochromatosis (p. 140) the increased absorption of iron can be balanced through regular venesection or the use of desferrioxamine, the drug that is also used to prevent the accumulation of iron in transfusion-dependent β-thalassaemia (p. 162).

Enzyme inhibition

Enzyme inhibitors can be used to prevent the synthesis of a compound that is toxic at high levels. Two examples have already been mentioned in this chapter: the use of NTBC to prevent the accumulation of high levels of the hepatotoxic metabolite fumarylacetoacetate in tyrosinaemia, and the widespread use of HMG-CoA reductase inhibitors to reduce cholesterol in familial hypercholesterolaemia.

Provision of enzyme cofactor

A small number of inborn errors are caused by either defects in the synthesis of an enzyme cofactor or reduced ability of a mutant enzyme to bind to its cofactor. In both situations small doses of the cofactor, which is often a vitamin, can increase residual enzyme activity sufficiently to restore normal metabolic function. As an example, thiamine provides an effective treatment for some forms of maple syrup urine disease and also for some of the mitochondrial disorders which present with the clinical picture of Leigh's disease.

Summary

Over 500 inherited metabolic disorders have been described. These are often referred to as inborn errors of metabolism. Most of these conditions show recessive inheritance. However, a few, such as most forms of porphyria, show dominant inheritance.

Inborn errors can occur in all metabolic pathways and include disorders of carbohydrate, amino acid, and lipid metabolism. Galactosaemia is caused by an inability to convert galactose to glucose and is treated by complete removal of milk from the diet. Glycogen storage disease type Ia is caused by a defect in the conversion of glycogen to glucose and is treated by regular feeding with glucose to prevent hypoglycaemia. Classic phenylketonuria is caused by reduced activity of phenylalanine reductase. Early treatment with a low-phenylalanine diet can prevent mental retardation. Maple syrup urine disease is a disorder of branched chain amino acid metabolism, which is treated by dietary restriction and, in some forms, supplementation with thiamine. The urea cycle disorders result in high circulating levels of ammonia. Treatment is by dietary restriction and with drugs such as sodium benzoate to promote the excretion of nitrogen. MCAD deficiency is a disorder of fatty acid oxidation which can present with sudden death. Treatment involves the avoidance of fasting and hypoglycaemia. Familial hypercholesterolaemia is caused by mutations in the gene which encodes the LDL receptor. Treatment is by low cholesterol diet and drugs, such as statins, which inhibit cholesterol synthesis.

Inborn errors can primarily affect specific cytoplasmic organelles. Lysosomal disorders are characterized by storage, often in association with neurodegeneration. Mitochondrial disorders affect organs with high energy requirements. Presentation can be with brain or muscle disease or with metabolic collapse associated with severe lactic acidosis. Peroxisomal disorders can involve a single peroxisomal enzyme or a failure of peroxisomal biogenesis.

Newborn screening for treatable metabolic disorders is carried out in most parts of the developed world. Screening for phenylketonuria and hypothroidism is universal. Screening for other conditions including galactosaemia, maple syrup urine disease, congenital adrenal hyperplasia, cystic fibrosis, and sickle-cell anaemia is available in some centres.

The treatment of metabolic disorders can involve substrate restriction (phenylketonuria), replacement of a missing product (hypothyroidism), the removal of toxic metabolites (urea cycle disorders), enzyme inhibition (tyrosinaemia), or the use of enzyme cofactors (maple syrup urine disease). New treatments are being developed based on recombinant replacement proteins and bone marrow transplantation (see Chapter 12).

Further reading

Benson PF, Fensom AH (1985) *Genetic biochemical disorders*. Oxford University Press, Oxford.

Holt IJ (ed.) (2003) *Genetics of mitochondrial diseases*. Oxford University Press, Oxford.

Nyhan WL, Sakati NO (1987) *Diagnostic recognition of genetic disease*. Lea and Febiger, Philadelphia.

Scriver CR, Beaudet AL, Sly WS, Valle D (ed.) (2001) *The metabolic and molecular bases of inherited disease*. McGraw-Hill, New York.

Multiple choice questions

1 A woman with phenylketonuria, who is married to her unaffected first cousin, requests genetic counselling. It would be correct to tell her that

(a) inheritance is autosomal recessive

(b) the probability that her partner is a carrier is 1 in 4

(c) if affected, her child should go on a low-phenylalanine diet from the age of 6 months

(d) when she becomes pregnant she should go on a low-phenylalanine diet from 20 weeks gestation onwards

(e) phenylketonuria is screened for routinely in the neonatal period by enzyme assay

2 An ill infant is found to have hypoglycaemia. Possible explanations for this would include

(a) glycogen storage disease

(b) alkaptonuria

(c) ornithine transcarbamylase deficiency

(d) medium chain acyl-CoA dehydrogenase deficiency

(e) Tay–Sachs disease

3 The following statements about metabolic disorders are correct:

(a) most show recessive inheritance

(b) most causal mutations exert a gain-of-function effect

(c) all inborn errors cause mental retardation

(d) all metabolic disorders are caused by enzyme deficiency

(e) all metabolic disorders are genetic

4 Neonatal screening is offered routinely for:

(a) familial hypercholesterolaemia

(b) galactosaemia

(c) hypothyroidism

(d) Tay–Sachs disease

(e) Zellweger syndrome

5 Examples of disorders which can be caused by mitochondrial mutations include

(a) Kearns–Sayre syndrome

(b) Leber's hereditary optic atrophy

(c) Hurler syndrome

(d) X-linked adrenleukodystrophy

(e) MELAS

Answers

1 (a) true—heterozygotes are unaffected

(b) true—this is explained in Chapter 13 (p. 249)

(c) false—a low-phenylalanine diet should commence as soon as the diagnosis is made

(d) false—she should go on a low-phenylalanine diet before becoming pregnant

(e) false—neonatal screening is based on assay of the level of phenylalanine

2 (a) true—this is a feature of von Gierke disease

(b) false—this condition has no effect on the blood sugar level

(c) false—this condition presents with elevated levels of ammonia

(d) true—this is because fatty acids cannot be oxidized to produce acetyl-CoA

(e) false—presentation in Tay-Sachs disease is with neurodegeneration

3 (a) true—most enzyme deficiencies show autosomal or X-linked recessive inheritance

(b) false—most causal mutations exert a loss-of-function effect

(c) false—many, such as alkaptonuria, do not affect the brain

(d) false—some are caused by failure of enzyme targeting or a defect in a cell receptor

(e) false—most are genetic but a few, such as porphyria cutanea tarda can be caused by exposure to toxins

4 (a) false—'cascade' screening is being considered in some populations

(b) true

(c) true

(d) false—population screening is carried out for carriers

(e) false—there is no effective treatment for this very rare disorder

5 (a) true—Kearns-Sayre syndrome is caused by mitochondrial DNA deletions

(b) true—Leber's optic atrophy is caused by mutations in mitochondrial DNA which are involved in oxidative phosphorylation

(c) false—Hurler syndrome is a lysosomal disorder

(d) false—X-linked adrenoleukodystrophy is a peroxisomal disorder

(e) true—MELAS is caused by a mutation in a mitochondrial gene for tRNA

Genes, drugs, and treatment

While developments in molecular biology have made a major impact on almost all branches of medicine, advances in genetically based therapy have lagged disappointingly behind. Although the recipients of much publicity, progress in gene therapy and the use of embryonic stem cells has been slow. Many inherited disorders still cannot be treated, and in those for which treatment is available it is usually based on conventional approaches.

However, as will become apparent later in this chapter, there is cause for cautious optimism. Genetically based treatment in the form of recombinant enzymes and proteins is now available for several disorders and over 800 trials based on gene therapy have been proposed in various parts of the world. As this book goes to press, scientists in Korea have just announced the successful culture of the first totipotent embryonic stem cell line to be derived from a therapeutic human clone. The potential benefits for treatment from these developments would have been unimaginable in the premolecular era.

Pharmacogenetics

Pharmacogenetics can be defined as the study of genetic variation in the body's response to drugs. For almost all agents there is increasing evidence that multiple genes are involved in the various processes through which a drug's effects are mediated (Table 12.1).

Pathogenic mutations in single genes involved in any of these processes can lead to significant changes in the

TABLE 12.1 Pharmacogenetic pathways in which genetic variation can occur

Pharmacokinetic	Pharmacodynamic
Absorption	Receptor binding/synthesis
Distribution	Ion channels
Metabolism	Signalling pathways
Excretion	Transcription regulation

potency of a drug or in its unwanted side effects. It is the study of these changes that constitutes the subject of pharmacogenetics. The study of minor changes in multiple genes falling within the spectrum of polymorphisms and rare variants (p. 141) is referred to as **pharmacogenomics**, as discussed later in this chapter.

Pharmacokinetic variation

Pharmacokinetics refers to the processes involved in the delivery of a drug to, or removal from, its target site. These processes include absorption, distribution, metabolism, and excretion. Large numbers of drug-metabolizing enzymes in which genetic variation occurs have been identified. Usually these have been recognized by measuring blood levels of the active drug or its metabolite. If these show a normal (Gaussian) distribution, it is assumed that metabolism is under polygenic regulation. However, if the blood levels show a bimodal or trimodal distribution, this suggests that a major single-gene effect is superimposed on a polygenic background.

The cytochrome P450 (CYP) family of enzymes, encoded by 55 known *CYP* genes, play a major role in the processing of most known drugs together with many environmental toxins and carcinogens. Four of these P450 enzymes—CYP2C9, CPY2C19, CYP2D6, and CYP3A4—are involved in the metabolism of over 80% of the most commonly used medications. Curiously, mutations in one *CYP* gene, *CYP1B1*, can cause congenital glaucoma (MIM 601771). The causal mechanism is not understood.

CYP2C9

This is responsible for the metabolism of various drugs including tolbutamide, verapamil, and the anticoagulant warfarin. Individuals with alleles that confer slow metabolizing activity are at risk of warfarin toxicity manifesting as potentially life-threatening haemorrhage.

CYP2C19

This enzyme is actively involved in the metabolism of several drugs with effects on the CNS including diazepam, imipramine, amitryptiline, and one form of phenytoin, together with other drugs including propranolol and the proton pump inhibitor omeprazole. This latter drug is used to treat *Helicobacter pylori* infection in patients with peptic ulcers. Several variant alleles that confer low enzyme activity have been identified. Enzyme activity shows a correlation with toxicity and efficacy. For example, the success rate of omeprazole in clearing *Helicobacter pylori* infection is 100% in slow metabolizers but only 28% in rapid metabolizers.

CYP2D6

This is involved in the metabolism of over 20% of all prescribed drugs including atenolol, codeine, haloperidol, and tamoxifen. Variation in CYP2D6 activity was first identified on the basis of the response to an antihypertensive drug known as debrisoquine, which is no longer in use. Patients were divided into 'extensive' (normal) metabolizers and 'poor' (slow) metabolizers. Subsequently an ultra-rapid metabolizer phenotype was identified and shown to be caused by the presence of multiple copies of the *CYP2D6* gene. Recently it has emerged that increased CYP2D6 activity shows an association with addiction to cigarette smoking. Another polymorphism in *CYP2D6* is associated with an increased risk of leukaemia because of reduced ability to detoxify ingested carcinogens.

CYP3A4

This enzyme is involved in the metabolism of 50–60% of all prescribed medications and is the most abundantly expressed P450 in human liver. It plays a central role in the metabolism of several drugs used in the treatment of cancer, and there is evidence that a specific polymorphism in *CYP3A4* is associated with an increased risk of secondary tumours such as treatment related leukaemia. As with other members of the cytochrome P450 family, polymorphic variation in the ability of CYP3A4 to metabolize environmental carcinogens and toxins is likely to play an important role in determining an individual's susceptibility to cancer. *CYP3A4* activity can be induced by various drugs including barbiturates and glucocorticoids.

Pharmacodynamic variation

Pharmacodynamics refers to the processes whereby the effects of a drug are mediated. Variable pharmacodynamic effects can reflect variation in binding of drugs to receptors, in signalling pathways, or in transcription pathways. As an aid to understanding, pharmacokinetics can be considered as encompassing the ways that the body handles a drug, whereas pharmaco-

TABLE 12.2 Examples of pharmacodynamic variation

Variation/polymorphism	MIM	Characteristics	Importance
β-1-Adrenergic receptor	109630	Altered response to β-agonists and β-blockers	Reduced therapeutic response. Also gain-of-function mutations predispose to heart failure
Phenylthiourea taster/non-taster	171200	Only 70% of population detect phenylthiourea as bitter	None known
Suphonylurea receptor	600509	Altered response to oral hypoglycaemic agents	Homozygous mutations cause familial hyperinsulinaemic hypoglycaemia
Thymidylate synthase activity	188350	Normally involved in synthesis of DNA	Over-expression causes resistance to 5-fluorouracil in treatment of cancer

Details of other known examples of pharmacokinetic and pharmacogenetic variation can be accessed at the Pharmacogenetics and Pharmacogenomics Knowledge Base website (*http://www.pharmgkb.org*).

dynamics refers to the effects that the drug has on the body.

Many examples of pharmacodynamic variation have been identified. Some of these result in failure of response to endogenous hormones. Androgen insensitivity (p. 179), in which the testosterone receptor is defective, represents a good example. Others involve defects in receptors which encode important ion channels. For example, defects in the ryanodine receptor gene, which encodes a calcium release channel, cause malignant hyperthermia, as discussed in the next section. Several other examples of historical interest and practical importance are indicated in Table 12.2.

> **Key Point**
>
> Genetic variation in drug response can be due to factors involved in the processing of a drug (pharmacokinetics) or to factors involved in the way that a drug's effects are mediated (pharmacodynamics).

Drug-induced genetic disease

In this section a number of inherited disorders which are unmasked by exposure to a specific drug or group of drugs are considered. The body's response to many external factors is determined by the interaction of several genes and metabolic pathways. However, the conditions discussed in this section all show single-gene inheritance. They are all important in that failure to diagnose and treat appropriately can have a fatal outcome.

Glucose-6-phosphate dehydrogenase deficiency (MIM 305900)

Glucose-6-phosphate dehydrogenase (G6PD) is an enzyme in the hexose monophosphate shunt, which, through the oxidation of glucose-6-phosphate, generates the only source of NADPH available in red blood cells for the prevention of damage by oxidation (Fig. 12.1). In individuals who are G6PD deficient, oxidant drugs, such as dapsone, phenacetin, primaquine (Box 12.1), and sulphonamide antibiotics, cause acute haemolysis. Other agents which can exert the same effect include naphthalene, the active constituent of mothballs, and fava beans. These are a popular vegetable in Mediterranean regions, where it has been recognized for over 2000 years that their ingestion can cause haemolysis. In these regions the condition is referred to as *favism*.

The gene that encodes G6PD is located on the X chromosome, and G6PD deficiency shows X-linked recessive inheritance. The worldwide distribution is similar to that of malaria, indicating that deficiency of G6PD somehow confers protection against infection by the malaria parasite. The incidence in males originating from areas in which malaria is endemic varies from 5 to 20%, with over 500 million males affected throughout the world.

Over 400 different allelic forms of G6PD have been described. The common variant in African-Americans, designated as G6PD^A-, results from two missense substitutions. This causes mild to moderate deficiency with self-limited intermittent haemolysis (Table 12.3). The common variant found in the Mediterranean, designated G6PD^Mediterranean, is caused by a single sub-

BOX 12.1 LANDMARK PUBLICATION: THE CAUSE OF PRIMAQUINE SENSITIVITY

The history of G6PD deficiency dates back over 2500 years to around 500 BC when the Greek philosopher and mathematician, Pythagoras, alerted his followers to the dangers of eating fava beans. However, it was not until 1956 that the basic enzyme defect was elucidated. This discovery was prompted by the experiences of large number of African-American troops who developed acute haemolytic anaemia when taking primaquine for malaria prophylaxis during the Korean war. Troops originating from western Europe were rarely affected, but, curiously, small numbers of troops from Mediterranean regions were more severely affected than their African-American counterparts. The basic defect was established by the Army Malaria Research Project group at the University of Chicago, who undertook conventional biochemical assays on samples from affected and unaffected men. As it was known that the level of reduced glutathione was low in affected men, attention was focused on enzymes involved in the glutathione reduction pathway (Fig. 12.1). The level of glutathione reductase was found to be normal, whereas the level of G6PD was very low in samples from all of the affected men.

In later studies it was shown that the level of G6PD is around 15% of normal in affected African-Americans and less than 10% in affected men of Mediterranean origin. In African-Americans young red blood cells are relatively resistant to oxidant damage. Thus when their red cells are exposed to an oxidant drug, such as primaquine, the haemolysis is usually short-lived and self-limiting as the marrow quickly responds by producing new red blood cells. In contrast, the Mediterranean mutation renders all red blood cells sensitive to oxidizing agents so that the haemolysis is much more severe and prolonged, to the extent that blood transfusion may be required. Severe G6PD deficiency can also lead to kernicterus (p. 147) in the neonatal period.

It has been estimated that worldwide over 500 million people have G6PD deficiency. Fortunately most of these will never experience haemolysis, but in poorer

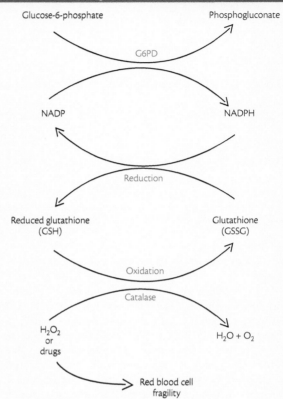

Fig. 12.1 Diagrammatic representation of how glucose-6-phosphate dehydrogenase (G6PD) deficiency results in failure to produce NADPH for the conversion of oxidized glutathione (GSSG) to reduced glutathione (GSH) which in turn is required to prevent oxidizing damage to the red blood cell.

parts of the Mediterranean and Middle East the condition remains an important cause of morbidity and occasional mortality.

Reference

Carson PE, Larkin Flanagan C, Ickes CE, Alving AS (1956) Enzymatic deficiency in primaquine-sensitive erythrocytes. *Science*, **124**, 484–485.

stitution resulting in severe deficiency with more prolonged haemolysis. Female carriers are unaffected. The diagnosis in males is confirmed by specific enzyme assay using commercially available kits. Treatment in severe haemolytic crises is by blood transfusion.

Malignant hyperthermia (MIM 145600)

Malignant hyperthermia, also known as malignant hyperpyrexia, is a rare but extremely important complication of general anaesthesia. If untreated, it conveys a high risk of mortality. It presents with a rapid rise in

TABLE 12.3 Common types of glucose-6-phosphate dehydrogenase

Type	Amino acid change	Ethnic distribution	Enzyme activity	Clinical consequences
A+	Asn126Asp	African-Americans (20%)	100%	None
A−	Asn126Asp	African-Americans (10%)	10–60%	Self-limited intermittent haemolysis
	+ Val68Met			
	or Arg227Leu			
B (= normal)	None	Worldwide	100%	None
Mediterranean	Ser188Phe	Mediterranean and Middle East populations (10–20%)	<5%	Severe prolonged haemolysis

temperature following the combined use of anaesthetics, particularly halothane, with muscle relaxants such as succinylcholine. The pyrexia is associated with muscle rigidity, tachycardia, lactic acidosis, and ultimately metabolic collapse. Treatment with cooling and the drug dantrolene is usually successful.

Malignant hyperthermia is caused by a rise in the intracellular level of calcium in muscle. In up to 50% of families this results from a mutation in the ryanodine receptor gene, *RYR1*, which encodes a calcium release channel. In these and other families that do not show linkage to the *RYR1* locus, inheritance is autosomal dominant. There is increasing evidence that other general so-called *modifier loci* also play a role.

At present molecular analysis is not usually deemed sufficiently reliable to determine whether someone from a malignant hyperthermia family can be given routine anaesthetic agents without precautions. Instead, susceptibility is determined by carrying out electrical stimulation studies on a muscle biopsy sample and by analysis of the effects of its exposure to halothane and caffeine. Individuals with a confirmed diagnosis of malignant hyperthermia or with a strong family history should be advised to carry details to that effect which can be readily accessed in the event of an emergency.

Pseudocholinesterase deficiency (MIM 177400)

Succinylcholine is a very powerful muscle relaxant which is often used when general anaesthesia is being induced to facilitate endotracheal intubation. Normally its effects wear off within a few minutes, but in approximately 1 in 3000 white North Americans the action is prolonged. This is because of reduced activity of pseudocholinesterase, the enzyme which normally hydrolyses succinylcholine to an inactive form.

Pseudocholinesterase is encoded by the gene *BCHE* (butyrylcholinesterase). Deficiency for pseudocholin-

esterase shows autosomal recessive inheritance. The most common 'atypical' variant form is caused by an A to G transition in the 70th codon resulting in the substitution of aspartic acid by glycine. This leads to reduced affinity of the enzyme for choline esters. A laboratory test based on inhibition of the enzyme by the local anaesthetic dibucaine can be used to detect heterozygotes and affected homozygotes in families.

Porphyria

The porphyrias constitute a group of inherited and acquired disorders involving defects in the haem biosynthetic pathway. Unlike most inborn errors of metabolism, inheritance is usually autosomal dominant because the various enzyme deficiencies are expressed in heterozygotes. Drugs such as barbiturates and glucocorticoids, which induce the synthesis of haem-containing cytochromes P450, can induce severe life-threatening attacks of porphyria in otherwise asymptomatic individuals. The various forms of porphyria have been discussed in Chapter 11 (p. 217).

> **Key Point**
>
> In a small number of individuals variation in drug response as determined by mutation in a single gene can have serious and potentially life-threatening consequences

Pharmacogenomics

Pharmacogenomics refers to the study of how variation in molecular function, determined by polymorphisms and mutations in multiple genes, influences the body's response to drugs. This contrasts with pharmacogenetics, which is concerned more with the way in which changes in single genes exert noticeable effects on variation in drug response. The distinction between these two terms and their application is not absolute and often they are used interchangeably. An alternative interpretation is that pharmacogenetics relates to the traditional study of genetic variation in drug response, whereas pharmacogenomics embraces the use of new genomic analytic techniques to predict an individual's susceptibility to beneficial and adverse effects.

Unfortunately, adverse drug reactions are common. Approximately 6–7% of hospital patients in the UK and the USA experience one or more adverse reactions, and in the UK these account for around 10% of all NHS bed occupancy. The pharmaceutical industry has risen to the challenge of trying to improve these figures through major investment in pharmacogenomics research. The preferred method at present is to undertake whole genome single nucleotide polymorphism (SNP—p. 141) analysis so that each individual can be assigned an SNP *profile* or *fingerprint* (note that this is not the same as a minisatellite *DNA fingerprint*).

The underlying rationale is that if large numbers of SNPs are analysed (>200 000) in sufficiently large numbers of individuals (including controls and patients who have experienced adverse events), then it will be possible to identify a profile that confers a high risk for an adverse event. This in turn will enable clinicians and pharmacologists to modify an individual's drug therapy according to their SNP susceptibility profile. To facilitate SNP profiling, automated high-throughput assays are anticipated, which will allow individual susceptibility to be determined before a drug is prescribed.

The mechanism that conveys genetic susceptibility could involve a single gene of major effect modified by several genes of minor effect, or multiple **polygenes** interacting with environmental factors such as diet. Hypertension is an example of a multifactorial disorder in which SNP profiling could be extremely helpful. At present treatment revolves around the use of four groups of drugs: angiotensin-converting enzyme (ACE) inhibitors, beta-blockers, calcium channel blockers, and diuretics. It can be difficult to predict how an individual with unexplained 'essential' hypertension will respond to a particular class of drugs. Younger patients and white people generally respond better to ACE inhibitors than do older patients and African-Americans, but there are no hard and fast rules and treatment is sometimes a matter of trial and error. The development of reliable pharmacogenomic susceptibility profiles would enable an individual's drug regime to be tailored to ensure maximum benefit.

It would be wrong to assume that SNP profiling is going to provide an accurate indication of drug susceptibility in every instance. The logistic challenges in terms of automated analysis and bioinformatic interpretation are considerable and the development costs are enormous. One of the most endearing characteristics of the human species is individual idiosyncrasy, and no doubt this will also apply at a pharmacogenomic level. However, an adverse drug reaction rate of 7% is unacceptably high, and all developments that could lead to an improvement are to be welcomed.

> **Key Point**
>
> It is hoped that technology will be developed to enable an individual's susceptibility to adverse drug reactions to be determined by molecular profiling

Treatment of genetic disease

Despite the remarkable advances of the last few decades, therapeutic options for genetic disorders are both limited and unsatisfactory. For many inherited conditions, particularly those characterized by neurodegeneration and neuromuscular disease, there is no effective treatment or cure. In many instances this reflects lack of knowledge or understanding of the basic molecular and protein defects, together with the enormous challenges posed by altering inherent patterns of gene expression.

There is cautious optimism that research in gene therapy will revolutionize this depressing scenario in the not too distant future, as outlined later in this chapter. Meanwhile, conventional medicine provides the mainstay of treatment for most genetic disorders. This has been considered in the context of cancer in Chapter 10 and of biochemical disorders in Chapter 11. Against this background there are several newly developed therapeutic strategies aimed at modifying the effects of genetic disease and in some situations possibly curing them. Generally these can be viewed as stopgap or interim measures, which bridge the existing era of conventional medicine and the brave new evolving world of gene therapy.

Replacement therapy with recombinant protein

The development of molecular genetic techniques for synthesizing recombinant human proteins has been exploited very successfully for the treatment of several human disorders. The basic principle is based on *expression cloning*. This involves the insertion of the human gene, together with promoters and regulatory elements, into an expression vector. Bacterial cells have the advantages that they grow rapidly and are relatively easy to manipulate. However, the proteins produced by transformed bacteria are not always identical to the normal human proteins because of different patterns of post-translational modification. Consequently, mammalian expression systems are now preferred. These can be laboratory based, as in the use of industrial bioreactors, or animal based, as with transgenic animals which have been engineered to produce a human protein in their milk.

Since the 1980s a large number of recombinant human proteins have been manufactured for the treatment of human disease (Table 12.4). Although more expensive than simply extracting the relevant protein from blood donations or cadaver organs, as used to be done to obtain factor VIII and human growth hormone respectively, the artificial synthesis of recombinant proteins has the major advantage that their use avoids any risk of contamination with agents such as hepatitis B or HIV. Both of these agents infected large number of males with haemophilia. Similarly, several children treated with growth hormone developed Creutzfeldt–Jakob disease, as a result of contamination of the

TABLE 12.4 Recombinant proteins used in the treatment of human disease

Protein	Disease	MIM
Lysosomal enzymes		
α-Galactosidase	Fabry disease	301500
α-Glucosidase	Pompe disease	232300
α-Iduronidase	Hurler disease	607014
Glucocerebrosidase	Gaucher disease	230800
Other proteins		
α1-Antitrypsin	α1-Antitrypsin deficiency	107400
Erythropoietin	Anaemia	
Factor VIII	Haemophilia A	306700
Factor IX	Haemophilia B	306900
Insulin	Diabetes mellitus	
Interferon	Multiple sclerosis	
Leptin	Obesity	

BOX 12.2 CASE HISTORY: TREATMENT OF HAEMOPHILIA A

I was diagnosed when I was 8 months old, after my parents had struggled to find out why I kept bruising. I started my home treatment when I was 8 years old and I was far from easy to inject. I have always had access to a high purity product. When I was between 9 and 10 I developed problems with my veins and eventually had to have a port-a-cath for 1 year. After my veins recovered I progressed to self-infusion and now I am totally self-sufficient and can inject into both arms.

During this time I have always been active and sporty. My parents' attitude throughout was not to wrap me in cotton wool. At the age of 5 I was spotted by a local tennis coach and began to play 'short' tennis. I played in a local club and won some tournaments and soon I progressed to tennis. Today I play doubles and singles for two clubs and I have won local and county tournaments in both. I have always played football and now at secondary school I also play basketball, badminton, and cricket. I've recently started shooting and joined a club in order to give my ankle a rest!

I tried playing the trumpet and got to grade 3 but breath control and the serving action of tennis gave me repeated stomach muscle bleeds which lead to periods of inactivity and bed rest and lots of treatment and pain. This is not what I like, so I gave up the trumpet and I'm very careful now with my serve!

All of this activity is only possible with good treatment—prophylaxis is supposed to be three times a week.

When I was 8 years old my ankle kept bleeding and wasn't getting any better—and they found I had inhibitors (inhibitory antibodies to factor VIII). This led to damage and arthritis. I received a high treatment regime and got rid of the inhibitors. Today I use a genetically manufactured factor treatment. I had an allergic reaction to the first one I tried but the one I use today is great.

My headmaster learned to inject me and he is my back-up for many school trips abroad. At school and with my family I have visited 10 countries and in July I visited Malaysia with the school and my big bag of factor VIII.

The future today looks great. My life is very full and active. I hope to do four A-levels and go on to do medicine. I have an active social life at church and school and recently passed my driving test.

Haemophilia rarely stops me. I fully appreciate the difference that regular high-purity injections and the support of the centre and the doctors have made to the sort of life I can lead compared to what it would have been if I had been born 10 or 20 years earlier. So I'm going to make the most of my life, live it to the full and hopefully, if I become a doctor, be sympathetic to those who need lots of injections and suffer pain.

This case history is reproduced with permission from 'DNA delivers. 50 years of progress told through the eyes of patients and their doctors', published by the Genetic Interest Group (GIG), London 2003.

growth hormone with an abnormal form of the prion protein. An added advantage of human recombinant proteins is that they are much less likely to induce antibody formation, as used to occur with the use of porcine-derived insulin in diabetes mellitus.

The availability of pure recombinant protein products has transformed the management of several conditions in which patients previously had to accept treatment with potentially immunogenic and contaminated therapeutic agents. Haemophilia A is an excellent example of such a condition (Box 12.2). This approach has also made it possible to treat some lysosomal storage disorders with regular infusions of recombinant enzyme. Lysosomal storage disorders can be subdivided on the basis of whether or not the central nervous system is involved (p. 214). For conditions in which the central nervous system is not involved, regular infusion offers the prospect of a cure if the recombinant enzyme can be modified to enable it to be taken up by the lysosomes. The non-neuronopathic form of Gaucher disease can now be treated successfully using recombinant gluococerebrosidase which has been modified so that it can be targeted to lysosomes in macrophages. This condition shows a general population incidence of 1 in 50 000 but a much higher incidence of around 1 in 500 in Ashkenazi Jewish populations. Regular infusions reverse almost all of the effects of the disease. The main drawback is the expense, which can exceed £100 000 per patient per year. Over 3000 patients have been treated successfully worldwide.

Similar treatment for lysosomal disorders which involve the central nervous system can have a beneficial effect on the physical consequences. However, it is unlikely that this will alter the disease course in the brain unless methods can be devised to enable the enzyme to cross the blood–brain barrier. Trials have been conducted for the treatment of Hurler syndrome (mucopolysaccharidosis Type I—p. 215) by regular infusions of recombinant α-iduronidase. These trials have shown that there is an improvement in lung function, joint mobility, and abdominal distension, and this form of treatment has now been approved by the Food and Drug Administration in the USA. As with all forms of recombinant enzyme therapy, this treatment for Hurler syndrome is very expensive.

Altering gene expression

One of the long-term goals of gene therapy is to develop methods for altering gene expression. The treatment of sickle-cell disease with hydroxyurea or butyrate (p. 159) illustrates the potential value of this approach. Both of these drugs lead to an increase in the synthesis of fetal haemoglobin. This not only compensates for the anaemia in sickle-cell disease, but also reduces the polymerization of Hb S, leading to a reduction in sickling. This in turn reduces the incidence of serious complications associated with increased blood viscosity.

Bone marrow transplantation

Bone marrow transplantation has proved to be extremely effective for the treatment of several genetic disorders. These include β-thalassaemia, as discussed in Chapter 8, and the severe immunological disorder known as severe combined immunodeficiency, in which both cellular and humoral immunity are deficient. Affected infants present with severe recurrent infection at the age of 3–6 months, and without treatment death ensues in early childhood. Bone marrow transplantation is curative.

Bone marrow transplantation has also been shown to be surprisingly effective for the treatment of several lysosomal storage disorders, including Hurler disease in which early transplantation can modify subsequent neurological degeneration as well as respiratory function and hepatosplenomegaly. This is achieved by the spread of bone marrow cells to various parts of the body including the brain. Successful bone marrow transplantation before the age of 2 years can lead to long-term survival with diminished neurodegeneration. However, the procedure conveys high risks of subsequent infection due to immunosuppression and graft-versus-host disease, together with a mortality rate of up to 30%, so many parents are reluctant to pursue this option.

> **Key Point**
>
> Several new genetically based strategies have been developed for the treatment of inherited disorders. The use of recombinant proteins to treat conditions such as haemophilia has proved to be particularly successful.

Gene therapy

Gene therapy can be defined as the treatment of human disease by the transfer of foreign DNA into a patient's cells. In germline gene therapy the genetic change involves the gonads and is therefore transmitted to future generations. This is widely regarded as unethical, partly because of the risks and uncertainties associated with contemporary technology, and partly because of unease at the notion of tinkering with an individual's genome. In somatic cell gene therapy the goal is to modify the genetic constitution of an organ or tissue for therapeutic purposes. Research into somatic cell gene therapy is generally regarded as being acceptable as long as all protocols have received formal approval and the disease being treated is sufficiently serious to justify a new and potentially hazardous therapeutic approach.

Gene therapy has promised much but, to date, has delivered little. Over 800 trials have been approved throughout the world and, although in many instances preliminary results have been encouraging, long-term success has been disappointing. With the possible exception of severe combined immunodeficiency, attempts to cure single-gene disorders have met with little success. Similarly, most research into treating cancer by gene therapy is still at a very early stage and of little proven benefit. However, the potential gains to be derived from successful gene therapy are so enormous, and alternative approaches so limited, that research and investment continue on a major scale.

Prerequisites for gene therapy

Given that gene therapy remains an experimental and largely untried approach, there is universal agreement that a number of important conditions must be met before gene therapy-based research can be undertaken.

The disease

This must be serious and essentially incurable using conventional treatments. Suitable examples include cystic fibrosis, Duchenne muscular dystrophy, transfusion-dependent β-thalassaemia, and neurodegenerative disorders such as Alzheimer disease. Chronic conditions requiring lifelong treatment such as haemophilia and immunodeficiency are also considered suitable. When (and if) safe and successful gene therapy strategies have been developed, it is likely that their use will be extended to conditions such as diabetes mellitus and cardiovascular disease.

The molecular basis

The causal gene must have been isolated and the molecular pathogenesis of the condition must be understood. This will directly influence the therapeutic strategy. If a disorder is caused by mutations with a loss-of-function effect, then a rational approach would be to try to insert a normal copy of the gene, ideally with adjacent control sequences if these are known. If a condition is caused by mutations with a gain-of-func-tion or dominant negative effect, then the insertion of a normal copy of the gene would probably be disastrous. In these situations an alternative strategy is required. This could involve either trying to replace the mutant gene with a normal copy by homologous recombination, or targeted inhibition of mutant gene expression using an antisense oligonucleotide. This consists of a short sequence of RNA which binds to complementary mRNA and prevents protein synthesis.

The target organ or tissue

Ideally this should be readily accessible with a long survival time and ability to replicate itself. Bone marrow stem cells represent an example of an ideal target for treating haematological and immune disorders. Experience with bone marrow transplantation for lysosomal storage disorders (p. 232) indicates that bone marrow stem cells could be used for treating other conditions, either because the modified cells would release an enzyme or hormone into the circulation or because some cells migrate to form the mononuclear-phagocyte system in other parts of the body.

Ethical approval

The ethical issues pertaining to gene therapy are particularly complex. All of the approaches are essentially untried and of unproven safety. However, in the absence of any alternative treatment, patients with life-threatening disorders may well feel that they have nothing to lose and therefore be willing to participate in what could be a risky procedure. As a safeguard to the abuse of vulnerable patients in this way, all gene therapy research protocols have to be reviewed and approved both locally by the host institution and by national regulatory bodies.

Transferring DNA into a cell

Two terms are used to describe the transfer of genes into animal cells. **Transduction** (Table 12.5) refers to virus-mediated gene transfer using as a vector a modified virus which naturally infects the host cell. **Transfection** refers

TABLE 12.5 Terms used in gene and stem cell therapy

Transdifferentiation	Change in tissue specificity shown by some somatic stem cells
Transduction	Virus mediated gene transfer
Transfection	Non-viral mediated gene transfer
Transformation	Malignant change in a cell; also refers to uptake of DNA by a bacterial cell
Transgene	Exogenous gene which has been transferred into a cell

TABLE 12.6 Advantages and disadvantages of different gene therapy modalities

Method of gene transfer	Advantages	Disadvantages
Adenoviruses	Can infect dividing and non-dividing cells	Very immunogenic
	Large DNA insert of up to 35 kb	Poor expression of exogenous DNA
	No risk of insertional mutagenesis	
Adeno-associated viruses	Inserted into chromosome 19, but no risk of insertional mutagenesis	Small DNA insert (<4.5 kb)
	Non-immunogenic	
Retroviruses	DNA integrated into host genome	Risk of insertional mutagenesis
	Non-immunogenic	Can only infect dividing cells
	Non-toxic	Relatively small insert (<8 kb)
Lentiviruses	Can infect non-dividing cells	Risk of insertional mutagenesis
	DNA is integrated into host genome	
Naked DNA	Technically straightforward	Inefficient DNA transfer
Liposomes	Can accommodate large amount of DNA	Inefficient DNA transfer
	Non-toxic	Poor expression of exogenous DNA
Receptor-mediated endocytosis	Efficient transfer of DNA	Poor expression of exogenous DNA
	Non-toxic	

to non-viral-mediated gene transfer by methods such as direct injection, binding to membrane soluble liposomes, and receptor-mediated endocytosis. **Transformation** refers to the change of a normal cell into a tumour cell, but in bacteria this term is used to describe the uptake and incorporation of foreign DNA.

Transduction

Several viral systems have been used in gene therapy trials. Most studies have focused on the use of adenoviruses and retroviruses, but unfortunately both of these have major disadvantages (Table 12.6).

Adenoviruses

These naturally occurring DNA viruses are a common cause of mild upper respiratory tract infection, particularly in young children. Their DNA does not become integrated into the host's genome. This has the advantage that there is no risk of the host DNA being damaged, a process referred to as **insertional mutagenesis**. However, this conveys the disadvantage of only short-term expression of up to a few weeks. A greater disadvantage is that adenoviruses generate vigorous host immune responses. In one instance this resulted in the death of the young man participating in a trial of gene therapy for the urea cycle disorder, ornithine transcarbamylase deficiency. This tragic event led to the

rapid cessation of the trial and a general reluctance to use adenoviruses in future studies.

The construction of a recombinant adenovirus for gene therapy involves a number of complex steps. The viral genes needed for replication are replaced by the desired therapeutic gene, known as the *insert*, by homologous recombination. The disabled adenovirus is then manufactured in a packaging cell line, which contains the genes and proteins necessary for viral replication. The disabled adenoviruses are shed from the packaging cell line and then purified before incorporation into the host cells by receptor-mediated endocytosis. The virus is then released and conveyed to the host cell nucleus where the therapeutic gene is transcribed. As the recombinant virus lacks the genes needed for replication, the viral infection cannot spread to adjacent cells.

Adeno-associated viruses are single-stranded non-pathogenic parvoviruses which can exist in isolation within a cell or become integrated into the host genome at a specific site on the long arm of chromosome 19. Subsequent co-infection with an adenovirus enables the adeno-associated virus to replicate. In contrast to adenoviruses, adeno-associated viruses are not immunogenic but they have the major disadvantage that, even after the removal of most of their genome, they can only carry inserts of foreign DNA of up to 4.5 kb, i.e. much less than the average-sized human gene. This is in contrast to adenoviral vectors, which can carry inserts of up to 35 kb of foreign DNA.

Retroviruses

These RNA viruses normally have the ability to transform cells. However, for the purpose of gene therapy they are disabled by the removal of most of their genetic material, which is replaced by the gene to be inserted. Retroviruses encode reverse transcriptase which converts RNA to cDNA. This is then integrated randomly into the host genome. Retroviruses have several advantages for gene therapy in that they are very efficient at entering almost all cells and they are both non-toxic and non-immunogenic.

Their major disadvantage relates to concerns about insertional mutagenesis, as there is no way of predicting or regulating the site of integration into the host genome. Thus the possibility exists that integration could damage an important host gene or generate a new chimaeric oncogene in a manner similar to that which occurs in chronic myeloid leukaemia or Burkitt's lymphoma (p. 190). Concerns about this possibility have proved to be justified, with the death from leukaemia of two of the first children to undergo successful gene therapy treatment for severe combined immunodeficiency.

Lentiviruses are a form of retrovirus which can infect non-dividing cells. They are easier to culture than retroviruses and provide greater capacity for the incorporation of regulatory sequences. However, as with other forms of retroviruses, they convey a risk for insertional mutagenesis.

Transfection

Various methods for non-viral transfer of DNA into cells have been considered but unfortunately none of these has proved to be effective. The attraction of these methods is that they avoid the use of potentially pathogenic viruses and possible risks of insertional mutagenesis. However, if foreign DNA is not integrated into the host genome then it is unlikely that it will be expressed over a sustained period.

Direct injection of 'naked' DNA into muscle cells has been assessed as a possible means of treating Duchenne muscular dystrophy. Transient expression of a specially constructed dystrophin minigene was observed in specific muscle cells, but with no obvious long-term benefit. It is difficult to envisage how an approach such as this would succeed for a generalized muscle disease, although direct injection of DNA into a small target

organ or gland might prove to be a more worthwhile option.

The use of DNA packaged with membrane-soluble synthetic lipid-based vesicles, known as liposomes, has been tried in the treatment of cystic fibrosis. Liposomes allow the transfer of large fragments of DNA, but in the studies carried out to date only transient expression has been observed.

Another approach to non-viral mediated transfer involves the conjugation of DNA with a protein which binds to a cell surface receptor and is then conveyed into the cell. This process is known as receptor-mediated endocytosis. The main problem with this technique is that the protein–DNA conjugate is usually taken up and degraded by lysosomes, so that very little free DNA is expressed.

The example of X-linked severe combined immunodeficiency (MIM 300400)

Severe combined immunodeficiency is a serious disorder of both cellular and humoral immunity, which results in susceptibility to bacterial, viral, and fungal infections. The disorder can be caused by several different genetic defects. The X-linked form, which accounts for approximately 50% of all cases, is caused by mutations in the gene *IL2RG*, which encodes the a chain of the interleukin-2 receptor. As a result, the growth and differentiation of both B cells and T cells are impaired.

In contrast to almost all other conditions, gene therapy for the X-linked form of severe combined immunodeficiency has proved to be very successful. In 11 boys for whom a histocompatible bone marrow transplant was not available, haematopoietic stem cells were incubated with a modified retrovirus containing *IL2RG* and then returned to their bone marrow. Nine were cured, and understandably the treatment was regarded as a great success. Unfortunately two of these boys subsequently developed leukaemia, probably as a result of insertional mutagenesis resulting in activation of a proto-oncogene, *LMO2*, which is known to be a cause of childhood leukaemia. This setback resulted in the temporary cessation of related gene therapy trials in both the UK and the USA.

> **Key Point**
>
> Despite extensive research, progress in gene therapy in a clinical setting has been very limited. Major problems and side effects include poor gene expression, viral immunogenicity and insertional mutagenesis.

Regenerative medicine

Two recent major developments, occurring within a year of each other, have raised the prospect of new genetically based therapeutic approaches, which can be considered under the heading of *regenerative medicine*. The first, announced in 1997, was the widely publicized cloning of Dolly the sheep (Box 12.3). The second, in 1998, was the successful isolation of human embryonic stem cells. Both of these events were heralded as major scientific achievements. Both also generated a major ongoing ethical and political debate. Although the potential therapeutic benefits are enormous, optimism has to be tempered by the reality that progress in this field is still very much at an early developmental stage.

Stem cells

Stem cells can be defined as undifferentiated cells, which are capable of both self-renewal and proliferation into specific differentiated cells in an organ or tissue. It has been customary to distinguish between embryonic stem cells with unlimited potential for differentiation (*totipotent*), and postnatal somatic stem cells which have a more restricted differentiation profile limited to cells in a particular organ (*multipotent* or *pluripotent*). However, recent discoveries indicate that some apparently organ- or tissue-specific stem cells, sometimes referred to as being *lineage specific*, can manifest different tissue specification. This is referred to as cellular or phenotypic **plasticity** or **transdifferentiation**.

BOX 12.3 CASE CÉLÈBRE: DOLLY THE SHEEP

Dolly the sheep led a short but celebrated life. She was born in the summer of 1996 close to the Roslin Institute in Edinburgh, where she soon found herself at the centre of a storm of publicity. Dolly's dubious claim to fame lay in the fact that she was the first mammal to be successfully cloned from adult tissue. In her particular case the cell originated from the mammary epithelium of an unknown donor ewe, an origin which was acknowledged by naming her in honour of the physical attributes of a famously well-endowed Country and Western singer. Dolly was cloned by transferring a single diploid nucleus into an enucleated unfertilized egg. Nuclear transfer and oocyte stimulation were induced using electrical impulses. The fused 'couplet' was then cultured in a ligated oviduct for 6 days before transfer into the uterus of a recipient ewe. Out of a total of 277 fused couplets, 29 blastocysts were implanted, of which only one, Dolly, survived to term. Confirmation that Dolly was a true clone was obtained by microsatellite analysis, which showed her to be identical to the donor mammary cells and not to be related to the recipient surrogate ewe.

Following news of Dolly's successful cloning the same technique was applied to other species, including goats, cattle, mice, pigs, and rabbits. The subsequent successful cloning of a cat prompted the fanciful notion that bereaved pet owners could clone their favourite family moggie. Inevitably attention began to focus on the possibility of human cloning, a prospect which was regarded, and continues to be regarded, with horror by all reputable scientists and government authorities. Concern about the ethics of cloning increased when news began to emerge that Dolly had developed premature arthritis, a complaint which contributed to the decision that she be 'put down' at the youthful age of 6 years. Postmortem examination failed to identify any sinister cloning-related pathology, and her medical problems of arthritis and obesity were attributed to an unhealthy celebrity lifestyle.

Dolly's successful cloning, although modestly announced, almost certainly raised unwarranted anxiety amongst the ethical community and unrealistic expectations amongst infertile couples. The techniques in existence for reproductive cloning are highly inefficient and also unsafe, as indicated by the high incidence of abnormalities in the small numbers of surviving embryos. In particular, there is concern that the pattern of methylation reprogramming in the developing embryo will be abnormal and that a clone derived from adult tissue will be susceptible to premature ageing. If there is a niche for reproductive cloning it probably lies in the preservation of endangered species rather than the procreation of human beings.

Reference

Wilmut I, Schnieke AE, McWhir J *et al.* (1997) Viable offspring derived from fetal and adult mammalian cells. *Nature*, **385**, 810–813.

Embryonic stem cells (ESTs)

These totipotent cells are derived from the inner cell mass of a blastocyst during gastrulation, the process whereby a trilaminar disc is formed during weeks 2 and 3 after fertilization. They can be cultured, with difficulty, by using a special culture medium layered with **feeder cells**, which provide essential growth factors and nutrients. In theory ESTs could have very widespread applications in medicine. These include the regeneration of a diseased tissue or organ, as has been proposed for the treatment of diabetes, Parkinson's disease, and spinal injuries, and as an adjunct to gene therapy. A recent example of this is the successful correction of a genetic immunodeficiency in mice. This was achieved by the genetic modification of histocompatible ESTs which were then stimulated to differentiate into haematopoietic stem cells before transplantation into the bone marrow of the immune deficient mice.

In practice there are two serious factors which limit the use of ESTs in humans. The first relates to the ethical issues surrounding the fact that ESTs can only be obtained by the creation and subsequent destruction of a human embryo, through the process known as therapeutic cloning (Fig. 12.2). This involves the transfer of a nucleus from the potential recipient into an unfertilized enucleated oocyte, which is then allowed to develop into a blastocyst from which the embryonic stem cells are removed. Attempts have been made to ban therapeutic cloning in the USA, although carefully regulated research is permitted elsewhere, notably in Australia, Singapore, and the UK.

The second concern surrounding the use of ESTs in humans is based on practical issues such as possible immunogenicity, safety, and technical limitations. Although therapeutic cloning should generate ESTs which are totally histocompatible, it is likely that they will contain mitochondria from the enucleated oocyte which will have a different immunogenetic profile. Safety concerns have emerged as a result of the discovery that transplanted ESTs can occasionally develop into teratomas and teratocarcinomas. These tumours contain multiple types of undifferentiated cells. (Ovarian teratomas can also be caused by autonomous development of an unfertilized diploid oocyte—p. 83.) Finally, despite extensive research, the technical limitations are substantial in terms of how EST growth and differentiation are regulated.

Somatic stem cells

The use of somatic stem cells from a healthy donor is much more acceptable ethically than EST-based treatments. Most research has focused on haematopoietic stem cells derived from bone marrow, which could in theory provide a source of normal haematopoietic cells including erythrocytes, platelet-producing megakaryocytes, granulocytes, and lymphocytes. The demonstration that cells which originate in bone marrow can migrate to other organs such as brain, heart, and liver has raised cautious optimism that stem cell transplantation into bone marrow could be used to treat multisystem disease. It has also been shown that haematopoietic stem cells are present in blood obtained from the umbilical cord and placenta after birth. This has prompted small numbers of parents to request that a pregnancy be 'genetically engineered' to provide an unaffected histocompatible sibling, whose cord blood could be used to treat an older sibling affected with a condition such as Fanconi's anaemia or severe β-thalassaemia. Concern on the part of society that this could represent the thin end of the wedge of a 'designer baby' culture has meant that some regulatory authorities, such as the Human Fertilisation and Embryo Authority in the UK (*www.hfea.gov.uk*) have been reluctant to grant permission for this to go ahead.

Unfortunately the factors which determine stem cell plasticity are not understood and it is likely to be many years before stem cell therapy based on successful transdifferentiation becomes a realistic option. An additional problem with all somatic stem cell therapy, apart from that involving identical twins, is that of immunogenicity. This can be overcome by bone marrow ablation when using haematopoietic stem cells, but this in itself conveys a significant mortality. Finally, it is worth noting that the use of histocompatible stem cells for the replacement of an organ destroyed by an autoimmune process is likely to be of only transient benefit as the immune system will probably attack the newly formed organ. Thus stem cell treatment for conditions such as type 1 diabetes (p. 127) could well prove to be disappointing.

Reproductive and therapeutic cloning

The distinction between reproductive and therapeutic cloning is crucial. Reproductive cloning involves nuclear transfer to an enucleated oocyte, which is then implanted in a surrogate mother and allowed to develop into an embryo and beyond (Fig. 12.2). As discussed in the context of Dolly (Box 12.3), this is a very unreliable procedure with a poor success rate. In addition, there is much to be learned about epigenetic aspects of reproductive cloning such as methylation and cellular ageing. Among both the scientific and the political

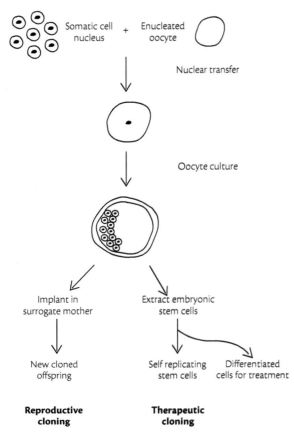

Somatic cell nucleus + Enucleated oocyte

Nuclear transfer

Oocyte culture

Implant in surrogate mother

New cloned offspring

Extract embryonic stem cells

Self replicating stem cells

Differentiated cells for treatment

Reproductive cloning

Therapeutic cloning

Fig. 12.2 The basic principle underlying reproductive and therapeutic cloning.

communities there is an almost universal consensus that reproductive human cloning is unacceptable. Although a few deluded older individuals might feel that their attributes justify cloning for posterity, it is extremely uncommon to encounter members of the younger generation who would wish to be a clone of one of their parents.

The initial procedures involved in therapeutic cloning are similar, but development does not proceed beyond the blastocyst stage, when the embryonic stem cells are removed. As already discussed, this procedure is being restricted in the USA despite a strong lobby from the scientific community. Elsewhere therapeutic cloning is being carried out on a limited and carefully controlled basis to foster research on gene regulation in early human development and the use of ESTs as potential therapeutic agents.

This research received a considerable boost in early 2004, when it was announced that scientists in South Korea had successfully cloned the first human blastocysts. To achieve this they used 242 ova from 16 donors resulting in a total of 30 blastocysts and a single line of totipotent embryonic stem cells. This low success rate indicates that the technical difficulties, although formidable, are not insurmountable. The next step is to exploit this development by finding ways to implement successful embryonic stem cell therapy.

> **Key Point**
>
> The use of embryonic stem cells obtained by therapeutic cloning could revolutionize the treatment of both genetic and non-genetic disease. Major technological obstacles have to be overcome before this approach can be implemented in clinical practice.

Summary

Pharmacogenetics refers to the study of genetic variation in the body's response to drugs. This variation can involve stages such as absorption, metabolism, and excretion whereby the body processes a drug (pharmacokinetics), or factors such as receptor sensitivity and signalling pathways through which a drug's effects are mediated (pharmacodynamics). The cytochrome P450 enzyme system provides several examples of how variation in the metabolism of a drug can influence its effects. Variation in the body's response to environmental carcinogens and toxins contributes to innate susceptibility to develop cancer.

A small number of potentially life-threatening single-gene disorders can be precipitated by exposure to specific drugs or dietary factors. These disorders include glucose-6-phosphate dehydrogenase deficiency, malignant hyperthermia, and pseudocholinesterase deficiency. The latter two conditions are precipitated by anaesthetic agents; careful inquiry should always be made about a personal and family history of adverse anaesthetic incidents before general anaesthesia is undertaken.

Pharmacogenomics refers to the use of new methods of genome analysis to identify molecular profiles which convey susceptibility to adverse drug reactions. The method most commonly proposed at present involves the study of large numbers of single nucleotide polymorphisms (SNPs). It is anticipated that in the future SNP profiling will be undertaken before drugs with serious side effects are prescribed.

New forms of genetically based treatment include replacement therapy with recombinant based protein, the use of drugs to alter gene expression, and bone marrow transplantation using histocompatible cells. This has proved to be very successful in the treatment of β-thalassaemia and severe combined immunodeficiency. Large numbers of gene therapy trials have been undertaken based on different methods of viral mediated (transduction) and non-viral mediated (transfection) DNA transfer for the treatment of single-gene disorders and cancer. Success has been disappointing because of problems resulting from viral immunogenicity and poor gene expression. Gene therapy has been successful in the X-linked form of severe combined immunodeficiency, but two of the first boys to be treated subsequently developed leukaemia as a result of insertional mutagenesis.

Stem cell therapy offers a new and potentially very powerful form of treatment for inherited disorders, degenerative disease and traumatic injury to the spinal cord. Embryonic stem cells are totipotent but their use is restricted at present by ethical concerns, safety issues and technical limitations. The use of somatic haematopoietic stem cells offers much greater prospects of success in the short term.

Reproductive cloning involves nuclear transfer to create a viable human embryo. This is condemned universally on ethical grounds. Therapeutic cloning involves nuclear transfer to obtain a pure line of embryonic stem cells for therapeutic purposes. This is permitted on a carefully regulated basis in some parts of the world. The first successful establishment of a totipotent human embryonic stem cell line from a therapeutic clone was announced in early 2004.

Further reading

Brooks G (ed.) (2002) *Gene therapy. The use of DNA as a drug.* Pharmaceutical Press, London.

Kolata G (1998) *Clone. The road to Dolly and the path ahead.* William Morrow and Company, New York.

Treacy EP, Valle D, Scriver CR ((2001) Treatment of genetic disease. In: Scriver CR, Beaudet AL, Sly WS, Valle D (ed.) *The metabolic and molecular basis of inherited disease*, 8th edn. McGraw-Hill, New York.

Weber WW (1997) *Pharmacogenetics.* Oxford University Press, Oxford.

Multiple choice questions

1 A patient admitted to hospital for surgery requiring a general anaesthetic recalls that her father almost died because of complications with a general anaesthetic many years ago. It would be correct to tell her that

(a) general anaesthesia is now much safer than it used to be so there is no cause for concern

(b) men are more susceptible to problems with general anaesthesia so there is no cause for concern

(c) her father's problems were probably due to pseudocholinesterase deficiency and as this shows autosomal recessive inheritance there is no cause for concern

(d) it would be very important to try to learn as much as possible about her father's anaesthetic problems before proceeding with the operation

(e) the anaesthetist must be alerted to this history before the operation is carried out

2 Replacement therapy with a recombinant product is available for

(a) haemophilia A

(b) Duchenne muscular dystrophy

(c) growth hormone deficiency

(d) Huntington disease

(e) Gaucher disease

3 Adenoviruses are useful in gene therapy because

(a) there is no risk of insertional mutagenesis

(b) they can accommodate a large insert

(c) they are non-immunogenic

(d) exogenous DNA is well expressed in the host

(e) they can infect both dividing and non-dividing cells

4 The following statements about gene therapy are correct:

(a) transfection refers to viral mediated gene transfer

(b) transduction refers to non-viral mediated gene transfer

(c) retroviruses convey a risk of insertional mutagenesis

(d) polygenic diseases will be easier to treat than monogenic diseases

(e) somatic cell gene therapy should avoid the possibility that a genetic change will be passed on to future generations

5 A man from Turkey, who is married to his first cousin, seeks genetic counselling because he has been diagnosed with G6PD deficiency. It would be correct to tell him that

(a) inheritance is autosomal recessive

(b) female carriers are not affected so there is no need for his daughters to be concerned

(c) the Mediterranean form of G6PD deficiency is more severe than the African-American form

(d) he should avoid fava beans

(e) sometimes severe G6PD deficiency can present in the neonatal period with jaundice

Answers

1 (a) false—it is true that general anaesthesia is safer now than previously, but this does not necessarily explain the cause of her father's problems

  (b) false—men are not more susceptible to problems with anaesthesia than women

  (c) false—her father could have had malignant hyperpyrexia which shows autosomal dominant inheritance

  (d) true—for the reason stated above

  (e) true—this is absolutely vital

2 (a) true

  (b) false—no method is available for targeting a recombinant protein to muscle

  (c) true

  (d) false—no method has been developed for targeting a recombinant protein to the brain

  (e) true—for the non-neuronopathic form

3 (a) true—the virus DNA is not incorporated into the host DNA

  (b) true—adenoviruses can take an insert of up to 35 kb

  (c) false—they are very immunogenic

  (d) false—expression is poor

  (e) true—this is an advantage over retroviruses

4 (a) false—this refers to non-viral gene transfer

  (b) false—this refers to viral mediated gene transfer

  (c) true—this is because the cDNA is incorporated into the host genome

  (d) false—the molecular basis of polygenic disorders is generally very poorly understood and this is a prerequisite for successful gene therapy

  (e) true—this is why it is much more acceptable ethically

5 (a) false—inheritance is X-linked recessive

  (b) false—it is true that carriers are not affected, but his daughters could be homozygous affected as his wife, who is his first cousin, could be a carrier

  (c) true—haemolysis is self-limited in the African-American form

  (d) true—these are a major precipitant of haemolysis in this condition

  (e) true—severe G6PD deficiency can cause kernicterus

Clinical skills and scenarios

In both the UK and the USA, the recommendations drawn up for the teaching of genetics to medical students emphasize the importance of acquiring specific clinical skills (Appendices 1 and 2). These include the ability to

- elicit a family history
- construct and interpret a family tree
- determine and convey information about genetic risks
- interpret genetic laboratory reports
- access written and electronic information resources.

In addition, by the completion of training, students are expected to 'understand approaches which can be used for the diagnosis of genetic disease' (UK), 'be able to recognize cases with abnormal development and dysmorphic features' (UK), and be able to 'formulate an appropriate plan for diagnostic evaluation and patient management' (US). Given the vast numbers of genetic disorders which exist, these are ambitious requirements and clearly it is not possible to cover every possible scenario in a book of this size. As a compromise, two relatively common presentations are considered here: a child with unexplained mental retardation and a child with multiple congenital anomalies. It is hoped that by providing an outline of the standard approach, students will be able to build on this to devise a plan of action when presented with a patient or family with a possible genetic disorder.

Construction of a family tree

When an individual or family is referred for genetic evaluation, the initial step almost always involves the construction of an accurate family tree, also referred to as a pedigree (from the French *pied à grue* = crane's foot). With the increasing recognition of the importance of genetic susceptibility to a wide range of multifactorial disorders (Chapter 6), a strong case can be made for including a detailed family history as a core component of every medical case history.

Although sophisticated computer packages have been developed, the essential tools consist of pen and paper together with knowledge of basic pedigree symbols. These were first introduced in Chapter 4 (p. 71) and are reproduced here as Fig. 13.1. A very detailed system of human pedigree nomenclature has been developed which caters for every eventuality, but in practice the symbols shown in Fig. 13.1 will cover most situations.

The logical starting point is with the nuclear family or individual who has been referred. As an example, consider a 2-year-old boy referred because of multiple congenital anomalies and developmental delay.

Step 1 First, enquire about his sibship. Have the parents conceived any other pregnancies, either as a couple or with different partners, and if so what was the outcome? Ensure that information about any pregnancy loss is included.

Step 2 Next, ask each of the parents about their own siblings and parents. Have their brothers and sisters had children or lost any pregnancies? Are their own parents alive and well? Did they lose any pregnancies? Have there been any unusual illnesses in any family members and are there any disorders known to show familial clustering?

Step 3 Next, try to extend the pedigree to include full details of at least three generations going back to the child's grandparents with enquiries about their own parents, their siblings, and their siblings' offspring. If possible, ages at death and known causes of death should be included.

Step 4 Finally, two key questions should be asked: (1) whether there is any possibility of consanguinity within the family, specifically between the parents of the affected child, and (2) whether there is anything else of note in more distant relatives.

By this point you should have constructed a 3–4-generation family tree showing all relevant family details (Fig. 13.2). It is customary to use a diagonal arrow to indicate the individual through whom the family has been ascertained. This person is known as the **proband** and can also be referred to as the **propositus** if male or **proposita** if female. Roman and Arabic numbers are added to indicate the generations and the individuals in each generation respectively. A number placed within an unaffected male or female symbol is a shorthand method for indicating multiple unaffected siblings.

A well-drawn family tree can provide very valuable information about the likely pattern of inheritance of a disorder present within a family. This is reviewed in the next section. However, it is important to remember that construction of the family tree can also lead to the recall of unhappy memories and raise some very sensitive issues such as non-paternity and concealed extramarital relationships. It is essential to exercise tact and diplomacy at all times and to recognize that it is better not to record confidential information that, if disclosed, could have a damaging effect on relationships within the family.

Distinguishing different patterns of inheritance

Having drawn up a family tree which indicates that several individuals are or have been affected with a particular disorder, the next step is to try to determine the underlying pattern of inheritance. In some instances this is quite straightforward, particularly if the diagnosis is clear and the relevant condition is known to

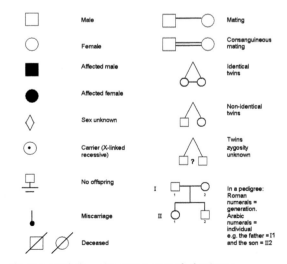

Fig. 13.1 Symbols used in construction of a family tree.

consistently show a specific mode of inheritance. However, many conditions show marked genetic heterogeneity in that they can be caused by mutations in different genes and show different patterns of inheritance (Table 13.1). Sometimes no clear diagnosis can be made. For example, there may be a poorly defined family history of 'congenital blindness' or 'muscular dystrophy' with no surviving affected individual in whom a precise diagnosis can be established. In these situations the only method available for determining the probable

pattern of inheritance is simple inspection of the pedigree.

The basic principles underlying the recognized patterns of single gene inheritance were outlined in Chapter 4 and are summarized in Table 13.2. By applying these it is usually possible to determine the most likely mode of inheritance in an individual pedigree as demonstrated in the following examples. The main points to consider when presented with a pedigree for analysis are summarized in Table 13.3.

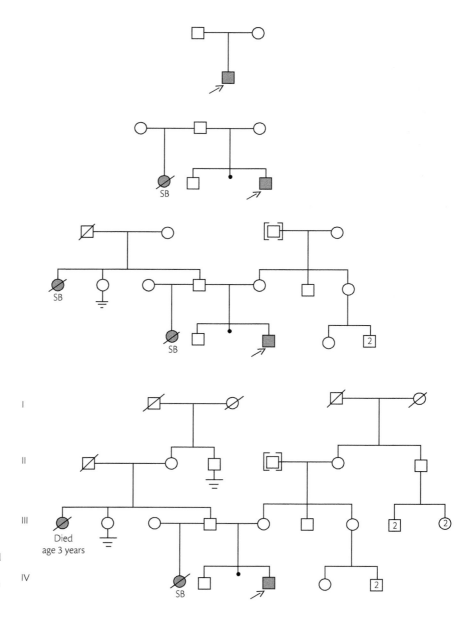

Fig. 13.2 Steps in pedigree construction. SB = stillborn; [ ] = adopted. The presence of three children with multiple congenital abnormalities would raise suspicion of a familial chromosome abnormality such as a reciprocal translocation.

TABLE 13.1 Examples of disorders which show different patterns of Mendelian inheritance

Disorder	Mode of inheritance
Alport syndrome	AD, AR, XR
Congenital cataract	AD, AR
Ehlers–Danlos syndrome	AD, AR
Epidermolysis bullosa	AD, AR
Hereditary motor and sensory neuropathy	AD, AR, XR
Ichthyosis	AD, AR, XR
Polycystic kidney disease	AD, AR
Retinitis pigmentosa	AD, AR, XR
Sensorineural hearing loss	AD, AR, XR, M

AD, autosomal dominant; AR, autosomal recessive; M, mitochondrial; XR, X-linked recessive.

TABLE 13.3 Approach to pedigree analysis

Observation	Implications
Affected members in different generations	Unlikely to be autosomal recessive
Affected members in only 1 sibship	Likely to be autosomal recessive
Parental consanguinity	Likely to be autosomal recessive
Only males affected	Consider X-linked recessive inheritance
More females affected than males	Consider X-linked dominant inheritance
Male to male transmission	Probable autosomal dominant inheritance
	Cannot be X-linked
Males have no affected descendants	Consider mitochondrial inheritance

TABLE 13.2 Characteristic features of different patterns of inheritance

Autosomal dominant	Vertical transmission through generations
	Male to male transmission possible
	Males and females equally affected
	Offspring risk is 1 in 2 for an affected parent
	Reduced penetrance and variable expression
Autosomal recessive	Usually only members of one sibship affected
	Males and females equally affected
	Positive association with parental consanguinity
	Offspring risk is 1 in 4 for carrier parents, very low for an affected parent
X-linked dominant	Transmitted by females to sons and daughters
	Transmitted by males only to daughters
	No male to male transmission
	Both sexes affected, but girls more often affected than boys
	Some conditions lethal in affected males
X-linked recessive	Generally only males affected
	Offspring risks for a carrier female are 1 in 2 for an affected son and for a carrier daughter
	Offspring risk for an affected male is that all daughters are carriers
	No male to male transmission
Mitochondrial inheritance	Only transmitted by females with offspring risk of up to 100% dependent on homoplasmy/heteroplasmy
	Both sexes usually equally affected
Chromosomal inheritance	Carriers of balanced rearrangements can have affected children due to unbalanced chromosome complements, so pedigree can contain apparently randomly affected individuals of either sex

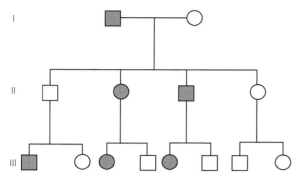

Fig. 13.3 Pedigree illustrating autosomal dominant inheritance with reduced penetrance.

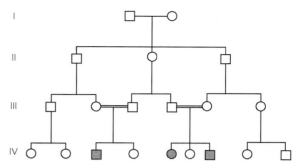

Fig. 13.4 Pedigree illustrating autosomal recessive inheritance with consanguinity.

Example 1 (Fig. 13.3). In this pedigree there are affected individuals in three generations, both males and females are affected, and there is an example of male to male transmission. These findings point to autosomal dominant inheritance. In passing from I1 to III1 the disorder has skipped a generation, which can be explained by non-penetrance in II1. Thus this pedigree shows autosomal dominant inheritance with reduced penetrance.

Example 2 (Fig. 13.4). This pedigree shows three affected individuals, all of whom are born to parents who are first cousins. Both males and females are affected. These findings point strongly to autosomal recessive inheritance.

Example 3 (Fig. 13.5). In this pedigree, four males are affected in two different generations. No females are affected, and there is no example of male to male transmission. These observations are consistent with

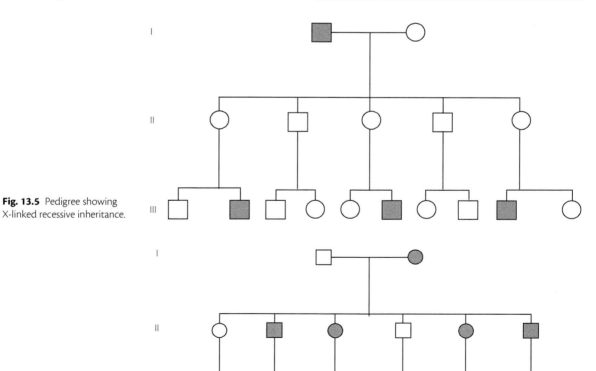

Fig. 13.5 Pedigree showing X-linked recessive inheritance.

Fig. 13.6 Pedigree showing X-linked dominant inheritance.

X-linked recessive inheritance. It could be argued that the pedigree is also consistent with autosomal dominant inheritance and reduced penetrance or sex-limited expression. However, the absence of male to male transmission and the fact that all three daughters in the second generation have transmitted the disorder to sons makes X-linked recessive inheritance more likely.

Example 4 (Fig. 13.6). On first inspection this pedigree could easily be misinterpreted as showing autosomal dominant inheritance. However, careful analysis indicates that there are no examples of male to male transmission and the affected males have passed the condition on to all of their daughters. Thus although autosomal dominant inheritance is a possibility, the pedigree is very suggestive of X-linked dominant inheritance.

Example 5 (Fig. 13.7). This pedigree is also consistent with autosomal dominant inheritance but once again there are no examples of male to male transmission. In addition it is notable that all of the children of affected females are affected whereas none of the children born to affected males are affected. These findings should raise suspicion of mitochondrial inheritance.

Risk calculation

This is one of the most important components of a genetic consultation and—unfortunately for students—it is also one of the most common subjects to arise in genetic examinations. Sometimes estimation of the probability that an individual will be a carrier or be affected is relatively straightforward, but in many situations a little mathematics is required.

> ### Key Point
>
> Questions on pedigree analysis are very common in genetic examinations. Sometimes, as illustrated above in Examples 3 and 4, the pedigree is consistent with more than one form of inheritance, and it may not be absolutely clear which answer is expected. In theory either of the correct answers should be acceptable, so unless the examiner is too idiosyncratic, fairness should prevail. In general it is wise to be wary of concluding that autosomal dominant inheritance is the correct answer unless there is at least one example of male to male transmission. The presence of parental consanguinity should always raise suspicion of autosomal recessive inheritance.

Autosomal dominant inheritance

The probability that each child of an affected individual will inherit the relevant condition is 1 in 2. Thus in Fig. 13.8, which shows a nuclear family in which a father and son have achondroplasia (p. 173), individual II1 can be informed that the probability that each of his children will be affected is close to zero. For II2, the probability that each of his children will be affected is 1 in 2.

This straightforward interpretation applies when the condition in question shows full penetrance, so that it is absolutely clear who is heterozygous and who is homozygous unaffected. However, if the condition shows reduced penetrance (p. 72) then the assessment of risk becomes more difficult. Figure 13.9 shows a three-generation family in which several members have otosclerosis (MIM 166800), a late-onset form of hearing loss which shows autosomal dominant inheritance with a pene-

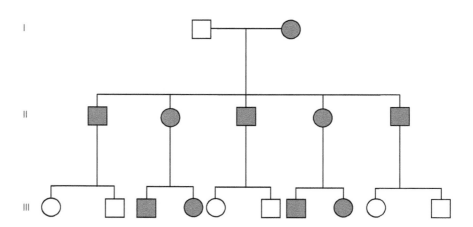

Fig. 13.7 Pedigree showing mitochondrial inheritance.

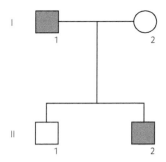

Fig. 13.8 Pedigree of a father (I1) and son (II2) with achondroplasia.

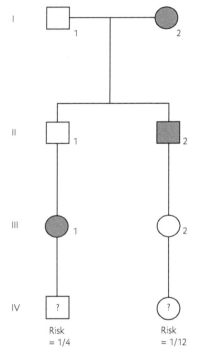

Risk
= 1/4

Risk
= 1/12

Fig. 13.9 Pedigree of a family with otosclerosis which shows autosomal dominant inheritance with reduced penetrance.

trance of 50%. For each child born to III1 there is 1 chance in 2 that the mutation causing otosclerosis will be inherited. However, only 50% of individuals with the mutation will become affected. Thus the probability that each child born to III1 will be affected equals $^1/_2 \times ^1/_2 = ^1/_4$. Essentially this means that for each four children born to III1, on average two will not inherit the mutation and therefore be unaffected. The other two will inherit the mutation but only one will be affected because of reduced penetrance. Thus on average one child in four will be affected, and of the other three one will have inherited the mutation and the other two will not.

This is important when considering children born to III2. This woman is not affected, but as her father (II2) is affected, it is possible that III2 will have inherited the disease mutation but not become affected because of reduced penetrance. As shown above, the probability for this is 1 in 3 (i.e. 1 in 3 unaffected children born to an affected individual will be an example of non-penetrance). This for III2, there is 1 chance in 3 that she is heterozygous for the disease mutation. Therefore there is 1 chance in 6 ($^1/_3 \times ^1/_2$) that each of her children will inherit the mutation and there is 1 chance in 12 ($^1/_6 \times ^1/_2$) that each of her children will be affected. Fortunately reduced penetrance is relatively uncommon amongst autosomal dominant disorders, so in practice calculations in which this has to be taken into account are uncommon. Unless stated otherwise it should be assumed that penetrance is complete when assessing autosomal dominant risks.

One other scenario which can rise is that of homozygosity for an autosomal dominant disorder (p. 72) although in practice this is very uncommon. The risks to offspring for someone who is homozygous for an autosomal dominant disorder are different to those for an individual who is heterozygous, in that all offspring will inherit a mutation so the risk for being affected is 1 (or 100%).

Autosomal recessive inheritance

As explained in Chapter 4 (p. 73) the probability that the sibling of someone with an autosomal recessive disorder will also be affected is 1 in 4. The probability that an unaffected sibling will be a carrier is 2 out of 3.

Risk assessment becomes more difficult when other family members are considered. The starting point is to determine the probability that each parent of a child for whom a risk is sought could transmit a mutant allele. This is considered for two possible situations:

Example 1. Both parents could be carriers (Fig. 13.10). In this situation the unaffected sister of a boy with phenylketonuria (p. 210) wishes to know the probability that she and her healthy unrelated partner will have an affected child. To calculate the risk it is first necessary to determine the probability that each of these prospective parents is a carrier. The probability that the sister (III3) is a carrier is $^2/_3$ (see Fig. 4.6 on p. 73). The probability that her partner is a carrier equals the frequency of carriers in the general population. This can be determined from a knowledge of the incidence of phenylketonuria in the general population using the formula for the Hardy–Weinberg distribution (p^2, $2pq$, q^2—see p. 137). In this distribution

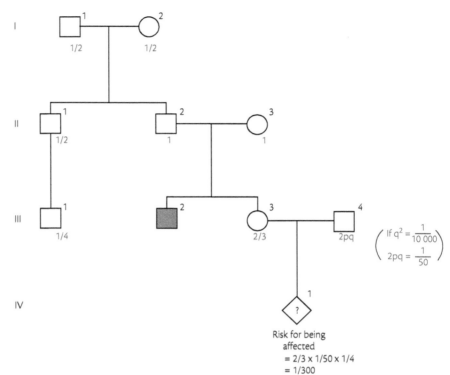

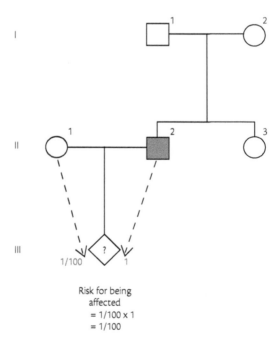

Fig. 13.10 Pedigree showing a boy affected with the autosomal recessive disorder phenylketonuria. Figures in red show the probability for being a carrier.

p^2 represents the frequency of unaffected homozygotes, $2pq$ equals the carrier frequency, and q^2 is the frequency of affected homozygotes. The incidence of phenylketonuria is approximately 1 in 10,000. Thus $q^2 = 1$ in 10,000, $q = 1$ in 100, and $2pq = 2 \times {}^{99}/_{100} \times {}^{1}/_{100}$, which is approximately 1 in 50.

Using this information the prospective parents, III3 and III4, can be told that the probability that their first child will be affected is equal to the probability that they are both carriers, i.e. ${}^{2}/_{3} \times {}^{1}/_{50}$, multiplied by the probability that two carrier parents will have an affected child (${}^{1}/_{4}$). This gives an overall risk of ${}^{2}/_{3} \times {}^{1}/_{50} \times {}^{1}/_{4} = 1$ in 300.

Note that if III3 had chosen to marry her first cousin, III1, then the risk that their first child would be affected would be ${}^{1}/_{4} \times {}^{2}/_{3} \times {}^{1}/_{4} = 1$ in 24. The probability that this first cousin, III1, will be a carrier, can be calculated simply by working backwards from the affected boy (III2) and halving the carrier risk for each generational step as shown in Fig. 13.10.

Example 2. One parent is affected (Fig. 13.11). In this pedigree, a man affected with phenylkenonuria (individual II2) wishes to know the probability that he and

Fig. 13.11 Shows the probability that the child of a man with phenylketonuria will be affected. Figures in red indicate the probability that the allele transmitted by each parent of III1 will be abnormal.

his healthy unrelated partner will have an affected child. The simplest way to approach this is to determine the probability that each parent will transmit a mutant allele. The affected father, II2, has two mutant alleles so he will automatically transmit one to all of his children. The probability that his partner, II1, will transmit a mutant allele equals the probability that she is a carrier, i.e. 1 in 50, multiplied by 1 in 2, as there is one chance in 2 that a carrier will transmit his or her mutant allele. This gives an overall risk of $^1/_{50} \times {}^1/_2 = 1$ in 100 that III1 will be affected. Note that in this situation there is no need to multiply by 1 in 4 as the parents are not both carriers.

Key Point

The probability that any two healthy parents will have a child with an autosomal recessive disorder equals the probability that the father is a carrier multiplied by the probability that the mother is a carrier multiplied by 1 in 4. If one of the parents is affected then the probability that a child will be affected equals the probability that the other parent is a carrier multiplied by 1 in 2.

X-linked recessive inheritance

Figure 13.12 shows the carrier risks for a female in a large pedigree showing X-linked recessive inheritance. Females indicated as a circle with a central dot can be deduced as being obligatory carriers on the basis of the

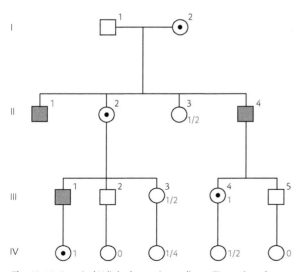

Fig. 13.12 A typical X-linked recessive pedigree. Figures in red indicate the probability for being a carrier.

pedigree structure. The probability that a son born to an obligatory carrier will be affected equals 1 in 2. The probability that a son, born to a woman who may be a carrier, will be affected equals the probability that she is a carrier (e.g. 1 in 4 for IV3) multiplied by 1 in 2.

Other X-linked recessive pedigrees can present more difficult calculations in which it is necessary to modify the probability that a woman is a carrier based on the observation that she has already had several unaffected sons. For any woman whose carrier status is unknown, the more unaffected sons she has then the less likely it becomes that she is a carrier. Bayes' theorem, named after the Reverend Thomas Bayes (Box 13.1), provides a method for calculating a precise carrier risk.

Essentially this involves determining the initial background (prior) probability that the woman is or is not a carrier and then modifying this by taking into account the unaffected sons (conditional probabilities) to arrive at a joint probability that the woman is or is not a carrier. These joint probabilities are then compared to determine the final posterior probability that the woman is or is not a carrier. This is a complex concept which is not easily understood. Fortunately it is generally regarded as being too difficult for students of genetics to cope with. To show that this insulting assessment of the average student's intellect is incorrect, an example of a typical Bayesian genetic calculation is provided.

Figure 13.13 shows part of a large pedigree in which several males, including I3 and II1, have non-specific mental retardation. (Sadly it is a well-established fact that there are many more retarded males than females because of the large number of X-linked recessive forms of mental retardation.) The daughter in the third generation, III5, wishes to know the probability that she is a carrier of this condition. If the fact that she has four unaffected brothers is ignored then her pedigree carrier risk is 1 in 4. However, if her four unaffected brothers are taken into account—and they certainly should be, as their existence makes it unlikely that their mother is a carrier—then the daughter's carrier risk will be much lower. Exactly how much lower is determined by a simple Bayesian calculation as shown in Table 13.5. This indicates that the posterior probability that II2 is a carrier has fallen from her prior carrier risk of 1 in 2 to a figure of 1 in 17. This means that the daughter (III5) can be reliably informed that her risk has fallen from 1 in 4 to 1 in 34 (i.e. half of her mother's posterior risk).

Readers should not be unduly concerned if Bayesian probability theory is not their favourite subject, as

BOX 13.1 LANDMARK PUBLICATION: BAYES' THEOREM

Thomas Bayes (1702–61) was a non-conformist (Presbyterian) clergyman who lived in Tunbridge Wells in Kent in the south of England. Little is known about his life other than that he was an enthusiastic mathematician who was honoured by being elected to the Royal Society, possibly because he published a tract in which he defended the views and thoughts on calculus promoted by Sir Isaac Newton. At the time these were regarded as somewhat heretical.

Thomas Bayes' main claim to fame is that he devised a method for calculating the probability of occurrence of a future event, based on previous events and relevant observations. Specifically, he set out to 'find out a method by which we might judge the probability that an event has to happen, in given circumstances, upon supposition that we know nothing concerning it but that, under the same circumstances, it has happened a certain number of times and failed a certain other number of times'.

As illustrated in the example on X-linked recessive inheritance, Bayes' theorem can be applied in genetics, or in any other situation, by (1) determining prior probabilities for each hypothesis, (2) modifying these on the basis of relevant observations, (3) obtaining joint probabilities for each hypothesis, and (4) arriving at the posterior probabilities which can be used for predictive purposes. As another simple example, consider a twin pregnancy in which an ultrasound scan at 20 weeks gestation shows that the twins are of the same sex. To what extent does this make it more likely that the twins are monozygotic (MZ) as opposed to dizygotic (DZ)? Approximately one third of all twin pairs are MZ and two thirds are DZ. Thus the prior probabilities for being MZ or DZ are $1/3$ and $2/3$ respectively. The conditional probabilities for being of the same sex if the twins are MZ or DZ are 1 and $1/2$ respectively. These values are incorporated into a simple Bayesian table to obtain a posterior probability of 1 in 2 that the twins are identical (Table 13.4).

The great value of Bayes' theorem in genetics is that it can be used to determine risks in many different situations without any need for a significance test or confidence interval. Thus, in the above example, in a large series of like-sex twin pairs, half will be MZ and half will be DZ. Bayes himself had doubts about the validity of his approach and his essay was not published until after his death by a friend who found it amongst his papers. Thus Thomas Bayes shares with Gregor Mendel the dubious distinction of posthumous rather than contemporary recognition of his genius.

TABLE 13.4 Bayesian table

Probability	Monozygotic	Dizygotic
Prior	$1/3$	$2/3$
Conditional (same sex)	1	$1/2$
Joint	$1/3$	$1/3$
(Odds	1	1)

Posterior probability for twins being monozygotic = $(1/3)/[(1/3) + (1/3)]$ = $1/2$.

Reference

Bayes T. An essay towards solving a problem in the doctrine of chances. Originally published in the *Philosophical Transactions of the Royal Society* in 1763; republished in 1958 in *Biometrika*, **45**, 296–315.

most examining bodies do not include it in formal examinations. However, it is a useful concept, which finds widespread application not just in genetics, but also in many other branches of medicine and science when information needs to be incorporated that modifies an existing possibility, making it more or less likely.

Interpreting a laboratory report

Clinicians are expected to be able to interpret reports issued by cytogenetic and molecular genetic laboratories. Often this is straightforward, and most laboratories qualify abnormal findings by providing a written explanation of their significance. If there is ever any doubt, then before conveying the results to a patient or family, it is essential to discuss the report with scientists in the laboratory from which it was issued.

Cytogenetics reports

The cytogenetics community has devised a standardized shorthand system of nomenclature for reporting the results of chromosome analysis. This builds on the system used for describing chromosomes and rearrangements as outlined in Table 2.3 (p. 29). Examples of abnormal reports are given in Table 13.6.

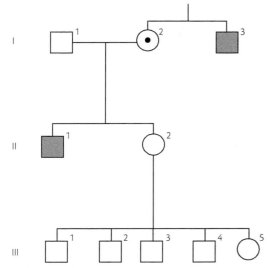

Fig. 13.13 Part of a pedigree showing two males with non-specific X-linked mental retardation.

TABLE 13.5 Bayesian calculation for family shown in Figure 13.13

	II2 carrier	II2 not a carrier
Prior probability	$^1/_2$	$^1/_2$
Conditional probability		
4 unaffected sons	$(^1/_2)^4 = 1/16$	$1^4 = 1$
Joint probability	$^1/_{32}$	$^1/_2$
Odds	1	to 16

The posterior probability that II2 is a carrier = $(^1/_{32})/[(^1/_{32}) + (^1/_2)]$ = 1 in 17.
Therefore the prior probability that III5 is a carrier = 1 in 34.

All chromosome reports start with the number of chromosomes in each cell, followed by a description of the sex chromosomes. This is followed by a description of any abnormality that has been identified. Gain or loss of a chromosome is indicated by a plus or minus sign, followed by the relevant chromosome. Note that

TABLE 13.6 Examples of chromosome reports and their interpretation (see also Table 2.3, p. 29)

Report	Interpretation
46,XX	Normal female
46,XY	Normal male
*47,XX,+18	Female with trisomy 18 (Edwards syndrome)
*47,XYY	Male with XYY syndrome
*46,XX,del(4)(p16)	Female with deletion of band 4p16 (Wolf–Hirschhorn syndrome)
*46,XY,dup(4)(q22q26)	Male with duplication of bands 4q22 to 4q26
*46,X,i(X)	Female with a long arm X isochromosome (a cause of Turner syndrome)
46,XY,inv(2)(p21q21)	Male with a pericentric inversion in one number 2 chromosome with breakpoints at p21 and q21
46,XX,inv(3)(p13p24)	Female with a paracentric inversion in one number 3 chromosome with breakpoints at p13 and p24
46,XY,t(1;2)(p21;q21)	Male with a reciprocal translocation with breakpoints in the short arm of chromosome 1 and the long arm of chromosome 2
*46,XX,der(2)t(1;2)(p21;q21)pat	Female with derivative number 2 chromosome inherited from her father instead of normal number 2 chromosome
45,XX,der(14;21)(q10;q10)	Female who is a balanced carrier of a 14;21 Robertsonian translocation
*46,XY,der(14;21)(q10;q10)mat,+21	Male with an unbalanced karyotype causing Down syndrome due to inheritance of a 14;21 Robertsonian translocation from his mother
*47,XY,+21/46,XX	Male with mosaicism for trisomy 21 (normal cell line is always listed last)
*46,XX.ish del(7)(q11.23q11.23)(ELN-)	Female with Williams syndrome due to interstitial microdeletion at 7q11.23 identified by *in situ* hybridization using an elastin gene (ELN) probe

* indicates an unbalanced karyotype.

in the past, plus or minus preceded by a chromosome arm (e.g. 46,XY,4p-) was sometimes used as a shorthand designation for deletion of part of that arm. Some textbooks still refer to the cri-du-chat syndrome as the 5p- syndrome. The correct designation should be 46,XX,del(5p) or 46,XX,del(5)(p13).

The nomenclature used to describe translocations can be confusing (Table 13.6). A balanced reciprocal translocation is represented by 't' followed by the breakpoints. If 'der' is included, this indicates that only one of the translocation chromosomes is present and that the rearrangement is therefore unbalanced. For Robertsonian translocations, a balanced carrier will be designated as having 45 chromosomes with breakpoints at q10 and q10 representing the centromeres. An individual with 46 chromosomes and a Robertsonian translocation, indicated as 'rob' or 'der', has an unbalanced karyotype.

Molecular genetic reports

A standardized system for describing mutations has been developed, based either on the genomic change or on the protein substitution.

Genomic changes are indicated by g, followed by the number of the nucleotide(s) involved, followed by a description of the change. Thus replacement of an adenine base by a thymine base at nucleotide 17 would be described as g.17A→T. A deletion of four nucleotides would be indicated as g.510 → 513del, and an insertion of two adenine nucleotides as g.307 → 308insAA. For mutations involving an intron, the intron number is indicated as IVS (intervening sequence), followed by the number, followed by details of the mutation. Thus a G to C mutation at the second nucleotide of intron 3 would be designated as IVS3+2G→C.

A change in the protein structure is indicated by p, followed by a description of the proteins involved using either single letters or three letter codes (Table 13.7). X or STOP indicates a stop or termination codon. Numbers refer to the codons involved. Thus p.P250R or p.Pro250Arg indicates that proline has been replaced by arginine at codon 250. A description of the protein change gives more information about the functional effect, whereas a description of a genomic mutation simply indicates which nucleotides are involved. An example of how mutations should be reported is illustrated by achondroplasia (p. 173), which is caused by a G→A or G→C mutation at nucleotide 1138 in *FGFR3*. This should be reported as g.1138G→A or g.1138G→C. All of these point mutations result in a substitution of glycine by arginine at codon 380. This is designated as p.G380R or p.Gly380Arg.

Mental retardation

Mental retardation is generally defined as the onset before the age of 18 years of impaired intellectual skills resulting in an IQ test score of less than 70, in association with functional impairment. It is customary to subdivide it into two groups, mild (IQ of 50–70) and severe (IQ less than 50). In the past it was felt that mild mental retardation largely represented the tail end of the normal distribution with social deprivation being an important contributory factor, whereas severe men-

TABLE 13.7	Amino acid abbreviations				
Amino acid	Single-letter abbreviation	Three-letter abbreviation	Amino acid	Single-letter abbreviation	Three-letter abbreviation
Alanine	A	Ala	Leucine	L	Leu
Arginine	R	Arg	Lysine	K	Lys
Asparagine	N	Asn	Methionine	M	Met
Aspartic acid	D	Asp	Phenylalanine	F	Phe
Cysteine	C	Cys	Proline	P	Pro
Glutamic acid	E	Glu	Serine	S	Ser
Glutamine	Q	Gln	Threonine	T	Thr
Glycine	G	Gly	Tryptophan	W	Trp
Histidine	H	His	Tyrosine	Y	Tyr
Isoleucine	I	Ile	Valine	V	Val
			Termination	X	Stop

tal retardation was more likely to have an underlying pathological cause. This is now recognized to be an oversimplification.

Mental retardation is common, with a worldwide prevalence of around 3.8 per 1000 for severe mental retardation and between 20 and 30 per 1000 for the mild form. Genetic factors are known to play a significant role, as discussed below. Consequently, children with mental retardation represent one of the most common referrals to genetics clinics.

Causes of mental retardation

Both genetic and environmental factors make a major contribution, although in many instances no clear cause can be established. This is the case for approximately 40% of children with severe mental retardation and up to 70% of children with mild mental retardation (Table 13.8). New genetic diagnostic techniques, such as multitelomeric FISH analysis (p. 30), have helped identify a causal factor in a small proportion of previously unexplained cases, and new developments in neuroimaging can identify subtle cerebral malformations such as neuronal migration disorders and defects in myelination.

There is also increasing recognition that many genes on the X chromosome are implicated, which probably explains why most studies have consistently found that there are more retarded males than females. The prevalence of X-linked mental retardation has been estimated to be approximately 1 in 500 males, and mutations in X chromosome genes are thought to account for 25–50% of all cases of mental retardation in males.

TABLE 13.8 Causes of mental retardation

Type of disorder	Mild (IQ 50–70) (%)	Severe (IQ <50) (%)
Chromosomal		
Down syndrome	5	30
Fragile X syndrome	3	
Other (including microdeletions)	2	5
Single-gene disorders (e.g. mucopolysaccharidoses)	5	5–10
Malformation syndromes	5	10–20
Acquired		
Pre- or perinatal	5–10	10
Postnatal	5	5–10
Unknown	70	30–40

Percentages are based on average prevalence figures taken from several different studies.

To date, 15 genes on the X chromosome have been identified which can cause non-specific mental retardation. These do not include fragile X syndrome (p. 112) which represents the most common inherited form of mental retardation in humans. The X chromosome genes which have been identified are believed to be involved in signalling pathways important in neuronal morphogenesis and in chromatin remodelling, errors in which probably lead to deregulation of the expression of genes required for normal cognitive function.

Many children with learning difficulties also show features of autism. These include poor verbal communication, impaired social responsiveness, and an adherence to ritualistic behaviour. Autism is approximately three to four times more common in boys than in girls, and a small proportion of affected boys are found to have fragile X syndrome. Non-syndromal autism is believed to show multifactorial inheritance, and despite extensive research the polygenes which contribute to this condition have not been identified. Thus at present there are no diagnostic genetic tests for non-syndromal autism, other than chromosome and fragile X mutation analysis.

Evaluation of a child with mental retardation

The approach to most situations in clinical practice consists of three key steps:

◆ taking a history

◆ carrying out an examination

◆ initiating appropriate investigations.

This tripartite approach is entirely applicable when evaluating a child with mental retardation.

Medical history

This should include a detailed review of both the family history and the child's own medical history from conception onwards. As always, careful but discreet enquiries should be made about other family members who might have been affected in previous generations. Sometimes this information is withheld from younger family members because of concern about stigmatization. Parental consanguinity and parental performance at school should also be noted.

Careful enquiry should also be made about possible adverse events in pregnancy, such as severe antepartum haemorrhage or maternal exposure to teratogens including alcohol and anticonvulsants. If possible, maternity and neonatal records should be reviewed to establish whether the child was in good condition after delivery and whether special care was needed in the neonatal period. Details of the child's own medical history should be sought, with particular reference to serious illnesses such as meningitis or encephalitis, which can cause damage to the developing brain.

Finally, particular attention should be given to the child's developmental history and possible neurological problems which can be associated with mental retardation. Specifically, it should be established whether the child has always been delayed or whether there has been evidence of recent regression which would point to a possible neurodegenerative condition. It should also be noted whether the child has had convulsions and whether or not there is evidence of neurological dysfunction which would be consistent with a diagnosis of cerebral palsy (e.g. non-progressive spasticity or athetosis). Information about unusual behavioural characteristics, such as autism, should be recorded together with whether there are any characteristic features which would point to a diagnosis such as Angelman syndrome (p. 84).

Clinical examination

This should be thorough with particular reference to growth parameters, neurological deficits, and dysmorphic features. The head circumference should be measured, as both small head size (microcephaly) and large head size (macrocephaly or megalencephaly) can be associated with mental retardation. Neurological abnormalities could be a manifestation of cerebral palsy or of a particular disorder such as ataxia telangiectasia (p. 66) or Duchenne muscular dystrophy (p. 109), both of which can be associated with learning disability. Abdominal palpation should be carried out to establish whether the liver or spleen are enlarged (i.e. hepatosplenomegaly). This would point to a neurodegenerative storage disorder (p. 214). The child's skin should be examined carefully for abnormalities of pigmentation, which occur in conditions such as neurofibromatosis type 1 and tuberous sclerosis. Finally, careful note should be made of whether the child has a normal appearance and whether there are any congenital abnormalities such as extra digits (polydactyly) or marked body asymmetry.

Investigations

The decision as to which investigations should be undertaken will depend to some extent on the relevant clinical findings. However, as a general rule, a standard set of investigations, summarized in Table 13.9, should be considered in all children with unexplained mental retardation. More specialized investigations are indi-

TABLE 13.9 Baseline investigations in unexplained mental retardation

Karyotype analysis	Standard chromosome analysis. Also multi-telomeric FISH if standard chromosome analysis is normal and clinical features or family history are suggestive of a 'cryptic' chromosome rearrangement (p. 29)
Molecular analysis	Fragile X mutation analysis. Also tests for specific mental retardation syndromes if clinical features or family history are suggestive, e.g. X-linked recessive inheritance
Metabolic analysis	Serum and urine amino acids
	Urine organic acids
	Urine glycosaminoglycans (mucopolysaccharides)
	Serum uric acid (Lesch–Nyhan syndrome)
	Creatine kinase (Duchenne muscular dystrophy)
Other investigations	Cranial neuroimaging—CT scan and/or MRI
	Ophthalmological assessment
	Thyroid function tests

cated in specific situations. These can include FISH for microdeletion analysis when a specific syndrome is suspected (p. 29), and more comprehensive metabolic studies if there is any evidence of an acid/base disturbance or of neurodegeneration.

Genetic counselling and empiric risks

The provision of genetic counselling should be an integral component of the management of any family in which one or more individuals has a learning disability. This is relatively straightforward if a clear underlying diagnosis can be established. Unfortunately, as already indicated, it is often not possible to establish a clear cause or diagnosis and in these situations recourse has to be made to empiric recurrence risks derived from large family studies. These have shown that the recurrence risk for the future sibling of a boy with severe unexplained mental retardation is around 5%, although some studies have yielded higher risks of up to 10%. The recurrence risk for siblings of a girl with severe mental retardation is between 2 and 5%. Recurrence risks for siblings of a child with unexplained mild mental retardation tend to be higher.

Dysmorphic features

Dysmorphology can be defined as the study of abnormal form, although the term is used in a more general context to describe the study of children with birth defects and/or unusual features (p. 183). A dysmorphic child or a child with dysmorphic features is one who looks unusual or has one or more congenital malformations.

As discussed in Chapter 9, approximately 1 in 40 of all newborn infants has at least one major malformation apparent at birth and a similar proportion have a serious abnormality which first becomes apparent later in childhood. Thus the study of dysmorphic children takes on an important dimension, given that many such children suffer chronic ill health. Often this is in association with neurodevelopmental problems such as convulsions and learning disability.

In dysmorphology, unlike most medical disciplines, the emphasis is on diagnosis rather than treatment. Nevertheless, the establishment of a diagnosis is important for several reasons. First of all, a child may benefit by being spared the discomfort and embarrassment of continuing diagnostic investigations. In addition, a diagnosis can provide information about possible complications, such as lens dislocation and aortic aneurysm in Marfan syndrome (p. 106), so that appropriate surveillance measures can be put in place. Parents can also benefit by gaining information about their child's prognosis and by being relieved of the guilt that they were in some way to blame for their child's unusual problems. Contact with other families through well-organized support groups can be extremely helpful. Finally, the establishment of a clear diagnosis enables the provision of accurate information about recurrence risks so that parents, and when appropriate, other family members can be offered precise genetic counselling.

Medical history

As in almost every other clinical situation, the assessment of a child with dysmorphic features should start with a detailed medical history. This should include full details of both the family history and the child's own medical history from conception onwards. Particular points of importance in the family history are possible instances of previous pregnancy loss or early childhood death. As always, sensitive enquiries should be made about possible parental consanguinity. Advanced maternal or paternal age should raise suspicion of either possible non-disjunction (p. 38) or a new dominant mutation (p. 37) respectively.

Review of the pregnancy history must include enquiry about possible teratogenic exposure (p. 181). Fetal movement is reduced in some neuromuscular conditions, such as arthrogryposis, in which the joints become fixed, and in disorders causing hypotonia such as the Prader–Willi syndrome. The pregnancy and delivery records should be scrutinized to establish if any abnormalities were noted at birth, either in the baby or in the placenta. For example, a single umbilical artery is a common finding in the VATER association (p. 180).

Enquiries should also be made about possible problems in the neonatal period. Babies with the fetal alcohol syndrome are often jittery and irritable as a result of withdrawal from alcohol exposure. Babies with the Prader–Willi syndrome are floppy and often require tube feeding. Infants with the DiGeorge syndrome have transient hypocalcaemia, whereas infantile hypercalcaemia is a feature of Williams syndrome.

Finally, special attention should be paid to the child's behaviour as it is now recognized that many of the more common dysmorphic syndromes are associated with particular behavioural characteristics. As an example, boys with fragile X syndrome are often aggressive with occasional autistic tendencies. Other examples of characteristic behavioural patterns are given in Table 13.10.

Clinical examination

This represents the most important step in the evaluation of any child with unusual physical features. Fortunately for the child, it should not be too unpleasant as the examination relies mainly on observation. As the child's confidence is gained it should become possible to carry out a detailed examination of every part of the body without causing embarrassment or discomfort. The assessment of a dysmorphic child requires patience and experience, and it would be unreasonable to expect a medical student to have acquired the necessary skills. However, there are a few general principles which always apply.

◆ A careful assessment should be made of the child's overall appearance including stature, physique, head

TABLE 13.10 Assessment of a dysmorphic child

History	
Family history	Miscarriage, stillbirth, and childhood deaths
	Parental ages and consanguinity
Pregnancy	Exposure to teratogens
	Fetal movements
	Abnormalities noted at birth in baby or placenta
Infancy and childhood	Neonatal problems, e.g. hypotonia, hypo- or hypercalcaemia
	Behavioural characteristics, e.g. extrovert personality (Williams syndrome)
	Inappropriate laughter (Angelman syndrome)
	Hyperphagia (Prader–Willi syndrome)
Examination	
General physique and growth parameters	
Facies	Unusual features. Should include careful inspection of hair, ears, eyes, nose, philtrum, lips and mouth (palate, tongue and teeth)
Neck	Neck webbing (Turner syndrome)
Trunk	Shield chest (Turner syndrome)
	Pectus excavatum (Marfan syndrome)
Limbs	Long or short and in proportion
	Digits and nails
	Brachydactyly (short digits)
	Other prefixes: poly- (extra), arachno- (long), clino- (curved), campo- (flexed)
	Palmar creases (single in Down syndrome)
Genitalia (in males only)	Hypospadias
	Undescended testes
Skin	Café-au-lait patches (neurofibromatosis)
	Depigmentation (incontinentia pigmenti)
	Birth marks, e.g. haemangiomata
Investigations	
Karyotype analysis	Standard or specific, e.g. breakage studies or FISH for microdeletions or telomeric rearrangements
Molecular analysis	Fragile X or other developmental syndromes in which gene defect is known
Biochemical analysis	Mitochondrial (blood lactate)
	Peroxisomal (very long chain fatty acids)
	Urinary mucopolysaccharides
Diagnostic imaging	Neuroradiology
	Skeletal survey
	Abdominal ultrasonography
Reaching a diagnosis	
Literature resources	Textbooks and journals
	Computerized dysmorphology databases
	Consultation with colleagues

circumference, and demeanour. It is important to avoid using derogatory or hurtful terms such as 'funny looking' or 'odd'. Descriptions such as 'slightly unusual' or 'characteristic' are much less likely to cause offence.

♦ Having gained the child's confidence, it is sensible to begin by examining whichever part of the body the child is willing to expose. This may mean starting with hands or feet, to gradually reassure the child.

♦ Aspects of the examination that may seem unpleasant or frightening to the child should be kept to the end of the assessment. Most children are nervous about strangers looking in their mouth, and curiously many young children resent having their head circumference measured so this is also best delayed.

♦ Finally, it is important that findings should be documented carefully with measurements checked against reference tables. If parental permission is granted, photographs can be taken for future reference and discussion. It is quite likely that the diagnosis will not yet be clear, and a review of photographs with colleagues can be extremely helpful.

Investigations

As with unexplained mental retardation, the choice of investigations will be governed by the clinical findings. In most situations the investigations undertaken will fall under one or more of the following headings.

Karyotype analysis

Standard chromosome analysis should be undertaken on every dysmorphic child. Additional chromosome investigations can be initiated as appropriate. These may include microdeletion analysis by FISH when a microdeletion syndrome (p. 63) is suspected, chromosome breakage studies (p. 66) for conditions such as Fanconi anaemia and Bloom syndrome, and multitelomeric FISH analysis when the clinical features suggest a chromosome abnormality but standard chromosome analysis is normal.

Molecular analysis

The underlying molecular basis of a large number of dysmorphic syndromes has now been established (p. 169), and when one of these conditions is suspected, a diagnosis can sometimes be confirmed by appropriate mutation studies.

Biochemical analysis

Several inborn errors of metabolism are associated with dysmorphic features and if one of these conditions is suspected then appropriate biochemical studies should be undertaken. Examples include peroxisomal disorders such as Zellweger syndrome (p. 217), storage disorders such as the mucopolysaccharidoses (p. 215), and disorders of cholesterol synthesis such as the Smith–Lemli–Opitz syndrome (p. 175).

Diagnostic imaging

X-rays should be used sparingly in children, but if there is any suggestion of an underlying skeletal abnormality then localized radiography or a whole-body skeletal survey should be requested. This certainly applies to any child in whom an underlying disorder of bone development, or *skeletal dysplasia*, is suspected. Ultrasonography is both painless and risk free and can provide very useful information about soft tissue organs such as the kidneys and spleen. Specific neuroimaging by CT scan or MRI is indicated in children with unexplained mental retardation or other neurodevelopmental abnormalities.

Reaching a diagnosis

When all of the diagnostic information is available an attempt should be made to arrive at a diagnosis. Sometimes this will be straightforward so that parents can be given full details of the condition which affects their child, together with information about genetic risks and support groups. If no clear diagnosis is apparent, then recourse can be made to a large number of excellent information resources, several of which are listed in the next section.

Unfortunately, despite exploring all of these avenues, it is often not possible to reach a confident diagnosis. In these situations honesty is the best policy. If an incorrect diagnosis is made, this will eventually become apparent when, for example, the parents joint an inappropriate support group or the child evolves into a different syndrome. There is no disgrace in admitting defeat. Parents usually recognize and respect the truth and some are quietly pleased and relieved that their child will not be stigmatized with a syndrome label. The unfortunate 'Elephant Man' represents an extreme example of dysmorphism for which a precise diagnosis has yet to be established with absolute confidence (Box 13.2).

Key Point

The evaluation of a child with dysmorphic features should include a full family and medical history, a thorough clinical examination and appropriate investigations which will usually include chromosome analysis. For many dysmorphic children it is not possible to reach a diagnosis.

BOX 13.2 CASE CÉLÈBRE: THE ELEPHANT MAN

Joseph Merrick was born in 1862 into a poor family living on the outskirts of Leicester in England, where he appears to have enjoyed a happy childhood until the death of his mother when he was 10 years old. Thereafter his increasing disability and disfigurement slowly led to social isolation, unemployment, and homelessness, culminating in admission to the local workhouse in his late teens. Here the conditions were so appalling that Joseph opted instead to become a 'freak' exhibit in a travelling show, a precarious occupation which he pursued until rescued by a surgeon, Frederick Treeves, who found him refuge at the London Hospital. There he was to spend the rest of his short life in comfort, well cared for until his death from either asphyxiation or spontaneous dislocation of the neck at the age of 27 years.

Joseph Merrick's diagnosis remains unclear, so it is worth pursuing the recommended tripartite approach of history, examination, and investigations to see if any light can be shed. He is said to have indicated that no one in his family was similarly afflicted, although his mother was known to have been a 'cripple'. This was probably as a result of polio or a poorly treated injury. During Joseph's pregnancy she had been frightened by an elephant in a visiting circus. It was this, coupled with his facial disfigurement, which led to him being described as the 'Elephant Man'. No other untoward events are recorded in his pregnancy or in his early medical history.

For details of Joseph's clinical features we have to rely on Frederick Treeves' detailed description. The major findings were massive thickening of the subcutaneous tissue with pendulous folds of skin, numerous wart-like growths or papillomata, and extensive bone thickening in the skull, right arm, and both legs (Fig. 13.14). Joseph's skull was noted to be particularly abnormal with a circumference of 91 cm (normal range 54–59 cm), and a very irregular outline due to huge rounded bony lumps. Examination of Joseph's skeleton, which has been preserved, indicates that these lumps were benign tumours known as osteomata.

Obviously it was not possible to carry out any relevant investigations during Joseph's lifetime, so retrospective attempts at diagnosis have been based mainly on his clinical features. The first suggestion was that Joseph had neurofibromatosis type 1, in which the most notable findings are brown café-au-lait marks on the skin and multiple soft tumours of neural origin known as neurofibromata. However, Joseph's features were much more severe than those

Fig. 13.14 A photograph of Joseph Merrick, age 26 years. Reproduced with the permission of the Trustees of the London Hospital pathology museum.

that have ever been seen in this condition. Other diagnoses which have been considered include various skeletal disorders including Paget's disease, fibrous dysplasia, and Maffuci's syndrome. Again, none of these is entirely satisfactory. At present Proteus syndrome is considered the most likely diagnosis. This is characterized by asymmetrical overgrowth, with multiple lipomas, naevi, and exostoses. Some cases are caused by mutations in the gene *PTEN*, which is a tumour suppressor. To date no mutation has been identified in *PTEN* in DNA extracted from Joseph Merrick's skeleton, so the diagnosis of Proteus syndrome remains unproven.

The story of Joseph Merrick has been well documented in several books and in the film *The Elephant Man* starring John Hurt. Further details of his medical history can be obtained from the references listed below.

References

Howell M, Ford P (1980) *The true history of the Elephant Man*. Allison & Busby, London.

Seward GR (1994) Did the Elephant Man have neurofibromatosis 1? In: Huson SM, Hughes RAC (eds) *The neurofibromatoses*. Chapman & Hall Medical, London.

Tibbels JAR, Cohen MM (1986) The proteus syndrome: the Elephant Man diagnosed. *BMJ*, **293**, 683–685.

Accessing information resources

The rapid expansion of knowledge in medical genetics means that it is not possible for anyone to keep abreast of all the developments that are taking place. Consequently, every department of medical genetics rightly values its library as one of its most important assets. The preclinical evaluation of most families referred to a genetics centre will involve a detailed literature review. This is particularly important in an era when reliable, up-to-date information about medical disorders is readily accessible through the Internet, to the extent that many families now arrive at a clinic clutching large reams of information printed off on their home computer. This means that the onus is on everyone involved in the provision of genetic services to ensure that they have access to the most up-to-date information available.

Despite the profusion of web sites offering genetic information, textbooks continue to provide a useful starting point when undertaking a genetic literature review. These can be supplemented by computerized databases which are invaluable when searching for a diagnosis. Some of the Internet sites dealing specifically with genetic disorders are excellent, but it has to be remembered that these are not always subject to peer review or any other form of policing. This means that patients with rare disorders, who are particularly vulnerable to the lure of expensive charlatan remedies, should be alerted to the importance of caution in believing all that they read.

General reference books and textbooks

Gorlin RJ, Cohen MM, Hennekam RCM (2001) *Syndromes of the head and neck*, 4th edn. Oxford University Press, Oxford.

Jones KL (1997) *Smith's recognizable patterns of human malformation*, 5th edn. WB Saunders, Philadelphia.

King RA, Rotter JI, Motulsky AG (2002) *The genetic basis of common diseases*, 2nd edn. Oxford University Press, New York.

McKusick VA (1998) *Mendelian inheritance in man*, 12th edn. Johns Hopkins University Press, Baltimore.

Rimoin DL, Connor JM, Pyeritz RE, Korf BR (ed.) (2002) *Emery and Rimoin's principles and practice of medical genetics*, 4th edn. Churchill Livingstone, London.

Scriver CR, Beaudet AL, Sly WS, Valle D (2001) *The metabolic and molecular basis of inherited disease*, 8th edn. McGraw-Hill, New York.

Spranger JW, Brill PW, Poznanski A (2002) *Bone dysplasias. An atlas of genetic disorders of skeletal development*, 2nd edn . Oxford University Press, Oxford.

Computerized databases

Human Cytogenetic Database. Schinzel A. Oxford University Press, Oxford.

London Dysmorphology Database (2003) Winter RM, Baraitser M. London Medical Databases, London.

London Neurogenetics Database (2003) Winter RM, Baraitser M. London Medical Databases, London.

Pictures of Standard Syndromes and Undiagnosed Malformations (POSSUM) Bankier A. Murdoch Institute for Research into Birth Defects, Melbourne.

Radiological Electronic Atlas of Malformation Syndromes and Skeletal Dysplasias (REAMS) (1999) Hall C, Washbrook J. Oxford University Press, Oxford.

Internet web sites

American Society of Human Genetics *www.ashg.org*

British Society of Human Genetics *www.bshg.org.uk*

Gene Clinics *www.geneclinics.org*

Gene Tests *www.genetests.org*

Genetic Alliance. Online Directory of Genetic Resources *www.geneticalliance.org*

Human Gene Mutation Database *www.hgmd.org*

Online Mendelian Inheritance in Man (OMIM) *http://www3.ncbi.nlm.nih.gov/entrez/query.fcgi?db=OMIM*

US Department of Energy Genome Programs (including Human Genome Project) *www.doegenomes.org*

Summary

By the completion of medical training, medical students are expected to have acquired a number of important basic clinical skills. These include the construction and interpretation of a pedigree, the calculation of simple genetic risks, the interpretation of genetic laboratory reports, the management of certain clinical scenarios, and the ability to access relevant information resources.

Pedigree construction is best approached by beginning with the nuclear family and then moving outwards to encompass both sides of the extended family. When interpreting a pedigree, autosomal dominant inheritance cannot be concluded with confidence unless there is an example of male to male transmission. Parental consanguinity points strongly to autosomal recessive inheritance.

To calculate the probability that unaffected parents could have a child with an autosomal recessive disorder, it is first necessary to determine the probability that each parent is a carrier. The product of these values is then multiplied by $1/4$ to obtain the risk that their child will be affected. Bayes' theorem can be used to determine risks when prior and conditional information have to be taken into account. This is particularly useful in X-linked recessive inheritance.

Standardized methods have been developed for reporting cytogenetic abnormalities and DNA sequence changes. If a clinician is in any doubt about the significance of a genetic report, then scientists in the relevant laboratory should be contacted.

An outline approach is suggested for the investigation of mental retardation and of a child with dysmorphic features. This involves taking a detailed family and medical history, carrying out a careful examination and undertaking appropriate investigations, which will usually include both chromosome analysis and, for mental retardation, fragile X mutation analysis. Details of information resources which can be consulted for further information are provided.

Further reading

Pedigree construction

Bennett RL, Steinhaus KA, Uhrich SB, *et al.* (1995) Recommendations for standardized pedigree nomenclature. *American Journal of Human Genetics*, **56**, 745–752.

Risk calculation

Bridge PJ (1997) *The calculation of genetic risks*, 2nd edn Johns Hopkins University Press, Baltimore.

Young ID (1999) *Introduction to risk calculation in genetic counselling*, 2nd edn. Oxford University Press, Oxford.

Interpreting laboratory reports

Den Dunnen JT, Antonarakis E (2001) Nomenclature for the description of human sequence variations. *Human Genetics*, **109**, 121–124.

Mitelman F (ed.) (1995) *ISCN 1995. An international system for human cytogeneic nomenclature.* Karger Basel.

Investigation for mental retardation and/or dysmorphic features

Aase JM (1990) *Diagnostic dysmorphology*. Plenum, New York.

Cohen MM (1997) *The child with multiple birth defects*, 2nd edn. Oxford University Press, New York.

Jones KL (1997) *Smith's recognizable patterns of human malformation*, 5th edn. WB Saunders, Philadelphia.

Stevenson RE, Schwartz CE, Schroer RJ (1999) *X-linked mental retardation*. Oxford University Press, New York.

Multiple choice questions

1 When assessing a family tree with more than one affected individual

(a) male to male transmission is consistent with mitochondrial inheritance

(b) parental consanguinity is suggestive of autosomal recessive inheritance

(c) if more females are affected than males this points to X-linked dominant inheritance

(d) if more females are affected than males this points to mitochondrial inheritance

(e) father to son transmission is consistent with X-linked dominant inheritance

2 A healthy man whose sister has cystic fibrosis (incidence 1 in 2500) seeks genetic counselling. It would be correct to tell him that

(a) the probability that he is a carrier is 1 in 2

(b) if his partner is unrelated with no family history of cystic fibrosis the probability that they will have an affected child is 1 in 150

(c) if his partner is an unaffected first cousin the probability that they will have an affected child equals 1 in 24

(d) if his affected sister has a child with her healthy unrelated partner the probability that they will have an affected child is 1 in 50

(e) if his father has a child with a different healthy unrelated partner the probability that they will have an affected child is 1 in 100

3 Chromosome analysis is undertaken on an unborn baby. The following reports would indicate that the baby is likely to be abnormal:

(a) 46,XY

(b) 46,XX,inv(4)(q22;q32)

(c) 47,XY,+21

(d) 45,XX,rob(13;21)(q10;q10)

(e) 46,XX,rob(13;21)(q10;q10)mat,+13

4 The following molecular genetic reports have been interpreted correctly:

(a) g.46T→A = substitution of thymine for adenine

(b) g.45_50del = deletion of six nucleotides

(c) g.45_46ins ATA = insertion of three nucleotides

(d) IVS4+1T→C = substitution of thymine by cytosine at the first nucleotide in the fourth intron

(e) p.G567T = replacement of glutamine by threonine at codon 567

5 Steps in the evaluation of a child with mental retardation would include

(a) a detailed family history

(b) a full clinical examination

(c) chromosome analysis

(d) molecular analysis for non-syndromal autism

(e) blood and urine metabolic analysis

Answers

1 (a) false—males do not transmit any mitochondria to offspring

(b) true—rare autosomal recessive disorders are more common in the offspring of consanguineous parents

(c) true—roughly twice as many women are affected as men, because women have two X chromosomes

(d) false—generally mitochondrial disorders affect males and females equally

(e) false—a father cannot transmit his X chromosome to his son

2 (a) false—the probability that he is a carrier is $^2/_3$

(b) true—the probability that both parents are carriers equals $^2/_3 \times ^1/_{25}$ and the probability that two carriers will have an affected child equals 1 in 4 (multiplying $^2/_3$ by $^1/_{25}$ by $^1/_4$ gives 1 in 150)

(c) true—the probability that an unaffected first cousin is a carrier is 1 in 4. This gives an overall probability for having an affected child of $^2/_3 \times ^1/_4 \times ^1/_4$, which equals 1 in 24

(d) true—the probability that his sister will transmit a mutant allele is 1. The probability that her partner is a carrier equals $^1/_{25}$, and if he is a carrier there is 1 chance in 2 that he will transmit the mutant allele, i.e. an overall risk of 1 in 50

(e) true—his father is an obligatory carrier. The probability that his partner is a carrier equals $^1/_{25}$, and the probability that they will have an affected child is $^1/_{25} \times ^1/_4$ which equals 1 in 100

3 (a) false—this represents a normal male

(b) false—this represents a balanced paracentric inversion

(c) true—this represents trisomy 21 (Down syndrome)

(d) false—this represents a balanced Robertsonian translocation

(e) true—this represents an unbalanced Robertsonian translocation resulting in Patau syndrome

4 (a) false—this represents substitution of adenine for thymine

(b) true—the deletion involves nucleotides 45 to 50 inclusive

(c) true—three nucleotides (ATA) have been inserted between nucleotides 45 and 46

(d) true—thymine has mutated to cytosine. IVS stands for intervening sequence (intron)

(e) false—G represents glycine, not glutamine. Otherwise the interpretation is correct

5 (a) true—this is an essential step

(b) true—this is also an essential step

(c) true—this should be carried out on all children with unexplained mental retardation

(d) false—there are no molecular tests for nonsyndromal autism

(e) true—many inborn errors of metabolism can cause mental retardation

Applied clinical genetics

Clinical genetics is the branch of medicine which deals with the management of inherited disorders in families. In the USA, the American Board of Medical Genetics was set up in 1979 and was recognized as a member of the American Board of Medical Specialties in 1991. In the UK, departments of clinical genetics have been established at all major medical centres together with formal training programmes for clinical geneticists and genetic counsellors.

The responsibilities of a department of clinical genetics are summarized in Table 14.1. Dysmorphology and syndrome diagnosis have already been considered in Chapters 9 (p. 183) and 13 (p. 257). The management of families with a predisposition to cancer is reviewed in Chapter 10 (p. 201). The remaining topics listed in Table 14.1 are outlined in this chapter, which concludes with consideration of some of the many difficult ethical issues which arise in clinical genetic practice.

TABLE 14.1 Responsibilities of clinical genetics

Genetic counselling

Carrier detection

Predictive testing

Prenatal diagnosis

Maintenance of genetic registers

Dysmorphology and syndrome diagnosis

Coordination of cancer genetic services

Genetic counselling

The provision of genetic counselling constitutes a cornerstone of clinical genetics. Formally, **genetic counselling** is usually defined as a process of communication or education, whereby individuals or families are given information about the nature of a condition, the likelihood of developing and/or transmitting it, and the options available for treating or preventing it. Implicit in this definition is the provision of information, so in some respects genetic counselling can be viewed as an information service that enables individuals to make informed choices. Implicit in the concept of genetic counselling is the provision of psychological and emotional support, to enable those who receive bad news to come to terms with their new situation and to adjust to the possible implications for their health and that of their family.

The key components of genetic counselling can be considered under four headings:

◆ *Construction of a family tree*. An approach for how this can be undertaken has been outlined in the previous chapter.

◆ *Confirmation of the diagnosis*. This is obviously crucial if correct information is to be given. Sometimes the diagnosis is well established. On other occasions it is necessary to review relevant medical records, as in the case of a family history of cancer, or to undertake appropriate diagnostic investigations, as when presented with a non-specific diagnosis such as mental retardation (p. 254).

◆ *Conveyance of information*. The range of information which can be discussed at genetic counselling is wide (Table 14.2). Given that the goal of genetic counselling is to enable the recipient, who is sometimes referred to as the **consultand**, to make his or her own informed decisions, it is necessary to provide all the information necessary to facilitate this process. The consultand needs to know not only the numerical value of a risk, but also whether the risk is for a minor or a more serious medical condition and whether effective treatment is available. Sometimes this will involve the provision of written information about a condition in the form of pamphlets or articles reproduced from journals or medical textbooks.

◆ *Provision of short- and long-term support*. Occasionally a consultand may receive very distressing news at a genetics clinic. For example, a young mother may discover that she is the carrier of a serious X-linked recessive condition, such as Duchenne muscular dystrophy, which will ultimately lead to the death of her son(s). Alternatively, a young couple may have to come to terms with the fact that there is a risk of 1 in 4 that each of their future children will have a serious and untreatable autosomal recessive disorder. In these situations the complex psychosocial issues can require professional support, which is often best provided by genetic counsellors who have received appropriate training in this field.

Information or advice?

Although the distinction is subtle, there is an important difference between providing information to enhance autonomous decision-making and giving advice as how best to proceed. As a general rule in medicine, and in genetic counselling in particular, counselling should be *informative* rather than *advisory*. In this context the term *non-directive* is often applied. Some consultands do seek guidance as to how best to proceed, but before acquiescing it is important to remember that it is the consultand and not the counsellor who has to live with the consequences of the outcome.

There will also be occasions when a consultand selects an option about which the counsellor is uneasy. In these difficult situations it is reasonable to try to ensure that the consultand is aware of the possible consequences of his or her decision, but ultimately it is the duty of the counsellor to respect the consultand's autonomy. Thus, in addition to being non-directive, counselling should also be *non-judgemental*.

This issue of whether genetic counselling involves the provision of information or advice is fundamental to the role of clinical genetics services. It may well be attractive to health care planners to perceive the role of genetics services as bringing about a reduction in the numbers of individuals born with serious, and thus

---

TABLE 14.2 Range of information conveyed at genetic counselling

Clinical details of the relevant conditions, including:
  Age of onset
  Range of severity
  Possible complications
  Long term prognosis
  Risk for carrying or developing the condition
  Risk for having a child affected with the condition
  Availability of carrier or predictive tests
  Availability of prenatal diagnostic tests
  Availability of effective treatment
  Details of local and/or national support groups

expensive, genetic disorders. In contrast, most professionals see their responsibility as ensuring that their counsellees are provided with full information in an appropriate, comprehensible, and sympathetic manner. In other words, their responsibility is to the individual or family and not to society or the holder of the health care budget. On occasions this may create a conflict, as when parents opt to continue a pregnancy knowing that their unborn child has a serious condition, the treatment of which will incur great expense. In reality clinical genetics services are widely deemed to be cost-effective in that their overall effect is to bring about a reduction in the burden of genetic disease in society. However, this is not their purpose. The aligning of clinical genetics services to any governmental policy for reducing genetic disease can be viewed as the first step along the pathway to the eugenics nightmare which blighted the history of genetics in the first half of the twentieth century (Box 14.1).

> **Key Point**
>
> Genetic counselling involves the provision of information in a non-directive and non-judgemental manner so that individuals can make their own informed decisions.

Carrier detection

Many referrals to genetics clinics involve family members who wish to establish whether they carry a particular inherited disorder. The approaches which can be used to determine carrier status are as follows.

BOX 14.1 CASE CÉLÈBRE: THE EUGENICS MOVEMENT

Eugenics can be defined as the improvement of the human species through selective breeding. The term was originally coined by Francis Galton, a cousin of Charles Darwin, who had a long-standing interest in quantitative inheritance and how it relates to characteristics such as height and intelligence. Galton and his contemporaries were concerned that the British upper classes were being out-bred by people of poorer 'stock' and they were anxious that this should be redressed by large families among the gentry.

This suggestion that those with a superior genetic endowment should be encouraged to reproduce represents an example of *positive eugenics*. The corollary, that the less genetically gifted should be discouraged from reproduction, falls within the domain of *negative eugenics*. Sadly, these ideas became fashionable during the first half of the twentieth century, leading to the introduction in parts of the USA and Europe of legislation which permitted involuntary sterilization of the mentally handicapped. Between 1907 and 1960 over 60 000 people were sterilized in the USA without their consent. Disturbing though this is, these events pale into insignificance when compared with the atrocities committed in Nazi Germany, where it has been estimated that over 1 million genetically 'inferior' individuals underwent forced involuntary sterilization. This included large numbers of patients with Huntington disease who had been ascertained through hospital-based disease registers.

There is now universal agreement that the eugenics practices of the twentieth century were both scientifically unsound and morally unacceptable. The genetic contribution to mental retardation is complex (p. 254). If autosomal recessive genes play a significant aetiological role, most of these will reside in unaffected heterozygotes. Thus the sterilization of the mentally handicapped will have almost no effect on the incidence of retardation in future generations (p. 138). Ethically, the concept of compulsory sterilization is contrary to all accepted standards of human behaviour.

Nevertheless, both society at large and the genetics community remain concerned that eugenic measures could be introduced surreptitiously by a misguided or malevolent government. These concerns prompted the Board of Directors of the American Society of Human Genetics to issue a formal statement in 1999 reaffirming their commitment to the 'fundamental principle of reproductive freedom' and their opposition to 'coercion based on genetic information'. Among clinical geneticists and genetic counsellors there is a clear consensus that their role is to inform, rather than to advise or dictate, and that there is no place for directive counselling in modern practice.

Reference

Board of Directors of the American Society of Human Genetics (1999) Eugenics and the misuse of genetic information to restrict reproductive freedom. *American Journal of Human Genetics*, **64**, 335–338.

Chromosome abnormality

Most children with chromosome abnormalities represent isolated events within a family. This applies to almost all cases of trisomy (e.g. trisomy 13, trisomy 18, trisomy 21, 47,XXX, 47,XXY, 47,XYY) and monosomy (45,X). In these situations the parents of an affected child will be reassured that they are not at increased risk of being carriers of a chromosome abnormality and usually they will not be offered chromosome analysis.

However, when a child is known to have an unbalanced rearrangement, or when there is a confirmed family history of a balanced rearrangement in other relatives, then it is essential that chromosome analysis be offered to all relatives who could be carriers. This is particularly important for rearrangements such as balanced translocations (p. 39) and pericentric inversions (p. 41), which can be inherited in an unbalanced form leading to severe physical and developmental abnormalities. Because of this possibility it is generally recommended that, for all balanced rearrangements with a potential to generate viable imbalance, an attempt should be made to track the rearrangement within the family by offering chromosome analysis to all relevant relatives. An important exception to this rule is that testing is usually not offered to children, as discussed later in this chapter (p. 277).

> **Key Point**
>
> An attempt should be made to ensure that all possible carriers of a balanced chromosome rearrangement are provided with an opportunity to undergo chromosome analysis.

Single-gene disorders

The methods available for detecting carriers of single-gene disorders are summarized in Table 14.3. Until recently this list would also have included a category entitled 'clinical examination', but with very few exceptions this approach has been superseded by developments in molecular genetics. By definition, carriers of autosomal recessive disorders are entirely healthy. Some female carriers of X-linked recessive disorders show subtle abnormalities on careful examination, particularly if the disorder involves the eyes or skin so that mosaicism resulting from X inactivation can be visible. This applies to X-linked ocular albinism and retinitis pigmentosa (Fig. 14.1) in which many female carriers show a mosaic pattern of

---

TABLE 14.3 Methods for detecting carriers of single-gene disorders

Biochemical assay		
Direct	Glucose-6 phosphate dehydrogenase (G6PD) deficiency	G6PD activity in blood (not 100% reliable)
	Tay–Sachs disease	Hexosaminidase A activity in serum and white blood cells
Indirect	Duchenne muscular dystrophy	Creatine kinase in serum (elevated in 2 out of 3 carriers)
Haematological indices	Sickle-cell disease	Sickling test and haemoglobin electrophoresis
	α-Thalassaemia	Full blood count, red cell indices and blood film
	β-Thalassaemia	Full blood count, red cell indices and haemoglobin electrophoresis
Mutation analysis	Preferred method for all autosomal recessive and X-linked recessive disorders in which a mutation can be identified	
	Autosomal recessive	*X-linked recessive*
	Congenital adrenal hyperplasia	Duchenne muscular dystrophy
	Cystic fibrosis	Fragile X syndrome
	Freidreich's ataxia	Haemophilia A and B
	Haemochromatosis	Hunter syndrome
	Phenylketonuria	
	Spinal muscular atrophy	
Linkage/gene tracking	Using markers linked to the disease locus when specific mutation cannot be identified	

Fig. 14.1 View of the retina of a female carrier of X-linked recessive retinitis pigmentosa showing small isolated patches of abnormal retinal pigmentation.

retinal pigmentation although clinically they are asymptomatic.

In general the preferred method for carrier detection is specific mutation analysis, as this gives an unequivocal answer if the mutation in the proband is known. An alternative approach is to use linkage/gene tracking when no specific mutation has been identified (p. 91). Biochemical assay of enzyme activity is not useful for most autosomal recessive and X-linked inborn errors of metabolism (p. 207) as the enzyme levels in carriers and non-carriers overlap. An important exception is Tay–Sachs disease, in which a clear distinction can usually be made between carriers and non-carriers (p. 148).

Haematological indices and haemoglobin electrophoresis can be used to identify carriers of sickle-cell disease and both α- and β-thalassaemia. This is the preferred method for population screening (p. 149) as the technology is widely available and relatively inexpensive.

Predictive testing

The development of molecular tests for late-onset genetic disorders has prompted an increasing number of requests for what is referred to as *predictive* or *preclinical* genetic testing. This applies mainly to late-onset disorders which show autosomal dominant inheritance. These can be divided into two groups: those for which prevention or treatment is available, and those for which there is no effective means of intervention. The approach to the management of these two situations differs fundamentally.

Disorders that can be prevented or treated

When effective intervention is available, in the form of either preventive measures or early treatment, then a strong case can be made for offering predictive testing to all individuals deemed to be at increased risk because of a positive family history. Examples of conditions for which this is relevant are given in Table 14.4. Many other conditions involve an increased risk for developing cancer. Approaches to offering regular surveillance for the development of relevant tumours have been outlined in Chapter 10 (Tables 10.7 and 10.8). Preventive measures which can be employed for other conditions include dietary modification as in familial hypercholesterolaemia, the avoidance of cigarette smoking as in α_1-antitrypsin deficiency, and the avoidance of exposure to certain drugs as in porphyria. Early diagnostic surveillance can be offered for complications such as hypertension in adult polycystic kidney disease and aortic dissection in Marfan syndrome.

The potential health benefits which can be achieved by preclinical genetic testing are such that a case can be made for trying to ensure that this is offered to all individuals known to be at increased risk within a community. For this reason several centres have established what are known as genetic registers for late-onset disorders, through which contact can be maintained with all known relatives at high risk who express a willingness to participate. These potential health benefits have to be balanced against the almost inevitable

TABLE 14.4 Predictive testing: late-onset disorders for which prevention or treatment is available			
Disorder	**MIM**	**Inheritance**	**Prevention or surveillance**
Adult polycystic kidney disease	173900	AD	Blood pressure and renal function
α_1-Antitrypsin deficiency	107400	AR	Avoidance of smoking and excess alcohol
Familial hypercholesterolaemia	143890	AD	Low cholesterol diet and HMG CoA reductase inhibitors
Haemochromatosis	235200	AR	Iron indices and early treatment with venesection
Marfan syndrome	154700	AD	Regular echocardiography
Porphyria—many different types		AD	Avoidance of specific drugs (p. 217)
Pseudocholinesterase deficiency	177400	AR	Avoidance of succinylcholine. (p. 229)

AD, autosomal dominant; AR, autosomal recessive.

anxiety that will be generated by adverse results. This emphasizes the importance of fully informed participation and the value of the long-term contact and support provided by genetic counsellors. Because of concern about possible psychological ill-effects there is a universal consensus that predictive testing should not be offered to children aged less than 16 years unless there is clear evidence that early preventive measures and/or diagnostic surveillance are of proven benefit. This is discussed further towards the end of this chapter.

Disorders that cannot be prevented or treated

In contrast to the disorders listed in Table 14.4, there is no effective treatment or cure for any of the conditions included in Table 14.5. This means that the justification for offering predictive testing is based on enhancing personal autonomy rather than improving the pros-

TABLE 14.5 Predictive testing in late onset disorders for which there is no prevention or treatment

Alzheimer disease (early-onset forms, p. 125)
Facioscapulohumeral muscular dystrophy
Familial motor neurone disease. (Most motor neurone disease is not genetic)
Hereditary motor and sensory neuropathy (also known as Charcot–Marie–Tooth disease)
Huntington disease
Myotonic dystrophy (surveillance available for cataracts and cardiac arrhythmias)
Spinocerebellar ataxia
Torsion dystonia

TABLE 14.6 Key steps in predictive testing for Huntington's disease

Preliminary assessment
Visit by genetic counsellor to consultand to: obtain family details and confirm diagnosis provide information on HD and predictive testing confirm that consultand wishes to explore predictive testing.
Formal clinic visit
Consultand is seen with a companion by genetic counsellor and clinical geneticist. Clinical and genetic aspects of Huntington's disease fully explained
Implications of testing discussed.
Neurological and psychiatric assessments
Consultand is offered referral for baseline neurological and psychiatric evaluation. Attendance is recommended but not essential
Formal clinical visit
Consultand confirms desire for predictive testing
Consent is obtained
Blood sample is taken
Appointment given for result disclosure
Disclosure appointment
Consultand confirms that he/she wishes to know the result
Result is disclosed—always in person and never by phone or letter
Follow-up arrangements agreed
Post-test counselling
Initial contact within 1 week
Further follow-up and counselling as required

This outline protocol is based on the recommended guidelines for predictive testing as drawn up by an international committee representing the International Huntington Association and the World Federation of Neurology Research Group on Huntington's Chorea as published in the *Journal of Medical Genetics* (1994), **31**, 555–559.

pects for an individual's health. Consequently, strong efforts are made to ensure that anyone requesting predictive testing for a disorder listed in Table 14.5 is fully informed of the progressive and untreatable nature of the condition and of the potential consequences of an adverse test result. This will usually involve a discussion of the natural history of the condition, together with an explanation of possible implications for issues such as employment, insurance, and reproduction. The importance attached to full counselling in these situations is illustrated by the standard recommended protocol for predictive testing in Huntington disease (Table 14.6). Experience to date indicates that serious adverse reactions are unusual among those who follow the full protocol through to completion. However, over 50% of individuals who express an interest in predictive testing for Huntington disease subsequently decide not to continue when the full implications of an adverse result are explained.

> **Key Point**
>
> Predictive testing for adult onset disorders should not be carried out in children unless effective early prevention and/or treatment are available.

Prenatal diagnosis

In the context of clinical genetics, prenatal diagnosis refers to the detection of a genetic or structural abnormality before birth. Inevitably this is a subject which arouses strong emotions, as termination of pregnancy is one of the options that arises when an abnormality is identified prenatally. However, it is important to emphasize that this is not the sole purpose of offering prenatal diagnosis. More often than not, the fetus is found to be unaffected and the prospective parents gain reassurance that all is well. Alternatively, if an abnormality is identified, some parents view this as an opportunity

---

BOX 14.2 LANDMARK PUBLICATION: ALPHA-FETOPROTEIN AND NEURAL TUBE DEFECTS

Anencephaly and open lumbosacral neural tube defects are serious malformations (p. 128). Anencephaly is invariably lethal. Open lumbosacral defects usually result in severe neurological disability, often accompanied by hydrocephalus because of an associated Arnold–Chiari malformation involving downward displacement of the brainstem and cerebellar vermis through the foramen magnum. In the 1950s and 1960s the prevalence of these conditions reached almost epidemic proportions in parts of the UK and elsewhere, particularly amongst lower socioeconomic groups and populations of Celtic origin. At that time there was no safe method available for detecting these conditions before birth.

This situation began to change in 1972 following the report by Brock and Sutcliffe, from Edinburgh, that AFP was elevated in the amniotic fluid of 1 spina bifida and 31 anencephalic pregnancies. AFP is produced in the fetal liver from around 6–7 weeks onwards and reaches a maximum level in fetal serum at 13 weeks. The probable explanation for the elevated levels in neural tube defect pregnancies is that AFP leaks from cerebrospinal fluid and open capillary beds into the surrounding amniotic fluid. Brock and Sutcliffe correctly surmised that as AFP crosses the placental barriers, elevated levels would also be found in maternal serum in pregnancies with an open defect.

Subsequent studies confirmed that there was a relatively clear distinction between amniotic fluid levels of AFP in open neural tube defect and unaffected pregnancies. This observation, in conjunction with developments in ultrasonography, meant that for the first time prospective parents with a family history of neural tube defects could consider prenatal diagnosis as an option. The levels of maternal serum AFP in affected and unaffected pregnancies show greater overlap than in amniotic fluid, so this approach alone does not provide a reliable diagnosis. However, maternal serum AFP screening is now widely practised to identify pregnancies in which further investigation is indicated.

This discovery of raised levels of AFP in neural tube defect pregnancies, together with the later observation of the preventive effect of folic acid (Box 6.3), led to a sharp decline in the number of babies born with severe and disabling defects. The ethics of prenatal diagnosis and termination of pregnancy will always be contentious. However, the fact that large numbers of parents have chosen to take advantage of this discovery, and that maternal serum AFP screening is practised worldwide, indicates that this observation represents a major milestone in the development of prenatal diagnostic services.

Reference

Brock DJH, Sutcliffe RG (1972) Alpha-fetoprotein in the antenatal diagnosis of anencephaly and spina bifida. *Lancet*, **ii**, 197–199.

to come to terms with their unborn child's diagnosis and to prepare for his or her birth. Medical services can also be prepared to deal with the child's abnormalities soon after birth. Finally, it is hoped that intrauterine treatment based on gene therapy will soon become an option for many single-gene disorders.

Nevertheless, prenatal diagnosis is an emotive topic and understandably many prospective parents and health care professionals struggle with the difficult ethical issues that it generates (p. 279). All prospective parents undergoing prenatal testing for a genetic disorder should be fully counselled in a non-directive manner, preferably well in advance of pregnancy. A decision to request prenatal testing must be entirely voluntary, with the explicit right to opt out at any point. It is totally unacceptable to offer prenatal diagnosis only on the condition that the parents pursue a particular course of action in the event of an adverse result.

Methods of prenatal diagnosis

There are four standard techniques currently used for prenatal diagnosis. Two of these, maternal serum screening and ultrasonography, are non-invasive. The remaining two, chorionic villus sampling and amniocentesis, both convey a small risk for causing miscarriage. A fifth category consists of novel and as yet largely research-based approaches such as preimplantation genetic diagnosis and the analysis of fetal cells obtained from the maternal circulation.

Maternal serum screening

In 1972 it was first realized that the level of alphafetoprotein (AFP) is raised in the amniotic fluid of pregnancies in which the fetus has a severe open neural tube defect (p. 128) such as anencephaly or lumbosacral spina bifida (Box 14.2). AFP is a fetal protein which

leaks through an open defect in the fetus into the surrounding amniotic fluid. Subsequently it was shown that the maternal serum level of AFP is also raised in many pregnancies with an open neural tube defect (Fig. 14.2). This led to the development of screening programmes in pregnancy based on the assay of AFP and other chemicals in maternal serum.

Serum screening for neural tube defects

In many parts of the world it is now standard practice to offer maternal serum screening to all pregnant women at 16 weeks gestation. An increased level of maternal serum AFP, arbitrarily designated as greater than 2.0 or 2.5 multiples of the median, prompts further assessment in the form of detailed ultrasonography. Sometimes an amniocentesis is also performed for assay of AFP in amniotic fluid. Serum screening followed by ultrasound scanning of pregnancies deemed to be at high risk will detect around 70% of severe open neural tube defects. Serum screening combined with ultrasound fetal anomaly scanning in all pregnancies will detect over 90% of severe neural tube defects.

The maternal serum level of AFP does not depend solely on the presence or absence of an open neural tube defect. Other causes of elevated maternal serum AFP include anterior abdominal wall defects, such as exomphalos and gastroschisis, a twin pregnancy, and fetal haemorrhage.

Serum screening for chromosome abnormalities

In contrast to pregnancies in which the fetus has a neural tube defect, the level of AFP in maternal serum at 16 weeks gestation is reduced in Down syndrome pregnancies (Fig. 14.2). This observation has led to the development of a 'triple' screening test for Down

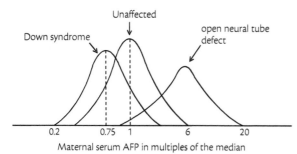

Fig. 14.2 Distribution of maternal serum AFP levels at 16 weeks gestation in normal pregnancies, pregnancies with Down syndrome, and pregnancies with open neural tube defects.

TABLE 14.7 Mean levels of maternal serum markers in the second trimester in trisomy 18 and trisomy 21

Trisomy 18	
Alpha-fetoprotein	0.6
Human chorionic gonadotrophin	0.3
Pregnancy-associated plasma protein A	0.1
Unconjugated oestriol	0.4
Trisomy 21	
Alpha-fetoprotein	0.75
Human chorionic gonadotrophin	2.0
Inhibin-A	2.1
Unconjugated oestriol	0.7

All levels expressed as multiples of the median in normal pregnancies.

syndrome based on analysis of the levels of AFP, unconjugated oestriol, and human chorionic gonadotrophin (Table 14.7). A formula, i.e. an *algorithm*, has been developed based on maternal age and the levels of these three chemicals at 16 weeks gestation to estimate the probability that any particular baby is affected with Down syndrome. When the risk is greater than 1 in 250 an invasive test such as amniocentesis is offered for definitive chromosome analysis. In theory this approach can lead to the identification of approximately 60% of all Down syndrome pregnancies, although in practice the actual proportion detected is lower because some women decline serum screening. The use of a fourth biochemical marker, inhibin-A, raises the theoretical detection rate to around 75%.

Serum screening at 16 weeks gestation based on the triple test and another marker, known as pregnancy-associated plasma protein A (PAPP-A), will detect almost 100% of cases of trisomy 18. All four markers show markedly reduced levels in trisomy 18 pregnancies (Table 14.7). The maternal serum levels of human chorionic gonadotrophin and PAPP-A are also significantly reduced at 12 weeks gestation in pregnancies in which the fetus has either trisomy 13 or trisomy 18. These pregnancies also show increased fetal nuchal translucency (see below). This information has been combined to develop a risk algorithm which can in theory identify 95% of trisomy 13 and trisomy 18 pregnancies at 12 weeks gestation. This facility is available at only a limited number of specialist centres.

> ### Key Point
>
> The mean level of maternal serum AFP is elevated in pregnancies with an open neural tube defect and reduced in pregnancies with Down syndrome. Serum screening based on assay of four biochemical markers at 16 weeks gestation can detect up to 75% of all Down syndrome pregnancies.

Ultrasonography

High-resolution ultrasound scanning provides a pain-free, safe, non-invasive method for imaging the developing fetus. Most pregnant women are offered routine ultrasound scanning in the first trimester to determine the exact gestation and to identify multiple pregnancy. Detailed scanning at this period can be used to determine the degree of *fetal nuchal translucency*. This reflects the amount of subcutaneous tissue fluid present at the back of the developing baby's neck (Fig. 14.3). This is increased in trisomy 13, 18, and 21. Screening

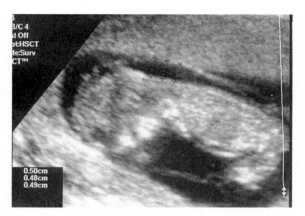

Fig. 14.3 Ultrasound scan at 12 weeks gestation showing increased nuchal translucency. Courtesy of Mr E. Howarth, Leicester Royal Infirmary, Leicester.

for increased maternal translucency followed by definitive chromosome analysis can identify approximately 80% of all cases of Down syndrome. Up to 90% can be identified at 12 weeks gestation if screening for nuchal translucency thickness is combined with serum screening for human chorionic gonadotrophin.

High-resolution ultrasound scanning at 18–20 weeks gestation is now offered widely as a means of *fetal anomaly scanning*. This will detect major structural abnormalities such as severe neural tube defects, major cardiac malformations, renal agenesis or hydronephrosis, and short or missing limbs (Table 14.8). Other more subtle abnormalities can be identified which raise suspicion of a more serious condition such as a chromosome abnormality (Fig. 14.4) or a multiple malformation syndrome. For example, the demonstration of duodenal atresia or a common atrioventricular defect could be indicative of Down syndrome (p. 54)

TABLE 14.8 Proportions of major structural abnormalities detectable by ultrasound at 18–20 weeks gestation

Abnormality	Proportion detectable (%)
Cardiac defects	25
Diaphragmatic hernia	60
Abdominal wall defects	90
Hydrocephalus	60
Limb malformations	90
Open neural tube defects	90
Renal abnormalities	85

Figures refer to major structural abnormalities only.

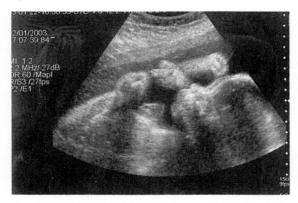

Fig. 14.4 Ultrasound scan at 18 weeks gestation showing showing clenched hands and small jaw (micrognathia) in a fetus subsequently shown to have trisomy 18. Courtesy of Dr M Khare, Leicester Royal Infirmary, Leicester.

and should lead to an offer of definitive chromosome analysis. Certain ultrasound findings are referred to as *soft markers* as they occur in many normal infants but show a slightly increased incidence in specific disorders. These include choroid plexus cysts, which show a weak association with trisomy 18, and echogenic foci in the heart, which are marginally more common in babies with Down syndrome than in unaffected babies.

Chorionic villus sampling (chorion biopsy)

Chorionic villi are derived from the extraembryonic part of the blastocyst, known as the trophoblast, and consist of cytotrophoblast with an outer layer of syncytiotrophoblast. Because of this shared origin with the embryo, they can be used to determine the genetic constitution of the developing pregnancy. **Chorionic villus sampling** (CVS) can be performed at 11–14 weeks, either transabdominally by inserting a needle through the anterior abdominal wall, or transcervically by passing a soft flexible catheter through the birth canal. The fronds of chorionic villi obtained by suction using a syringe are carefully dissected to remove maternal decidual cells before being used for genetic analysis.

Molecular analysis

Chorionic villi provide a rich source of DNA which can be used for specific mutation analysis or for the analysis of markers known to be linked to the disease locus. This is now the standard method used for the prenatal diagnosis of single-gene disorders such as cystic fibrosis and Duchenne muscular dystrophy or any other condition in which the basic molecular defect is known within the family. Later in pregnancy a transabdominal placental biopsy can be performed for the same purpose.

Chromosome analysis

Chromosome analysis using chorionic villi can be undertaken directly, using cells which are in mitosis, or following a 2–3-week period of culture after which the cells are harvested and synchronized in metaphase. A direct analysis can be undertaken within a few hours of the biopsy. This provides a rapid indication of the fetal sex and the presence or absence of an additional chromosome 21. Other chromosome abnormalities can only be diagnosed reliably following cell culture.

Biochemical analysis

Cultured chorionic villi can be used to detect most inborn errors of metabolism in which the basic biochemical defect has been established. However, there are a few biochemical disorders, notably phenylketonuria, in which the relevant enzyme is not expressed in chorionic villi or amniocytes (see below), so that not all inborn errors can be diagnosed using this approach.

Advantages and disadvantages of CVS

On the positive side, CVS provides a well-established method for obtaining a relatively early diagnosis for single gene and chromosome abnormalities. However, there are some important negative aspects. Experience worldwide indicates that there is an associated risk of miscarriage of 1 in 50 to 1 in 100. Early CVS before 9–10 weeks *may* convey a risk for causing limb defects in the developing embryo. This remains unproven, but there is substantial suggestive evidence. In 1–2% of cases CVS provides an unclear result because of mosaicism manifesting as two cell lines with different chromosome constitutions. This can be a true reflection of the fetal karyotype or it may be spurious in that the mosaicism is limited to the rapidly dividing trophoblast either *in vivo* or *in vitro*. Usually this can only be resolved by carrying out amniocentesis. Finally, CVS cannot be used to diagnose neural tube defects or other structural abnormalities. Thus a normal result can sometimes be interpreted by parents as providing reassurance, which may subsequently prove to be false.

Amniocentesis

Amniocentesis involves the removal of fluid from the amniotic cavity by the insertion of a needle through the anterior abdominal wall under ultrasound guidance. This technique is usually performed for diagnostic purposes at 16–18 weeks gestation although at some centres amniocentesis is performed from 12–14 weeks onwards.

The 10–20 ml of fluid obtained at amniocentesis consists of amniotic liquor containing a small number of cells, known as amniocytes, of diverse fetal origins.

TABLE 14.9 Indications for offering prenatal chromosome analysis

Advanced maternal age (35 years or greater)		
Previous child with a chromosome abnormality		
One parent carries a rearrangement such as an inversion or a translocation		
Risk of 1 in 250 or greater on triple test		
Abnormal findings on ultrasound, e.g.	Duodenal atresia	(Trisomy 21)
	Holoprosencephaly	(Trisomy 13)
	Clenched hands	(Trisomy 18)
	Hydrops fetalis	(Turner syndrome)

These include the fetal skin, the alimentary and urogenital tracts, and the surface of the placenta. The cell pellet obtained by centrifugation is cultured for a period of 2–3 weeks, following which there are usually sufficient cells for chromosome analysis or biochemical assay. A longer period of cell culture may be necessary to obtain sufficient cells for molecular analysis.

Amniocentesis is used widely for the prenatal diagnosis of chromosome abnormalities, the indications for which are summarized in Table 14.9. It can also be used to diagnose metabolic disorders in which the relevant enzyme is expressed in amniocytes. The supernatant fluid obtained from centrifugation can be used to detect the presence of abnormal metabolites, as in the mucopolysaccharidoses, and for assay of AFP when ultrasound has raised suspicion of a neural tube defect.

Amniocentesis conveys a risk of 1/100 to 1/200 for causing miscarriage. The lengthy wait of 3–4 weeks for a result is also, understandably, seen as a disadvantage by many parents. The analysis of cells in interphase by FISH (p. 29), using centromeric probes specific for chromosomes 13, 18, 21, X, and Y, provides a rapid screen for the common aneuploidy syndromes. It has been proposed that rapid interphase FISH provides a satisfactory chromosome analysis, although obviously this does not detect other chromosome abnormalities or rearrangements.

Using a technique similar to amniocentesis, a small sample of fetal blood can be obtained from the umbilical cord. This procedure, referred to as cordocentesis or fetal blood sampling, is more hazardous than amniocentesis with an associated risk for causing miscarriage of 3–5%. Consequently cordocentesis is only used sparingly. Indications include the investigation of hydrops fetalis, possible intrauterine infection, and unresolved chromosome mosaicism.

Novel approaches to prenatal diagnosis

All of the techniques outlined so far involve the diagnosis of an abnormality after pregnancy is established, with a subsequent option of termination. In addition, the two procedures which can be used to achieve a definitive diagnosis of a chromosomal or molecular abnormality, i.e. CVS and amniocentesis, both convey a risk for causing miscarriage. Two new approaches are being developed which could potentially avoid these drawbacks.

Preimplantation genetic diagnosis (PIGD)

This is the name given to two slightly different techniques which have been developed for the detection of genetic abnormalities by *in vitro* analysis before implantation. The first is based on the analysis of polar bodies expelled following the first and second meiotic divisions in the female (p. 35). These are obtained by micromanipulation (Fig. 14.5) and then subjected to

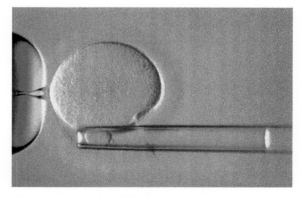

Fig. 14.5 Removal of a polar body by micromanipulation for preimplantation genetic diagnosis. Reproduced with permission from Verlinsky Y, Kuliev A (2000) *An atlas of preimplantation genetic diagnosis.* Parthenon Publishing Group, New York.

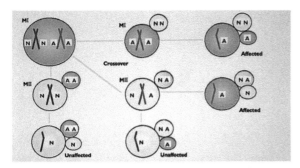

Fig. 14.6 How analysis of the genotype of polar bodies can be used to predict the genotype of an oocyte for preimplantation genetic diagnosis. Reproduced with permission from Verlinsky Y, Kuliev A (2000). *An atlas of preimplantation genetic diagnosis*. Parthenon Publishing Group, New York.

chromosome analysis by FISH or molecular analysis by PCR. The underlying rationale is that by analysing the genetic constitution of the polar bodies in a woman who is known to carry a genetic abnormality, it will be possible to predict the genotype of the oocyte from which the polar bodies were obtained (Fig. 14.6). If this is normal, the oocyte can be fertilized artificially *in vitro* and then implanted into the mother's uterus.

The second approach is based on the analysis of a single cell obtained from the eight-cell zygote, also known as a **blastomere** (Fig. 14.7), following *in vitro* fertilization. If this is found to have a normal genetic constitution, then the totipotent seven-cell blastomere is implanted. As with the first approach, the analysis of the single cell is achieved by FISH for a suspected chromosome abnormality or by PCR for a known molecular defect.

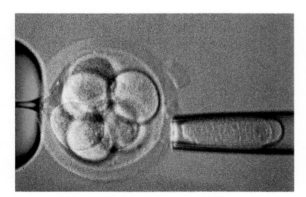

Fig. 14.7 The removal of a single cell from an eight-cell blastomere for preimplantation genetic diagnosis. Reproduced with permission from Verlinsky Y, Kuliev A (2000) *An atlas of preimplantation genetic diagnosis*. Parthenon Publishing Group, New York.

Both of these techniques have been employed successfully in specialist centres. However, both are extremely demanding technically, and therefore expensive, with limited availability. Most centres offering *in vitro* fertilization and PIGD achieve a success rate of one pregnancy in every 4–5 cycles.

The use of fetal cells in maternal blood

For many years attempts have been made to isolate the small number of fetal-derived cells which circulate in maternal blood. To date no reliable prenatal diagnostic tests have been developed based on this approach. However, some centres have achieved success using circulating trophoblastic cells or nucleated red cells of fetal origin for the determination of the fetal sex and the detection of a small number of single-gene disorders including sickle-cell disease and spinal muscular atrophy.

Ethical considerations

Ethical challenges can arise in all branches of medicine, and clinical genetics is no exception. In reality they tend to occur more often in clinical genetics than in other medical specialties, partly because of the sensitive nature of the information being conveyed and partly because genetic issues have implications for other family members and for the unborn.

The *Concise Oxford English Dictionary* defines ethics as 'the science of morals in human conduct' or as 'a set of moral principles'. Everyone has their own set of moral values which they try to apply in day-to-day life. The practice of clinical genetics can raise some particularly difficult and challenging ethical dilemmas for which there is often no precedent or easy solution. In these situations an acceptable approach is to try to apply the following set of well-established fundamental ethical principles which are deemed to form the cornerstone of good medical practice:

◆ *Autonomy*. Autonomy refers to the right to personal freedom, implying that an individual is entitled to make his or her own decisions without pressure from external influences. Patient autonomy in medicine demands fully informed consent.

◆ *Beneficence*. It is axiomatic that doctors should try to do good (beneficence) and strive not to do harm (non-maleficence).

◆ *Justice*. Justice implies fairness with equal access and opportunity for all, regardless of wealth, race or, religion. Essentially this means that everyone should have equal access to genetics services with an expectation of equal care and attention.

◆ *Privacy*. Respect for privacy and confidentiality can be a particularly contentious issue in clinical genetics, as will become apparent in the following discussion.

Ethical issues in genetic counselling

Much has already been made of the importance of being both non-directive and non-judgemental in genetic counselling. This is entirely in keeping with the principle of autonomy. But what happens if there is a potential conflict between autonomy and beneficence? Imagine a situation in which a counsellor has been contacted in advance by one member of a partnership with a plea that a high recurrence risk for offspring is not disclosed as this could jeopardize the relationship. How does the counsellor balance the need for autonomy and honesty with a desire not to do harm? Consider also how a counsellor should respond to a couple who have had a child with cystic fibrosis, when the counsellor knows that the molecular results have revealed non-paternity. This means that the recurrence risk is almost certainly much lower than 1 in 4, but how can this be conveyed without disclosing the molecular results?

The issue of beneficence has already been considered in the context of whether the counsellor's responsibility is to the counsellee or to society at large (p. 266). Here most would agree that the counsellee's interests are paramount. But the decision becomes much more difficult when the counsellor is confronted by the reality that telling the truth, and thereby respecting autonomy, will almost certainly be harmful.

Ethical issues and carrier detection

Carrier detection, when offered on an equitable, informed, and voluntary basis, satisfies the criteria for all of the four cardinal ethical principles. However, problems can arise in two particular situations: when testing is requested for children, and when individuals who are found to be carriers refuse to allow this information to be passed on to other family members.

Genetic testing in children is generally not recommended unless there is a clear medical benefit to be gained. Children cannot give informed consent. Carrier testing for children is sometimes requested by parents, but usually they accept that testing be delayed until the child reaches adulthood. A more difficult situation arises when a prenatal diagnosis is performed which reveals that the fetus is not affected with a chromosome abnormality or single-gene disorder but is instead a carrier. Some authorities have recommended that this should not be disclosed and that the parents should simply be informed that their unborn baby is not affected. If parents do wish to know the full test result, it can be difficult to convince them that they are not entitled to this information.

A much more difficult dilemma arises when a carrier of a translocation or an X-linked recessive disorder refuses either to pass this information on to other family members or to allow the counsellor to approach the relevant relatives. This raises a conflict between the counsellee's right to privacy and the relatives' rights to information. Opinion is divided as to how members of the clinical genetics team should respond. Most would tend to accept, reluctantly, the counsellee's right to confidentiality, although some would argue that the at-risk relatives' right to know outweighs the counsellee's right to confidentiality.

Ethical issues and predictive testing

As with carrier detection, predictive testing clearly enhances individual autonomy, and it is difficult to argue that informed consenting adults should not be allowed to learn about their genetic destiny. However, predictive testing can raise several very difficult ethical issues.

The first concerns children and whether it is acceptable that they should be subjected to predictive testing for adult-onset disorders. The universal consensus view is that this is not acceptable unless there is a clear medical benefit to be gained, such as might be achieved by implementation of a low-cholesterol diet in familial hypercholesterolaemia (p. 213) or early screening for tumours in familial cancer syndromes (p. 196). Most parents feel that this is reasonable, but a few argue cogently that they want their children tested so that their subsequent education can be geared to cope with anticipated medical problems. The child's right to make its own autonomous decision almost certainly outweighs the parents' right to determine its future career path.

A particularly difficult variation of this theme arises when prenatal diagnosis is undertaken at the parents' request for a late-onset disorder such as Huntington disease (p. 105). According to the fundamental principle of autonomy, the parents are fully entitled to decide how they wish to proceed when the result becomes available. If the result is positive and they decide to continue the pregnancy, then effectively the child has had a predictive test and will go through childhood and adult life with the unwanted knowledge that he or she will eventually be affected.

Huntington disease provides another illustration of a problem which can arise from predictive testing when the grandchild of an affected individual requests testing but the healthy, unaffected intervening parent does not want to know whether he/she will become affected. If the grandchild tests positive, then the intervening parent indirectly also receives a positive test result. In general, it is felt that the grandchild's right to know outweighs the parent's right not to know. Clearly this situation raises issues of autonomy (the parent's versus the grandchild's), beneficence (helping the grandchild may harm the parent), and also confidentiality if in due course the parent requests testing when the counsellor already knows that his or her test will be positive.

Finally, predictive testing raises the issue of who owns a DNA sample. An individual may provide a sam-

---

BOX 14.3 CASE CÉLÉBRE: WHO OWNS ABRAHAM LINCOLN'S DNA?

Abraham Lincoln (1809–1865) is widely regarded as one of the world's most respected and influential political figures. A man of humble origins, he became a lawyer in 1836 and slowly established a national reputation in politics to win the Republican Party nomination for President in 1860, being duly elected later that year as the 16th President of the USA. His presidency is remembered chiefly for his success in maintaining the Union during the American civil war and for his role in initiating the abolition of slavery. Amongst his many famous speeches he is particularly remembered for the Gettysburg address in which he coined the term 'government of the people, by the people, for the people'. Sadly, his life ended prematurely when he was assassinated at Ford's Theater in Washington by John Wilkes Booth, a Southern sympathizer who did not share Lincoln's tolerant views.

Contemporary descriptions and photographs (Fig. 14.8) indicate that Lincoln was a very tall man of around 193 cm (6 feet 4 inches) with long limbs and large extremities. These observations, coupled with the discovery that a distant descendant of Lincoln's great-great-grandfather had Marfan syndrome, prompted the suggestion that Lincoln might also have been affected.

Lincoln's features, as depicted in his majestic statue in the Lincoln memorial in Washington, would certainly be consistent with this diagnosis.

Following the discovery of the Marfan syndrome gene in 1991, it was suggested that some of Lincoln's macabre remains, notably fragments of his skull and bloodstained clothes, could be used to extract DNA for mutation analysis. A small committee was formed under the auspices of the US National Museum of Health and Medicine to determine whether this was justified both ethically and scientifically. It was concluded that the proposal to carry out posthumous analysis of Lincoln's DNA did not violate his privacy and that he himself would probably not have objected, particularly if the knowledge that he had been affected would be a comfort and inspiration to other Marfan syndrome sufferers. However, on scientific grounds it was felt that because Marfan syndrome shows such marked mutational heterogeneity, the small sample of DNA that could be obtained from Lincoln's remains might well prove insufficient for full mutation testing. Consequently it was decided not to release Lincoln's remains for DNA analysis at that time.

It can reasonably be argued that historical figures, especially those who occupied positions of power, have ceded their right to privacy if knowledge of their medical problems might be judged to be of historical importance. However, some of today's politicians might not be too enthusiastic about having their DNA samples analysed posthumously, particularly if genes for unusual personality traits or unattractive behavioural characteristics could be identified. As far as Lincoln's possible diagnosis of Marfan syndrome is concerned, this issue is still unresolved and is likely to remain so until reliable methods are developed for mutation screening using minute samples.

Fig. 14.8 President Abraham Lincoln with a group of officers. From the Library of Congress American Memory Collection.

Reference

Reilly PR (2000) *Abraham Lincoln's DNA and other adventures in genetics.* Cold Spring Harbor Laboratory Press, New York.

ple for testing but then decide that he or she does not wish to be told the result. This is perfectly acceptable. But what happens when this person dies and his or her children want to know the result? The law tends to the view that dead people have relinquished their rights to ownership. So the moral of the story is that if you don't want your children to discover your genetic secrets, then don't leave them a sample of your DNA for future analysis, even if inadvertently, as in the case of Abraham Lincoln (Box 14.3).

Ethical issues and prenatal diagnosis

There can be few subjects in medicine more contentious than that of prenatal diagnosis. The issue of termination of pregnancy is always emotive, and although fetal abnormality constitutes the grounds for termination in less than 2% of cases (1853 out of 170 145 terminations in England and Wales in 1997), this is a particular focus of attention because of concern that prenatal diagnosis is an attempt on the part of society to rid itself of the handicapped.

In reality most terminations on the grounds of fetal abnormality are carried out at the request of parents, who voluntarily decide not to continue the pregnancy when a serious abnormality is identified. Thus the debate revolves around parental autonomy and the ethical perception of whether termination for a serious abnormality constitutes an act of beneficence or maleficence. At this point ethical views become very personal. However, the issue becomes blurred in practice by what constitutes a serious abnormality. In the UK termination of pregnancy is legal until and beyond 24 weeks gestation for a 'serious' abnormality, but this is not defined.

So what does constitute a serious abnormality? A condition which leads to death in early infancy or childhood, such as anencephaly or Tay–Sachs disease, would generally be accepted as serious, as would a disorder associated with lifelong pain or disability such as severe lumbosacral spina bifida or Duchenne muscular dystrophy. But classification becomes more difficult for conditions such as sickle-cell anaemia or cystic fibrosis, both of which, if well treated, are compatible with a reasonably good quality of life well into adulthood. If quality of life is a major criteria, then individuals with Down syndrome score highly as they are often as healthy and sometimes much happier than their unaffected peers. Similarly most men and women with a sex chromosome aneuploidy syndrome, such as 45,X, 47,XXX, 47,XXY, or 47,XYY, also enjoy a good quality of life and normal life expectancy. Yet when prospective parents are confronted with one of these diagnoses, identified inadvertently by amniocentesis carried out for Down syndrome, many find it extremely difficult to know how to proceed.

Against this background, most medical professionals and members of the genetic team see their responsibility as providing information and support to enable parents to reach a decision with which they are comfortable. When fully counselled, around 10% of prospective parents elect to continue pregnancy when the baby has Down syndrome and approximately 50% choose to continue when the baby has a sex chromosome aneuploidy syndrome. This traditional nondirective, non-judgemental stance does not extend to termination because the baby is of the 'wrong' sex, as this constitutes a trait rather than an abnormality. Nor do geneticists support societal pressure for termination of all abnormal or handicapped individuals. Health care professionals are responsible to their patients and families, not to society or to a misplaced vision of perfection.

Summary

Clinical genetics is the branch of medicine which deals with inherited disorders in families. The main responsibilities include genetic counselling, carrier detection, predictive testing, and prenatal diagnosis.

Genetic counselling aims to enhance autonomous decision making by providing information in a nondirective and non-judgemental manner. The primary role is to inform rather than to advise, and to offer support both in the short and in the long term.

Carrier detection is offered to individuals who are deemed to be at increased risk of carrying a chromosome rearrangement or an autosomal/X-linked recessive disorder. Ideally this should include an attempt to offer testing to all relevant adult relatives of reproductive age who could be carriers.

Predictive testing is offered to individuals who wish to know if they will develop an inherited late-onset disorder known to be present in their family. The relevant disorders can be subdivided on the basis of whether or not they are treatable. When prevention or treatment is available, active programmes of intervention and screening can be initiated. When no effective prevention or treatment is available great care has to be taken to ensure that the individuals concerned are fully informed of the possible consequences of their decisions.

Prenatal diagnosis involves the detection of an inherited or structural abnormality before birth. Four standard approaches are used. These are maternal serum screening, ultrasonography, chorionic villus sampling (CVS), and amniocentesis. Maternal serum screening is generally offered at 16 weeks gestation for neural tube defects (serum AFP is raised) and Down syndrome (serum AFP is reduced). Ultrasonography in the form of fetal anomaly scanning is offered at 18–20 weeks gestation for the detection of structural abnormalities. CVS is usually performed at 11–12 weeks gestation for the detection of chromosomal abnormalities and single-gene disorders. Amniocentesis is performed at 16–18 weeks gestation for the detection of chromosome abnormalities and single-gene disorders. Both CVS and amniocentesis convey small risks for causing spontaneous pregnancy loss.

Ethical dilemmas arise in all aspects of clinical genetics and often there is no easy solution. An acceptable approach is to try to apply the cardinal ethical principles of autonomy, beneficence, justice and privacy. There is universal agreement that participation in all forms of genetic testing should be voluntary, and that genetic testing should only be carried out in children if a clear medical benefit can be gained.

Further reading

Clarke A (ed.) (1998) *Genetic testing of children.* Bios Scientific Publishers, Oxford.

Harper PS (1998) *Practical genetic counselling,* 5th edn. Butterworth Heinemann, Oxford.

Harper PS, Clarke AJ (1997) *Genetics, society and clinical practice.* Bios Scientific Publishers, Oxford.

Simpson JL, Elias S (2003) *Genetics in obstetrics and gynaecology,* 3rd edn. WB Saunders, Philadelphia.

Multiple choice questions

1 Carrier testing for autosomal recessive disorders is usually achieved by

(a) clinical examination

(b) enzyme assay

(c) mutation analysis or linkage analysis

(d) karyotype analysis

(e) protein product analysis

2 A man whose father had Huntington disease asks for a test to establish if he has inherited the condition. It would be correct to tell him that

(a) it is very unlikely that he will become affected as Huntington disease shows X-linked recessive inheritance

(b) testing can be carried out at his first visit to the genetic clinic

(c) testing will involve molecular analysis for a triplet repeat expansion mutation

(d) if he tests positive he should not have children

(e) if he already has children they will have to be tested

3 A woman with a strong family history of spina bifida presents in early pregnancy. She requests a test to show if her baby is affected. It would be correct to tell her that

(a) maternal serum screening will detect all cases

(b) neural tube defects can be detected by CVS

(c) most severe neural tube defects can be detected by ultrasound in mid-pregnancy

(d) prenatal chromosome analysis is indicated

(e) periconceptional folic acid is indicated in all future pregnancies

4 A young woman with a family history of Down syndrome presents in early pregnancy. It would be correct to tell her that

(a) Down syndrome cannot run in families and there is no increased risk that she will have an affected child

(b) there is a small chance that she is a carrier of an inherited form and that this could be clarified by testing her chromosomes

(c) maternal serum screening will detect all cases

(d) definitive prenatal chromosome analysis involves an invasive test

(e) amniocentesis conveys a greater risk for causing miscarriage than CVS

5 A woman who has two sons with Duchenne muscular dystrophy presents in the early weeks of pregnancy and asks about the availability of prenatal diagnostic tests. It would be correct to tell her that

(a) Duchenne muscular dystrophy can be diagnosed prenatally by ultrasound scanning at 20 weeks

(b) prenatal diagnosis is based on molecular analysis

(c) prenatal diagnosis will involve an invasive test such as CVS

(d) if the test shows that the baby is affected she should have a termination

(e) early treatment after birth can prevent development of the disease process

Answers

1 (a) false—carriers (heterozygotes) are clinically normal

(b) false—for most inborn errors of metabolism enzyme levels in carriers overlap with the normal range

(c) true—this is the only method available for most autosomal recessive disorders

(d) false—chromosome analysis is not relevant

(e) false—for many autosomal recessive disorders the protein product is either not known or cannot be measured

2 (a) false—inheritance is autosomal dominant

(b) false—testing would not be carried out until he has had an opportunity to reflect on how he wishes to proceed

(c) true—Huntington disease is caused by expansion of a CAG triplet repeat

(d) false—this is a decision that only he and his partner should make

(e) false—predictive testing for Huntington disease is never carried out on children

3 (a) false—maternal serum screening will detect around 70% of all open neural tube defects

(b) false—there are no chromosomal or molecular abnormalities detectable in most cases of neural tube defect, so they cannot be detected by CVS

(c) true—this is the most reliable method for detecting severe open defects

(d) false—chromosome analysis is normal in non-syndromal neural tube defects

(e) true—this prevents around 80% of all non-syndromal neural tube defects

4 (a) false—Down syndrome is usually caused by trisomy 21, which does not run in families, but a small proportion of cases are caused by unbalanced Robertsonian translocations which can be carried in a balanced form by relatives

(b) true—chromosome analysis would identify a balanced translocation

(c) false- maternal serum screening at 16 weeks will detect 60–75% of cases

(d) true—either chorion villus sampling or amniocentesis

(e) false—CVS conveys a slightly greater risk than amniocentesis

5 (a) false—there are no structural abnormalities in babies with Duchenne muscular dystrophy

(b) true—either specific mutation analysis or linkage analysis

(c) true—this is the only reliable way to obtain material of fetal origin

(d) false—this is up to the mother, and her partner

(e) false—unfortunately there is no treatment that is effective in preventing the development of the disease process

Medical school core curriculum in genetics

Association of Professors of Human and Medical Genetics/American Society of Human Genetics

Document dated 27 December 2001. Reproduced with permission of the American Society of Human Genetics.

Preamble

Medical genetics is one of the most rapidly advancing fields of medicine, and molecular genetics is now integral to all aspects of biomedical science. Every physician who practices in the twenty-first century must have an in-depth knowledge of the principles of human genetics and their application to a wide variety of clinical problems. The American Society of Human Genetics and the Association of Professors of Human and Medical Genetics have developed this Medical School Core Curriculum to provide guidance to deans and curriculum committees regarding medical genetics knowledge, skills, and behaviors that all current medical students will need during their careers as physicians. Each medical school must find the best way to incorporate genetics teaching into its own curriculum, but some generalizations are possible:

- Medical genetics provides a unique perspective on function of the human body in health and disease; it is both a clinical specialty and a basic science. Medical genetics teaching must span the entire undergraduate medical school curriculum and continue into the postgraduate years.

- Medical genetics must be explicitly included in the curriculum. Although some aspects of medical genetics overlap with and may be taught by other disciplines, specific learning objectives in medical genetics need to be established.

- A well-qualified medical genetics specialist (or small committee of medical geneticists) should be given the authority and responsibility for implementing the genetics curriculum at each medical school. This responsibility should extend throughout the under-graduate medical curriculum and include involvement in all courses that deal with genetic principles or disorders.

◆ Medical genetics can be taught effectively by a variety of methods and in various formats. Problem-based learning is particularly well-suited to medical genetics because it involves integration of skills and knowledge from many fields. Genetics can also be taught in various clinical contexts and at different points in clinical training, depending on the par-ticular circumstances at each school. Specific clinical examples are important, but the focus of the curri-culum must be on medical genetic principles illustrated by the examples.

Given the rapid advance of medical genetics, this Core Curriculum is a work in progress. (The previous version was published in the *American Journal of Human Genetics* (1995), **56**, 535–537.) The American Society of Human Genetics and the Association of Professors of Human and Medical Genetics welcome all comments on these objectives, which will be revised as necessary to reflect changes that occur in our understanding of genetics and its application to medicine.

General medical competencies essential to medical genetics

During their training, medical students must acquire many general skills and behaviors that are important in all aspects of clinical practice, including medical gen-etics. These general competencies include the ability of students to:

1.1 explain the importance of disease prediction and prevention;

1.2 understand the developmental stages of human behavior, maturation, and intelligence;

1.3 apply appropriate techniques for conveying difficult medical information;

1.4 understand how to respond appropriately to patients' defense mechanisms;

1.5 recognize the importance of reiterating informa-tion to patients who are anxious or unfamiliar with the concepts being presented;

1.6 recognize the importance of patient confiden-tiality;

1.7 make appropriate referrals to genetics support groups, community groups, or other resources that can benefit the patient and family;

1.8 respect the autonomy of all patients, but also provide guidance with decision-making when requested;

1.9 respect patients' religious, cultural, social, and ethical beliefs, even if they differ from their own beliefs;

1.10 interpret their own attitudes toward ethical, social, cultural, religious and ethnic issues and develop an ability to individualize each patient or family member;

1.11 cope emotionally with patient responses;

1.12 recognize the limitations of their own skills and seek consultation when necessary;

1.13 effectively use resources such as medical text-books, research articles, and computer-based sys-tems to obtain information necessary for good patient care;

1.14 apply the principles of evidence-based medicine to clinical practice;

1.15 understand how clinical observations can provide insight into human biology and disease patho-genesis and, through research, lead to improve-ments in health; and

1.16 undertake a program of life-long learning.

Specific knowledge requirements

The practice of modern medicine includes recognition of the role of genetic factors in health and disease. Students must know:

2.1 Structure and function of genes and the general organization of the human genome

2.1.1 what genes are, how they are organized and con-trolled, what they do, and how they segregate;

2.1.2 how gene expression is affected by differences in coding and non-coding regions, effects of trans-acting factors, and the structure of chromatin;

2.1.3 how protein function is influenced by mRNA and polypeptide processing and interactions;

2.1.4 how gene activity varies during development and in normal and pathological cell function;

2.1.5 what information can and cannot be predicted from the DNA sequence of a gene;

2.1.6 what information can be obtained from measuring RNA or protein levels that cannot be obtained from the DNA sequence alone;

2.1.7 how processes such as gene duplication and divergence, exon shuffling, and the activity of transposable elements help to explain genomic variability, redundancy, and plasticity;

2.2 Genes and disease

2.2.1 the patterns of inheritance characteristic of autosomal dominant, autosomal recessive, X-linked dominant, and X-linked recessive traits;

2.2.2 factors that affect development of the phenotype in single-gene disorders, including modifier genes, and stochastic and pleiotropic effects, which result in variable expressivity and incomplete penetrance;

2.2.3 the clinical manifestations of common Mendelian diseases;

2.2.4 the basic principles of inborn errors of metabolism and of pharmacogenetic variations and their general clinical manifestations;

2.2.5 the genetic basis of mitochondrial diseases and the expected inheritance patterns for mitochondrial traits;

2.2.6 the nature of mutations and premutations and how they contribute to human variability and disease;

2.2.7 the concepts and clinical importance of genetic imprinting and uniparental disomy;

2.2.8 how polymorphisms, human gene mapping, and gene linkage and association studies are used in medicine;

2.2.9 the multifactorial nature of most human traits, both normal and abnormal, and the principles of multifactorial inheritance;

2.2.10 how genes interact with other genes and with various environmental factors to produce disease, and how amelioration of non-genetic factors can prevent development of disease in a genetically-predisposed individual;

2.3 Chromosomes and chromosomal abnormalities

2.3.1 how genes are organized into chromosomes, how chromosomes replicate in mitosis and meiosis, and how they are transmitted from parent to child;

2.3.2 the clinical features of common numerical, structural, and mosaic chromosomal abnormalities;

2.4 Population genetics

2.4.1 how the principles of population genetics account for varying frequencies of particular mutations in populations, the effects of consanguinity, the continuing occurrence of new mutations, and the resistance of gene frequencies to change by medical intervention;

2.4.2 how evolutionary principles can be used to understand human biology and disease;

2.5 Genetics in medical practice

2.5.1 how knowledge of a patient's genotype can be used to develop a more effective approach to health maintenance, disease prevention, disease diagnosis, and treatment for that particular individual;

2.5.2 common molecular and cytogenetic diagnostic techniques and how they are applied to genetic disorders;

2.5.3 how constitutional and acquired genetic alterations can lead to the development of malignant neoplasms and how identification of these changes can be used in the diagnosis, management and prevention of malignancy;

2.5.4 the potential advantages, limitations, and disadvantages of presymptomatic testing for genetic disease;

2.5.5 the potential advantages, limitations, and disadvantages of predictive testing for genetic disease;

2.5.6 how appropriate applications of genetic medicine can improve public health, and how to determine whether such interventions are warranted in a particular population;

2.5.7 the alternative approaches and goals of screening programs for genetic diseases in newborn infants, pregnant women, and other adults, and the ethical issues involved in justifying each program;

2.5.8 the existence of and justification for screening programs to detect genetic disease, and the difference between screening and more definitive testing;

2.5.9 conventional approaches to treatment of genetic diseases and the general status of gene-based therapies;

2.5.10 what exposures are likely to be teratogenic in humans and how such exposures can be prevented;

2.5.11 how to recognize and classify congenital anomalies and multiple congenital anomaly syndromes;

2.5.12 the purpose of genetic counseling;

2.5.13 when and how to refer individuals with a genetic disease or congenital anomaly to medical genetics specialists, and why referral is beneficial to the patients;

2.5.14 how novel scientific discoveries are evaluated in a clinical context and applied appropriately to the care of patients;

2.5.15 how legal and ethical issues related to genetics affect general medical practice;

2.5.16 how organizational and economic aspects of the health care system affect delivery of clinical genetic services;

2.5.17 what lessons the history of use and misuse of human genetics teach about the proper application of contemporary medical genetic knowledge.

Specific skills

Students must learn to synthesize factual material related to genetic diseases and to use this information to formulate an appropriate plan for diagnostic evaluation and patient management. They need the ability to:

3.1 elicit a comprehensive family medical history, construct an appropriate medical pedigree, and recognize patterns of inheritance and other signs suggestive of genetic disease in the family history;

3.2 recognize features in a patient's medical history, physical examination or laboratory investigations that suggest the presence of genetic disease;

3.3 identify patients with strong inherited predispositions to common diseases and facilitate appropriate assessment of other at-risk family members;

3.4 recognize and classify common congenital anomalies and patterns of anomalies;

3.5 recognize and initiate the evaluation of patients with inborn errors of metabolism;

3.6 interpret the results of common cytogenetic, molecular genetic, and biochemical genetic diagnostic techniques efficiently;

3.7 estimate recurrence risks for Mendelian and multifactorial disorders in affected families;

3.8 use the information that a patient has a genetic predisposition for a particular disease to help reduce the risk of developing that disease or deal with it more effectively if it does develop;

3.9 describe appropriate techniques and approaches to providing genetic counseling for commonly-encountered genetic diseases;

3.10 communicate genetic information in a clear and non-directive manner that is suitable for individuals of different educational, socio-economic, ethnic and cultural backgrounds;

3.11 recognize and accept varying cultural, social, and religious attitudes in relation to issues such as contraception, abortion, parenting, and gender roles;

3.12 utilize community support services and agencies, in particular, support groups for genetic diseases, appropriately;

3.13 provide patients with access to diagnostic and predictive tests that are appropriate for the condition in their family and advise patients of the benefits, limitations, and risks of such tests;

3.14 work with a medical genetics specialist to develop a comprehensive plan for the evaluation and management of patients with genetic disease;

3.15 make available to patients with genetic diseases appropriate treatments, including dietary, pharmacological, enzyme-replacement, transplantation, and gene therapies, as well as anticipatory guidance regarding health screening practices specific to the diagnosis;

3.16 appreciate the important role of biomedical research and acquire skills that enable critical analysis of scientific developments.

Specific behaviors

Students must learn to be sympathetic, nonjudgmental, and non-coercive counselors who recognize their own limitations and seek consultation whenever necessary. Students should:

4.1 present all relevant options fairly, accurately, and non-coercively;

4.2 be aware of the dilemmas posed by confidentiality when relatives are found to be at risk for a serious disease;

4.3 appreciate the implications that information regarding a genetic abnormality can have for a person's self-image, family relationships, and social status and that patients' reactions may differ depending on factors such as gender, age, culture, and education;

4.4 when appropriate, encourage patient participation in medical research provided the patient and/or family is fully informed and understands the risks and benefits of participation in terms of their own disease, treatment, and social context.

Teaching medical genetics to undergraduate medical students

Recommendations of the UK Joint Committee for Medical Genetics and the British Society of Human Genetics

Essential core knowledge and skills

I Basic genetics

- General features of the human genome (amount of DNA, number of genes, organization into chromosomes, repetitive DNA, amount of inter-individual variation)

- Chromosomal basis of inheritance (mitosis and meiosis)

- Modes of inheritance (Mendelian and non-Mendelian) including penetrance and expressivity including mitochondrial and complex multifactorial disorders

- Mechanism of origin of numerical chromosome abnormalities

- Major types of structural chromosome abnormalities and their basic implications

- DNA as genetic material (outline of replication, transcription and translation)

- Use of DNA polymorphisms as genetic markers

- How mutations cause partial or complete loss of function or gain of function

- Types of DNA test (testing for a specific mutation vs scanning a gene for mutations)

- Gene frequencies of common recessive mutations

- Genetic heterogeneity

- Parameters governing population genetic screening

- Developmental genetics: selective transcription; differentiation; stem cells.

- The clinical embryology of human malformation syndromes
- Epigenetic events including imprinting
- Principles of teratogenesis
- Evolution, natural selection and selective advantage
- History of eugenics movement

Learning objectives

- To provide the basic knowledge required to underpin the learning objectives associated with clinical genetics which will be required for a newly qualified doctor to practice safely
- In order for a doctor to be able to ensure that they can achieve the learning objectives associated with clinical genetics, certain basic scientific and historic knowledge is required. For most students, much of this material will have been covered before entry into an undergraduate medicine degree. Where this is not the case it will need to be provided in the early years of any medical degree to facilitate an understanding of place of genetics in modern medical practice.

Justification

Not only is this material essential to understand the current practice of clinical genetics but it is also necessary to facilitate the continued medical development of a doctor as our understanding of the molecular basis of disease and its treatment progresses.

II Clinical genetics

Specific learning objectives for clinical genetics

At the end of your undergraduate teaching you will be expected to be able to:

- Take a family history
- Construct and interpret a family tree
- Recognize basic patterns of inheritance
- Appreciate the risk of individuals suffering simple Mendelian disorders
- Have a clinical knowledge of several Mendelian disorders
- Have a clinical knowledge of chromosomal disorders including translocations, micro-deletions and the methods used to detect them
- Have a clinical knowledge of the genetic factors associated with cancer predisposition
- Recognize the genetic and environmental contribution to multifactorial conditions, e.g. congenital heart disease, cancer, diabetes and psychiatric illness

- Understand approaches which can be used for the diagnosis of genetic disease and carrier detection
- Understand different forms of DNA testing: prenatal diagnosis, preimplantation diagnosis, predictive testing: and appreciate when such testing may not be appropriate.
- Be able to interpret a simple DNA report and chromosome report
- Be able to recognize cases with abnormal developmental and dysmorphic features
- Be aware of current population genetic screening programs and guidelines for the introduction of such programs
- Be familiar with the practice of the genetic counselling clinic, its motives and methods including the principles of non-directive, non-judgemental counselling and impact of genetic diagnosis on the extended family. Be able to communicate the concept of risk in a manner that can be understood by the patient
- Know when and where to get genetic advice and information
- Perceive major ethical issues in genetics

Learning objective

- To be able to take a family history and construct and interpret a family tree from a verbal description.
- To be able to draw an accurate family tree using standard symbols from a verbal description of a family structure as would typically occur in a clinic. Family structures may include up to 25 individuals over 3 generations and include siblings, half-siblings, cousins, and twins.

Justification

Inability to perform this skill is likely to result in misunderstanding of family structures by the doctor, inaccurate communication to other professionals, and subsequent incorrect risk assessments. This creates risk of clinical harm (e.g. wrong screening advice to cancer family) or poorly informed reproductive decisions (e.g. choices made by possible carrier woman from family with X-linked condition).

Learning objective

- To be able to recognize inheritance patterns.
- Based on the family history of disease and the family tree, to be able to identify all forms of Mendelian inheritance, consanguinity and founder effects.

Justification

In order to be able to diagnose genetic disease, it is necessary to know the type of inheritance and as a part of the management of the condition, identify others at risk, and prepare the family for relevant clinical issues.

Learning objective

◆ To have a clinical knowledge of several Mendelian and chromosomal conditions.

◆ To be able to state, for several genetic conditions, usual mode of inheritance, two major features, one major complication, usual diagnostic test, one source of further information. Conditions: e.g. Marfan syndrome, HNPCC, neurofibromatosis type 1, cystic fibrosis, Duchenne muscular dystrophy, Down syndrome, microdeletion syndrome.

◆ To be able to understand the implications of balanced and unbalanced chromosomal translocations and microdeletions and the methods that can be used to detect them.

Justification

Most doctors are likely to have contact with relatively common Mendelian conditions from time to time. Lack of basic knowledge of features and complications of these conditions creates a risk of clinical harm, lack of knowledge of inheritance leads to a risk of poorly informed reproductive choices. Lack of knowledge of sources of further information exacerbates both of the above risks.

Learning objective

◆ To have a clinical knowledge of the genetic factors associated with cancer predisposition.

◆ To be able to identify those characteristics of a family history which suggest the presence of a familial cancer syndrome.

Justification

As cancer is a common condition it will occur in most families. It is useful if a doctor can differentiate those factors which make it more probable that a family harbours a familial cancer syndrome or may simply have a greater genetic contribution to cancer risk than the general population

Learning objective

◆ To be able to recognize the genetic and environmental contribution to multifactorial conditions e.g. congenital heart disease, diabetes, and psychiatric illness.

◆ To be able to state for a few common conditions the contribution of genetic and environmental factors in their causation and the relevance of this to the management of the patient and to population health issues, if any.

Justification

As the Human Genome Project progresses and the relevance of its findings to health and disease are clarified, so will our understanding of the interactions between constitutive genetic contribution and environmental factors. This has relevance for the classification, management and prevention of disease.

Learning objective

◆ To be able to understand approaches which can be used for the diagnosis of genetic disease and carrier detection.

◆ To be able to understand the methods utilized in the diagnosis of genetic disease including family history, ethnic background, clinical phenotype, and the role of laboratory testing for the diagnosis of disease and for the identification of individuals who are carriers of genetic conditions.

Justification

In the diagnosis of genetic disorders a combination of methods are usually required. In addition to allowing the diagnosis of patients with symptoms or signs, carriers of a predisposition to the condition or individuals at risk of passing on the condition can also be detected. The effective use of appropriate approaches can facilitate diagnosis and often avoid unnecessary or even inappropriate investigation.

Learning objective

◆ To understand different forms of DNA testing including prenatal and preimplantation diagnosis, predictive and diagnostic testing.

◆ To understand that DNA tests can be used to diagnose a genetic disorder, identify individuals who are at risk of developing genetic disease both pre- and postnatally or to identify individuals who are at risk of having a child with a genetic disease. To understand that with the current molecular techniques, DNA diagnosis can be carried out on single cells such as can be obtained from a developing embryo.

Justification

Clinically relevant tests, performed with the consent of the patient, can not only diagnose genetic diseases which are already resulting in signs and symptoms, but can identify mutations in samples from individuals who are at risk of developing a genetic disease. The use of such tests requires considerable counselling to ensure that those requesting testing understand the full significance of a result.

Learning objective

♦ To be able to interpret a DNA report and chromosome report.

♦ To be able to interpret the type of information provided in reports from DNA diagnostic and cytogenetic laboratories for common conditions.

Justification

Analysis of an individual's chromosomes or their DNA can provide information which makes a definite diagnosis, alters the likelihood of a specific diagnosis, or produces a result, the significance of which is unknown and an incidental finding. The ability to appropriately interpret such results and apply them to patient management is essential for evidence-based medicine.

Learning objective

♦ To be familiar with the practice of the genetic counselling clinic, its motives and methods including the principles of non-directive and non-judgemental counselling, and the impact of genetic diagnosis on the extended family.

♦ To be familiar with the workings of genetic counselling clinics in regional genetics units. To understand the aims and methods of these clinics and where appropriate the use of non-directive counselling and evidence based methods of providing information and advice. To understand the impact of the diagnosis of a genetic condition on the extended family of the consultand.

Justification

Genetic counselling involves a team of professionals working together to allow the diagnosis of genetic disease and appropriate counselling of both the consultand and their extended family where appropriate. As genetic processes can also identify at risk situations, education and non-directive counselling is an important part of the process.

Learning objective

♦ To know when and where to get help and information.

♦ To know under what clinically relevant circumstances advice from a clinical geneticist should be sought and to know where to find further information such as guidelines and relevant literature.

Justification

As several thousand conditions can be inherited in a Mendelian fashion and many diagnosed using genetic techniques, non-specialist clinicians cannot be expected to have a detailed knowledge of these rare disorders or the guidelines relating to evidence based diagnosis and management. Hence, it is appropriate for these generalists to know where and when to obtain advice and seek clinical genetic referral for their patients. Knowledge of where to obtain appropriate guidelines and reviews of the literature can facilitate the management of these processes.

Learning objective

♦ To perceive major ethical issues.

♦ To know what are considered as major ethical issues relating to the use of genetic information and procedures.

Justification

There are many issues relating to the use of genetic information and techniques which can produce ethical dilemmas in certain circumstances or even more generally. It is important to realize that individuals will have their own opinion on these matters and that certain societies may regard some possibilities as inappropriate.

III Special study modules

Medical students should be offered the option of a series of student-selected components in both basic scientific and clinical aspects of medical genetics allowing students to study in depth, areas of particular interest to them.

Conclusion

This paper provides a recommended core basic science and clinical curriculum for the teaching of medical genetics to medical students. It is based on previous work carried out under the auspices of the Royal College of Physicians looking at the perceived top Curriculum and Skills required for a medical graduate. In addition, it takes into account the recommendations of the GMC documents *Tomorrow's Doctors*, published in 1994 and 2002.

Glossary

Acrocentric A chromosome with the centromere close to one end (i.e. 13–15, 21, 22, and Y)

Allele (adj. **allelic**) A different form of a gene at a locus

Allelic heterogeneity Different mutations in the same gene causing an identical phenotype (cf. **locus heterogeneity**)

Amniocentesis Withdrawal of amniotic fluid for prenatal diagnosis

Anaphase lag Loss of a chromosome in anaphase leading to monosomy

Aneuploid (n. **aneuploidy**) A chromosome constitution which is not a multiple of 23

Anticipation Increase in severity or earlier age of onset in succeeding generations

Anticodon A three-base sequence of transfer RNA which is complementary to a three-base codon of mRNA

Antiparallel Opposite orientation of the two strands in DNA

Apoptosis Programmed cell death

Association The occurrence of a particular allele with a disease more or less often than would be expected by chance

Assortative mating Non-random mating due to partner selection on the basis of phenotype

Autosome (adj. **autosomal**) One of the chromosomes numbered 1 to 22

Autozygous Homologous alleles identical by descent

Bacterial artificial chromosome (**BAC**) A recombinant plasmid vector which is propagated in a bacterial cell

Barr body The inactive X chromosome seen in the nucleus of normal female cells (also known as sex chromatin)

Base pair Two complementary nucleotide bases

Bivalent Two homologous chromosomes which have paired in meiosis

CentiMorgan (**cM**) A unit of genetic (linkage) distance. 1 cM = 1% chance of recombination between two linked loci

Centromere The primary constriction on a chromosome where the spindle attaches in mitosis and meiosis

Chiasma (pl. **chiasmata**) Crossover in meiosis

Chimaera (**chimaerism**) An individual with two different cell lines derived from different zygotes (cf. **mosaic**)

Chorionic villus sampling Removal of chorionic villi for prenatal diagnosis

Chromatid One of the two strands of a chromosome after DNA synthesis, which forms a new daughter chromosome after mitosis

Chromatin The material of which chromosomes are made. Consists of DNA and histone proteins

Chromatin remodelling Change in the conformation of DNA which allows gene expression

Chromosome Structure in which DNA is packaged in the cell nucleus

Clastogen An agent which induces chromosome breakage

Codominant The expression of both alleles in a heterozygote

Codon A three-base triplet in DNA or RNA which encodes an amino acid

Complementary Matching in DNA/RNA strands of base pairs (A with T or U, C with G)

Compound heterozygote An individual with different mutant alleles at a homologous locus (cf. **double heterozygote**)

Concordant The presence or absence of a condition in two relatives, usually twins (cf. **discordant**)

Congenital Present at birth

Consanguineous (**n. consanguinity**) Related by descent from a common ancestor

Consultand An individual who seeks genetic counselling

Contig A set of overlapping DNA fragments

Contiguous Closely adjacent. Usually refers to a series of genes lost in a microdeletion

Continuous Having normally distributed or overlapping phenotypes (cf. **discontinuous**)

Cosmid A phage–plasmid hybrid capable of accepting a large DNA insert

Coupling The occurrence of two alleles at linked loci on the same chromosome (cf. **repulsion**)

Crossing-over Exchange of material between homologous chromosomes at meiosis I

Cytogenetics The study and analysis of chromosomes

Deformation A birth defect resulting from an extrinsic mechanical force

Deletion Loss of a sequence of DNA or part of a chromosome

Developmental biology The study of molecular development and embryology

Diploid Possessing 46 chromosomes (cf. **haploid; polyploidy; triploidy**)

Discontinuous Having distinct rather than overlapping phenotypes (cf. **continuous**)

Discordant Different phenotypes in two relatives, usually twins (cf. **concordant**)

Disruption A birth defect resulting from extrinsic damage to an otherwise normal developmental process

Dizygotic Originating from two separate zygotes, i.e. non-identical twins (cf. **monozygotic**)

DNA (deoxyribonucleic acid) The molecule that encodes mRNA and transmits genetic information from cell to cell and from generation to generation

Dominant Expressed in the heterozygote (cf. **recessive**)

Dominant negative A mutational effect which results in disruption of the normal product from another allele

Double heterozygote An individual who is heterozygous for two mutations at different non-homologous alleles (cf. **compound heterozygote**)

Double minute A small supernumerary chromosome caused by gene amplification

Dysmorphic Having an abnormal shape or form

Dysmorphology The study of children who are dysmorphic

Dysplasia A birth defect caused by abnormal organization of cells into tissues or organs

Empiric risk A risk based on observation rather than theory

Epigenetic Alteration in gene expression not caused by a change in gene structure

Epistasis An alteration in the expression of one gene by another gene

Euchromatin Chromatin which is light staining and transcriptionally active (cf. **heterochromatin**).

Eugenics The improvement of a species through selective breeding

Exon A region of a gene which is expressed and is present in mature mRNA (cf. **intron**)

Expression (= expressivity) The extent to which the effects of a gene are manifest

FISH (fluorescence *in situ* hybridization) A technique for analysing chromosomes based on hybridization of a labelled probe and visualization under UV light

Founder effect The presence of a high gene frequency in a population derived from a small group in which one or a few individuals were heterozygotes

Fragile site A non-staining gap in a chromosome or chromatid

Frame-shift A change in the reading frame caused by a deletion or insertion which is not a multiple of three

Functional cloning The identification of a gene through analysis of its protein product

Gain-of-function The effect of a mutation which causes an increase in the normal activity or in a new activity of its product (cf. **loss-of-function**)

Gene A sequence of DNA which controls the structure and synthesis of a specific polypeptide chain

Gene conversion A change in the structure of a gene which makes it identical to another gene

Gene flow The movement of genes from one population to another as a result of migration

Gene therapy Treatment of a disease by the introduction of recombinant DNA sequences

Genetic counselling The provision of genetic information to facilitate informed decision making

Genetic drift Change in gene frequency by random fluctuation

Genetic polymorphism The occurrence of different alleles at a locus at frequencies greater than can be maintained by recurrent mutation

Genotype An individual's genetic constitution (cf. **phenotype**)

Germline The cell line which generates gametes (ova and sperm)

Germline (= gonadal) mosaicism The presence of cell lines containing two or more different genotypes in the germline

Haploid Possessing a single set of 23 chromosomes (cf. **diploid**)

Haploinsufficiency The phenotypic effect resulting from a loss-of-function mutation in one allele

Haplotype The genotype of a group of alleles in coupling at a group of closely linked loci. Commonly used to refer to alleles at the HLA locus

Hardy-Weinberg equilibrium The maintenance of constant proportions of different genotypes in a population

Hemizygous The presence of only a single copy of a chromosome or gene, e.g. the X chromosome in males

Heritability The proportion of total phenotypic variance due to genetic factors

Heterochromatin Chromatin which is dark staining and transcriptionally inactive (cf. **euchromatin**)

Heterogeneity (adj. heterogeneous) A general term implying variation, e.g. alleleic, locus, or genetic heterogeneity

Heteroplasmy (adj. heteroplasmic) The presence of more than one population of DNA in an individual's mitochondria (cf. **homoplasmy**)

Heteroploidy Any alteration in the normal diploid number of 46 chromosomes

Heterozygote (adj. heterozygous) An individual with different alleles at a homologous locus

Heterozygote advantage Increased biological fitness in heterozygotes as compared to homozygotes

Histones The proteins around which DNA is wound

Holandric Present on the Y chromosome

Homeosis The transformation of one body segment into another

Homogeneously staining region A region of a chromosome which stains uniformly because of the presence of multiple copies of a gene

Homologous (n. homologue) Member of a pair of chromosomes

Homoplasmy (adj. homoplasmic) The presence of a single population of DNA in an individual's mitochondria (cf. **heteroplasmy**)

Homozygote (adj. homozygous) An individual with identical alleles at a homologous locus (cf. **heterozygote**)

Imprinting Different expression of a gene dependent upon the parent of origin

Induction The process whereby one group of cells influences the behaviour of another group of cells in embryogenesis

Insertion An abnormality in DNA or a chromosome caused by the presence of additional material from another source

Insertional mutagenesis The creation of a mutation in a gene by the insertion of foreign material

Intron A region of a gene which is not expressed and which separates exons (cf. **exon**)

Inversion A chromosome abnormality in which a segment of the chromosome is reversed as compared to its normal orientation

Isochromosome A structurally abnormal chromosome consisting of two short arms or two long arms

Juxtacrine Adjacent, as in cell–cell induction (cf. **paracrine**)

Karyotype An individual's chromosome constitution. Also used to describe a photograph of an individual's chromosomes

Kilobase (kb) 1000 base pairs

Liability A collective term for environmental and genetic susceptibility

Linkage Co-inheritance of alleles at two or more loci more often than would be expected by chance

Linkage disequilibrium The presence of specific alleles in coupling at linked loci more or less often than would be expected by chance

Linkage phase Whether alleles at linked loci are on the same chromosome (in coupling) or on opposite chromosomes (in repulsion)

Linked The term used to describe two or more loci at which alleles show linkage

Locus (pl. **loci**) The position occupied by a gene on a chromosome

Locus heterogeneity Similar or identical phenotypes being caused by mutations at different loci (cf. **allelic heterogeneity**)

Loss-of-function The effect of a mutation which results in a reduction of the normal activity of the product (cf. **gain-of-function**)

Loss of heterozygosity Loss of an allele in a cell in an individual who is heterozygous at that locus. Commonly refers to loss of a tumour suppressor gene in a tumour

Lyonization The process of X chromosome inactivation in the female

Malformation A birth defect due to an intrinsically abnormal developmental process

Megabase (Mb) 1000 kilobases = 1 000 000 base pairs

Meiosis The process of cell division in gametes, which reduces the chromosome complement from diploid to haploid (cf. **mitosis**)

Metacentric A chromosome with the centromere located centrally

Microdeletion A deletion of less than 5–10 Mb which cannot be seen using a light microscope

Microsatellite A DNA polymorphism caused by a short tandem repeat made up of two, three, or four nucleotides

Microsatellite instability A marked increase in the number of microsatellites in a cell (usually from a tumour) due to an error in the DNA mismatch repair system

Minisatellite A DNA polymorphism caused by a variable number of a 10–100-bp core sequence present in a tandem (head to tail) arrangement. Also known as **variable number tandem repeats**

Missense A point mutation which results in a codon specifying a different amino acid

Mitosis The normal process of cell division which results in two daughter cells, each with a diploid chromosome complement (cf. **meiosis**)

Mitotic recombination Exchange of genetic material between homologous chromosomes in mitosis

Monogenic Determined by a single gene

Monosomy A chromosome abnormality caused by loss of a single chromosome

Monozygotic Originating from a single zygote, i.e. identical twins (cf. **dizygotic**)

Morphogen A protein which diffuses during embryogenesis to determine a developmental process

Morphogenesis The process of the development of human form and shape

Mosaic (mosaicism) An individual with two or more cell lines which have originated from a single zygote (cf. **chimaerism**)

Multifactorial Caused by the interaction of environmental and genetic (usually polygenic) factors

Mutation Change in the normal structure or sequence of a gene

Non-disjunction Failure of a pair of chromatids or homologous chromosomes to separate in meiosis or mitosis

Non-parametric Without defined characteristics, such as a specific inheritance pattern, in linkage analysis (cf. **parametric**)

Nonsense A point mutation which results in a stop (termination) codon

Northern blotting A diagnostic technique for detecting RNA fragments by hybridization (cf. **Southern blotting; Western blotting**)

Nucleic acid A chain of nucleotide bases

Nucleosome The basic structural unit of chromatin consisting of DNA coiled around histones

Nucleotide The basic unit of DNA and RNA made up of sugar and phosphate moieties and a nitrogenous base (A, C, G, T, or U)

Okazaki fragment A small sequence of nucleotides generated in the synthesis of the lagging strand in DNA replication

Oligogenic Determined by a small number of genes

Oligogenic inheritance Inheritance pattern resulting from the interaction of a small number of genes

Oncogene A mutated proto-oncogene which can cause uncontrolled cell growth (cf. **tumour suppressor**)

Pachytene quadrivalent The cluster which is formed by chromosomes involved in a translocation in pachytene in meiosis I

Paracentric Involving only one arm of a chromosome (cf. **pericentric**)

Paracrine At a distance, as in cell–cell induction

Paralogous The term used to describe genes which share a similar structure and which have probably originated from a common ancestral gene, e.g. the α- and β-globin genes

Parametric With known characteristics, such as specific pattern of inheritance, in linkage analysis

Penetrance A statistical parameter indicating the proportion of individuals with a mutation who show any phenotypic expression

Pericentric Involving both arms of a chromosome

Phage (= bacteriophage) A virus which can infect bacteria and be used as a vector

Pharmacogenetics The study of genetic variation in the body's response to drugs

Pharmacodynamics The processes whereby the effects of a drug are mediated

Pharmacogenomics The study of how molecular variation influences the body's response to drugs and their side effects

Pharmacokinetics The processes involved in the delivery of a drug to, or its removal from, its target site

Phenocopy A phenotype caused by environmental factors which closely resembles that caused by genetic factors

Phenotype The clinical outcome of an expressed gene or genes. Also refers to the outcome of the interaction of genes with environment

Plasmid A small circular molecule of DNA, used a vector, which can replicate independently within a bacterium

Plasticity The change in tissue specificity shown by some somatic stem cells. Also known as **transdifferentiation**

Pleiotropy Multiple and diverse effects of a single gene

Polygene A gene which makes a small additive contribution to the overall phenotype

Polygenic Caused by multiple polygenes

Polygenic inheritance The pattern of inheritance shown by conditions caused by multiple polygenes

Polymerase chain reaction (PCR) A technique for amplifying a short sequence of DNA or RNA

Polyploidy (adj. **polyploid**) The presence of more than two copies of the haploid number of chromosomes

Population genetics The study of the distribution of genes in populations

Positional cloning The isolation of a gene by mapping and subsequent mutation analysis of candidate genes

Premutation A change in a gene, such as a small expansion of a triplet repeat, which renders the gene unstable and susceptible to further change to become a full mutation

Proband The index case. Also the individual through whom a family comes to medical attention. Also known as the **proposita/propositus**

Probe A labelled sequence of DNA or RNA which will hybridize to complementary sequences

Proposita/propositus The female/male index case. Also known as the **proband**

Proto-oncogene A gene which is normally involved in the regulation of cell growth and which can mutate to become an oncogene

Pseudoautosomal region The region on the short arm of the X and Y chromosomes where pairing occurs during meiosis I in the male

Pseudogene An inactive mutated vestigial copy of a functional gene

Pseudohermaphroditism The presence of ambiguous genitalia or genitalia of the sex opposite to that of the individual's chromosomal sex

Quantitative The term used to describe the field of genetics which is involved with the study of continuous phenotypes

Quantitative trait loci Loci which contribute to a continuous phenotype

Reading frame The order of the codons in a series of nucleotides,

Recessive Expressed in the homozygote (cf. **dominant**)

Reciprocal translocation Exchange of material between two chromosomes

Recombination Crossover between homologous loci during meiosis I to form new recombinant chromosomes

Recombination fraction The probability that a crossover will occur between two linked loci

Replication error positive The presence of microsatellite instability in a tumour

Repulsion The occurrence of two alleles at linked loci on opposite chromosomes (cf. **coupling**)

Restriction enzyme A bacterial endonuclease which can cut DNA at a specific nucleotide sequence

Restriction fragment length polymorphism Polymorphic variation in DNA fragment length identified by Southern blotting, caused by a mutation in a restriction site

Ring An abnormal chromosome caused by breaks at both ends of a chromosome with loss of the terminal portions and union of the 'sticky' ends

RNA (ribonucleic acid) A nucleic acid containing ribose, a phosphate group and a series of bases (A, C, G, and U) into which DNA is transcribed

Robertsonian translocation A specific type of reciprocal translocation caused by fusion at the centromeres of the long arms of two acrocentric chromosomes with loss of their short arms

Semi-conservative DNA replication in which the daughter molecule consists of one parental strand and one new strand

Semi-discontinuous DNA replication in which one strand (the leading strand) is synthesized continuously while the other strand (the lagging strand) is synthesized in fragments

Sequence Multiple abnormalities resulting from a single defect or mechanical factor

Sex chromosomes The X and Y chromosomes

Slippage An error in DNA replication which results in loss or gain of one or more repeat units

Somatic Pertaining to all parts of the body other than the germline

Southern blotting A diagnostic technique for detecting DNA fragments by hybridization (cf. **Northern blotting; Western blotting**)

Stem cell An undifferentiated cell capable of both self-renewal and proliferation into differentiated cells

Submetacentric A chromosome in which the centromere is located closer to one end than the other

Substitution A point mutation causing a change from one nucleotide to another

Synapsis The pairing of homologous chromosomes in prophase of meiosis I

Synaptonemal complex The protein structure which holds homologous chromosomes together during synapsis

Syndrome A pattern of abnormalities thought to be causally related

Synergistic heterozygosity The interaction of a small number of genes at different loci to cause a disease phenotype

Telomere The terminal portion of each chromosome arm

Teratogen An environmental agent which can cause a congenital abnormality

Tetraploidy The presence of four copies of the haploid chromosome complement (92 chromosomes)

Tetrasomy The presence of four copies of a chromosome

Transcription The synthesis of RNA from DNA

Transdifferentiation The change in tissue specificity shown by some somatic stem cells. Also known as **plasticity**

Transduction Virus-mediated gene transfer

Transfection Non-virus-mediated gene transfer

Transformation Malignant change in a cell. Also refers to uptake of DNA by a bacterial cell

Transition Substitution of one nucleotide for a nucleotide of the same type, i.e. a purine for a purine or a pyrimidine for a pyrimidine

Translation The process of polypeptide synthesis from messenger RNA

Translocation The transfer of part of one chromosome to another chromosome

Transposon A mobile genetic element which can insert into a different part of the genome

Transversion Substitution of a nucleotide by a nucleotide of a different type, i.e. a purine for a pyrimidine or vice versa

Triploidy (adj. **triploid**) The presence of three copies of the haploid chromosome complement (69 chromosomes)

Trisomy The presence of three copies of a chromosome

Tumour suppressor A gene which exerts a negative regulatory effect on cell division and in which homozygous loss-of-function mutations can lead to tumour development

Uniparental disomy The inheritance of two homologous chromosomes from one parent

Variable number of tandem repeats (VNTR) A DNA polymorphism caused by a variable number of a 10–100-bp core sequence present in a tandem (head to tail) arrangement. Also known as **minisatellites**

Vector An agent, such as a cosmid, phage, plasmid, BAC, or YAC, which can accommodate recombinant DNA or RNA and replicate in a specific host

Western blotting A technique for detecting proteins, in which the proteins are probed with an antibody on a nitrocellulose membrane(cf. **Northern blotting; Western blotting**)

Wild-type The normal version of an allele

X-linked The pattern of inheritance shown by a gene on the X chromosome

Y-linked The pattern of inheritance shown by a gene on the Y chromosome

Yeast artificial chromosome (YAC) A constructed yeast chromosome which can accommodate a large amount of recombinant DNA and can be used as a cloning vector

Index